I0072567

Waste Management: A Modern Approach

Waste Management: A Modern Approach

Edited by **Victor Bonn**

R CALLISTO
REFERENCE

New York

Published by Callisto Reference,
106 Park Avenue, Suite 200,
New York, NY 10016, USA
www.callistoreference.com

Waste Management: A Modern Approach
Edited by Victor Bonn

© 2015 Callisto Reference

International Standard Book Number: 978-1-63239-604-4 (Hardback)

This book contains information obtained from authentic and highly regarded sources. Copyright for all individual chapters remain with the respective authors as indicated. A wide variety of references are listed. Permission and sources are indicated; for detailed attributions, please refer to the permissions page. Reasonable efforts have been made to publish reliable data and information, but the authors, editors and publisher cannot assume any responsibility for the validity of all materials or the consequences of their use.

The publisher's policy is to use permanent paper from mills that operate a sustainable forestry policy. Furthermore, the publisher ensures that the text paper and cover boards used have met acceptable environmental accreditation standards.

Trademark Notice: Registered trademark of products or corporate names are used only for explanation and identification without intent to infringe.

Printed in the United States of America.

Contents

Preface

A modern approach in the field of waste management has been extensively discussed in this profound book. Solid waste management poses significant challenges for society because of its great variability in composition and production, and also due to its environmental and sanitary effects. For the purpose of fighting against this situation, this book has been compiled of analyses of technical and conceptual advancements and experiences gained through research and development projects worldwide. In addition to decreasing generation, regarded as the most preferable practice, this book also presents alternatives of valuation for potentially recoverable waste and techniques to decrease final conditioning and disposition risks. The book also covers the political, technical, social and economic aspects for the purpose of finding solutions for solid waste management.

This book has been the outcome of endless efforts put in by authors and researchers on various issues and topics within the field. The book is a comprehensive collection of significant researches that are addressed in a variety of chapters. It will surely enhance the knowledge of the field among readers across the globe.

It is indeed an immense pleasure to thank our researchers and authors for their efforts to submit their piece of writing before the deadlines. Finally in the end, I would like to thank my family and colleagues who have been a great source of inspiration and support.

Editor

Solid Waste Management in Different Regions of the World. Strategies for the Sustainability

Waste Management Threats to Human Health and Urban Aquatic Habitats – A Case Study of Addis Ababa, Ethiopia

Elias Mazhindu, Trynos Gumbo and Tendayi Gondo

Additional information is available at the end of the chapter

1. Introduction

Over recent decades one of the commonest characteristics manifest in the developing nations has been the disparity between rapid urban population growth and sanitation infrastructure provision. This disparity is being worsened by the challenges of poor waste management practices impacting on the deteriorating ecosystems of the rapidly transforming cities in these countries. The product of this mismatch, described as 'urbanisation without health', is the catalogue of overcrowding, growth in illegal settlements, uncollected household waste, and the absence of water, sanitation and other basic facilities which are typical of many urban centres in Africa, Asia and South America. As a result many millions of the urban poor live in neighbourhoods typically hazardous to their everyday health and general well-being. The major concern is that despite advances in technology and innovative responses towards mitigating the threats to environmental health, notable deficiencies in the access, maintenance and management of sanitary facilities in the cities of most developing countries still persist. Despite these advances, the question is why are the environmental threats endangering human health and ecosystem welfare on the increase?

In explaining this question, some studies have argued that the rapid rates of urbanization in Third World countries – in both spatial and demographic terms, are urban growth and transformations in Third World countries in recent decades – which are the major drivers of the current environmental and public health problems [1, 2]. An environmental health problem has been defined as "either an inadequate supply of a resource essential to human health or urban production (e.g. sufficient fresh water) or the presence of pathogens and toxic substances in the human environment which can damage human health or physical resources such as forests, fisheries or agricultural land" [3]. Most studies, discussed in this chapter, concur that the health of the residents of Addis Ababa is imperiled by the pitiable physical environmental conditions that are presently characterized by poor shelter, overcrowding in squalid housing and neighbourhoods, unsafe drinking water, poor sanitation, water pollution, indoor and air pollution and poor waste management. This poor urban environmental fabric, worsened by the low priority accorded to sanitation, has been largely blamed for the high incidence of waterborne pathogens in the catchment interface of Addis Ababa that are responsible for the spread of communicable diseases such as cholera, typhoid, and amoebic infections, mainly dysentery.

The built-up area of Addis Ababa – featuring ultra-modern buildings adjacent to slums - lies within the Big Akaki and Little Akaki river basin which has a catchment area of about 540 square kilometres. The Big and Little Akaki rivers, with their dendritic tributaries, drain the city from north to south (Fig. 1). The inappropriate practices of dumping household and industrial wastes in the river catchments has resulted in the spread of anthropogenic diseases in the city. Some earlier studies lament that the biological pathogenic vectors in the hydrological cycles of urban centres in Ethiopia account for four-fifths of all diseases and the related high mortality rates [4]. These diseases have been closely associated with the high prevalence of urban poverty and weaknesses in municipal waste management interventions – thereby increasing the vulnerability of the majority low income households.

Recent studies have established that nearly two-thirds of the urban citizens of Ethiopia use pit latrines for sanitation while close to a third defecate in open fields and less than 5% of the population use flush toilets [5]. It is mainly the residents in the slum settlements, constituting an estimated 80% of Addis Ababa's estimated 4 million people, who live with the most insidious environmental problems due to poorly developed existing environmental infrastructure and services such as sewers, drains, or services to collect solid and liquid wastes and safely dispose of them [6]. This situation is comparable to other sub-Saharan African cities where the majority of the urban population – 65% in Dar es Salaam [7], 67% in Blantyre [8], and 80% in Luanda [9] lives in squatter settlements. The most recent household survey conducted in 1998, revealed that 11% of the households in Addis Ababa had private flush toilets, while 73% owned private or shared pit latrines and 16% of the households had no toilet facilities of any kind [10]. In the peripheral residential areas of the rapidly sprawling city, 40% or more of the households had no access to latrines. It is common practice that some of the city residents who have no access to both public and private latrines are forced to defecate in nearby open grounds hidden from public view or else they have to walk longer distances to ravines, ditches, or wooded areas. The sight of some residents resorting to relieve themselves in public along road pavements and in storm water drains is also common – not least, in the congested areas of the city such as Merkato, Piassa, Bole and Gurd Shola.

Not surprisingly, the informal vending and sub-letting of private latrines on house plots and in the concealed spaces is a growing trend in the poorly serviced slum areas of Addis Ababa. The bathing of limbs on street pavements and in the congested public spaces of the city centre is a common sight during the lunch hour breaks. The practice is most visible

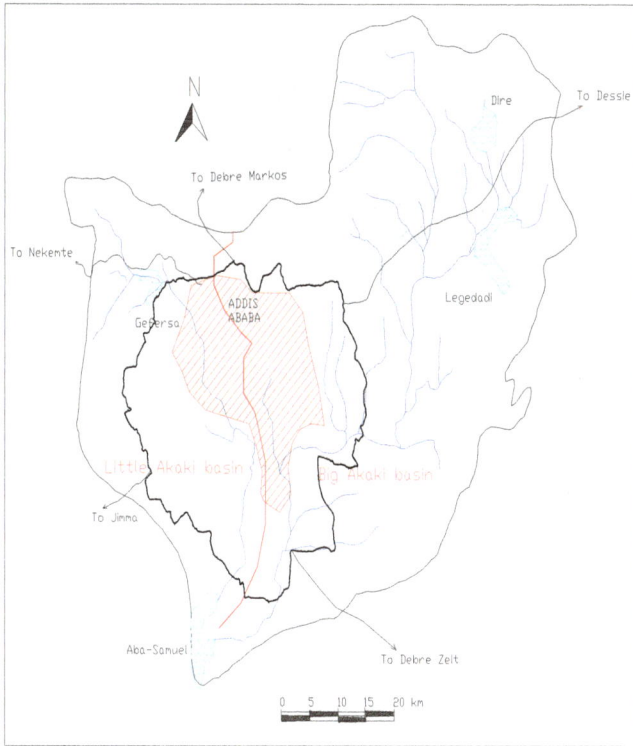

Akaki river catchment (in red built up area)

Figure 1. Addis Ababa catchments of the Little and Big Akaki River basin.

along the pedestrian alleys of the Merkato Open Air Market and the congested bus termini throughout the city. Urinating and defecating on open spaces scarcely hidden from public view - does not seem taboo (Fig. 2 a and b). These examples illustrate the rising importance of the environmental health risks inherent in the waste management challenges at the municipal, neighbourhood and household or personal levels. In addressing such challenges, both international and development aid communities have recognized the identification of waste management as an integral component in the conceptualization and implementation of city-wide development policy strategies – guided by the protection and enhancement of ecosystem services. Thus the United Nations Conference on the Environment and Development concluded that "...solid waste production should be minimized, reuse and recycling maximized, environmentally sound waste disposal and treatment promoted and waste service coverage extended" [11]. Not surprisingly, the UNCHS Habitat prioritizes "environmentally sound and resource efficient approaches in mitigating the problem of growing solid waste quantities, and considers waste management as a crucial component of human development policies and programmes" [12, 13].

(a)   (b)

Figure 2. a. Waste dumping site in Merkato; b. Human excreta in home surroundings in Merkato

The management of waste in the urban centres of Ethiopia is the responsibility of the Municipal Division of Health. All municipalities – except the chartered cities of Addis Ababa and Dire Dawa who have cabinet representation – exercise some autonomy in managing their own affairs. All the chartered cities and the certified smaller urban centres are mandated "to provide, maintain and supervise environmental health services along with other activities in their own areas" [14]. However worthy these objectives sound, most of the municipalities and urban centres do not seem to have efficiently run environmental planning and management institutions let alone sufficient resources for discharging their responsibilities effectively. This scenario is worsened, in part, by the sustained low priority accorded to essential sanitation activities in most of the country's urban centres largely attributing to insufficient local revenue bases. Besides their routine administrative duties, the sanitarians assigned to the regional health departments and environmental health centres can only afford attending to emergency cases especially water pipe outbursts and toilet flooding - considered imminent threat to the residents.

With this scenario in view, the chapter assesses some of the current waste management practices of domestic households in the slum areas of the city and the risks to both human health and ecosystems that these practices play out in the surroundings of homes and aquatic systems of Addis Ababa. In doing this, the chapter draws on the growing body of literature on waste management in urban Ethiopia in order to trace some of the important relationships between the current waste management practices and their impact on public health, especially in the congested parts of the central business district (CBD) of Addis Ababa. In the context of this chapter, solid waste management is taken to mean "the processes of controlling the generation, storage, collection, transfer and transportation, processing and disposal of solid wastes in accordance with the best principles of public health, economics, engineering…that is also responsive to public attitudes" [15]. Meanwhile, sanitation will be taken to extend further than physical access to latrines and toilet facilities such as hand washing basins, cleaning towels and lighting. Sanitation encompasses the whole process of enhancing the conditions of the living environment (both inside and outside the home), personal hygiene, as well as improving the physical infrastructure of [latrine and] toilet facilities, a safe and adequate water supply, and the safe disposal of domestic solid and liquid wastes [16][own emphasis]. The quality of water – for domestic or personal consumption - dependents on healthy ecosystems and sustainable land use management in watersheds.

The chapter examines some of the key underlying questions of improving the sustainability of ecosystems and the environmental health status of sub-Saharan African cities that continue to be threatened by the fragmented waste management policy responses using an ecohydrological perspective. As a new trend of thinking towards promoting livable urban settings, the theory and implementation of ecohydrology has been developed in the framework of the International Hydrological Programme of UNESCO (Zalewski *et al.*1997[17]; Zalewski and McCain 1998[18]). Zalewski *et al.* (2010: 102) have suggested that "an integrative approach, expressed by ecohydrology principles, should be helpful to distillate the general patterns of ecohydrological interplays, which confronted with social challenges [largely due to the rapid urban population growth rates and underlying transformations] should provide a dynamic framework for the formulation and implementation of realistic strategies for problem solving by focusing on ecological processes for enhancing sustainability [19][own emphasis]. But why is the ecohydrological perspective particularly useful for mitigating the negative effects of the current poor waste management on the environmental public health status of residents through the enhancement of ecosystem services to society? In answering this question, Zalewski *et al.* contend that "…given the conditions of the increasing demographic pressures, sustainability can be achieved through policy responses that regulate the whole range of water biota interactions with the human settlement activities towards enhancing the carrying capacity of the city – water resources, biodiversity and ecosystem services [20]. Due to the complexity of such interactions, an integrative understanding of the interactions between different biological and settlement activity patterns in urban ecosystems is essential. The integration for synergy in the basin scale and regulatory measures have been viewed as reducing the negative effects of the cumulative load of excessive effluents into the acquatic system significantly [21].

The essence of the ecohydrological perspective is rooted in the defining classification by Odum (cited in Zalewski et al 2010:102) that "ecology is the economy of nature" [22]. Thus the implementation of this strategy posits that the enhancement of the carrying capacity of urban ecosystems has to begin by quantifying the hydrological cycle (such as trends in the eutrophication of rivers through waste dumping) and the identification of threats to ecosystems and public health engendered by such waste disposal practices. The next step is the assessment of the ecosystems as they are modified by human settlement activities, their distribution in the catchment interface and their impact on the livability of the urban built environment. Finally, the regulation of water biota processes through interventions such as the reduction of point source pollution should be based on an understanding of the hierarchical complexity of ecological processes in the catchment area [23]. However, this approach contradicts the many environmental management policies that are top-down (command and control approaches) involving direct regulation along with monitoring and enforcement standards, permits and licences that have been criticized for being costly and difficult to enforce [24, 25]. The regulatory approaches have been perceived as contributing to the worsening of environmental health risks attributing to waste management policy strategies, mainly due in part to the lack of awareness of existing environmental instruments on the part of many residents [26, 27].

In this definitive context, the chapter examines the current domestic solid and liquid waste management practices in Addis Ababa with a view to suggesting possible policy options for mitigating the environmental health risks that are highlighted in the most recent literature (Kebbede 2004 [28]; Kuma 2004 [29]; Tadesse *et al.* 2004 [30]; Bihon 2008 [31]; Van Rooijen *et al.* 2009 [32]. A number of environmental health problems occurring at the varying spatial scales from the home through the neighbourhood to the city will be reviewed in light of the findings of case studies carried out mainly in the built-up area of Addis Ababa. The surveys by Abebe 2001[33], Kuma 2004 [34], and Tessema 2010 [35] attribute the proliferation of pathogens in the living areas of poor homes and neighbourhoods to a combination of inadequacies in the provision of sanitation facilities, inappropriate anthropogenic practices of sanitation at household level and the current waste management problems. The surveys reveal that the cramped living conditions and the presence of pathogens in the home environments due to the lack of basic infrastructure; the dangerous and unhealthy sites of some neighbourhoods due to the irregular or non-collection of garbage and the city-wide problems of toxic or hazardous waste disposal pose the major threats to the health of most residents in city.

The upsurge in the urbanization and industrialization following the structural reform programmes adopted by most sub-Saharan African countries is generating domestic waste in the form of raw sewage, untreated effluents with potential contaminant pollutants and toxic waste in the urban settlements. Current literature reveals that most of this waste ultimately finds its way into the clogged city streams and rivers ending up in inland water bodies such as the Aba-Samuel Dam, one of the main sources of water supply to Addis Ababa city. This trend persists as the standard practice by both the population and the practitioners. According to Alebachew *et al.* (2004), deficiencies of sanitary services, low capacity for urban waste management and the absence of regulations and scientific criteria for enforcement pose increasing environmental and public health hazards in the major towns of Ethiopia [36]. Arguably, there are many interconnected factors militating against the current top-down policy engagements in mitigating these problems. These factors include massive rural-urban migration fuelling rapid urban population growth, poor planning and ineffective development control measures, weak urban institutions, and insufficient institutional resources set aside for tackling the ever present environmental health risks prevailing in the city. Drawing on the diverse case study findings reviewed in this chapter, the empirical findings of our case study suggest that the integration and harmonization of the interactions of hydrological and ecological processes in urban settings, and striking the balance, can be the key for sustainable waste management in the "city-to-come" in sub-Saharan Africa - comparable to Addis Ababa.

Ligdi and Nigussie (2007) suggest that the presence of elements of ecohydrology in relevant policy and project documents, as well as the various capacity building efforts are good starting points to promote the use of ecohydrology in the country [37]. This approach will have the potential of bringing the fragmented approaches into one whole system that promotes sustainable waste management. Specific studies that can reveal the importance of the hydrological and ecological processes in managing waste should be carried out in selected "ecological hot spots" [38]. The results of such studies will help in convincing decision makers and practitioners to understand the value addition of ecohydrolgy as a tool for integrated waste management in the fragile ecological hot spots prone to widespread anthropogenic health risks. We first turn to the problematic of environmental waste management in Addis Ababa to help us map out the spread of pathogens and threats to the health of residents living in the overcrowded enclaves of the city particularly in Merkato and Kasanchis where the epidemiological footprint and its associated anthropogenic practices is most visible.

2. Research problem

The challenges brought about by inappropriate anthropogenic practices threatening the health of most residents and sustainability of the existing aquatic habitats are mostly visible along stream banks and public open spaces in Addis Ababa ranking the city as one of the dirtiest in the world. These threats and related land use imbalances have not scaled down for a long time in the city [39] owing mainly to unrelenting in migration into the city and the paucity of resources to manage the increasing quantities of waste accumulating in the living spaces. The excessive pressure of "unplanned" (by modernist planning standards) land uses including encroachments on the fragile aquatic systems through the dumping of all range of solid waste into the riverine network seems to continue unabated (Fig. 3 a and b). The impacts of poor waste management and disposal most visible in the slums of the city have associated with the endemic spread of communicable diseases affecting mainly the poor sections of the city residents.

Table 1 depicts the trends in the spread of the top ten diseases mainly attributed to the indiscriminate solid waste management practices in the city. As can be noted in the table, the number of cases and their frequencies over the three years was too high for a city of 3 million (in 1999) relative to other cities in developing countries.

However, even these figures obscure the true picture of the number of cases not reported to the health institutions and the widespread practices of self-treatment and traditional healers in the city.

(a)   (b)

Figure 3. a. Car washing contaminating groundwater; a. Quarrying and structures encroaching stream bank

Solid Waste Related Diseases	1997	1998	1999
1. Parasitic infection	57 887	36 827.	36 845
2. Bronchitis	38 100	28 849	28 780
3. Skin diseases	34 426	27 119	27 047
4. Broncho pneumonia	30 219	25 744	25 158
5. Dysentery	20 782	13 596	14 631
6. Bronchial asthma and allergic conditions	11 607	7 677	6 291
7. All other respiratory diseases	7 932	3 845	7 532
8. Typhoid	6 596	3 622	4 046
9. Influenza	3 593	1 905	1 858
10. Trachoma	1 619	1 015	1 346

Table 1. Solid waste related diseases and morbidity in Addis Ababa (Source: Annual Morbidity Report of Addis Ababa 1997 – 1999). *Please note that the data for 1997 include clinics whereas data for 1998 and 1999 cover hospitals and health centres only.*

The studies by Girma Kebbede (2004) and Kuma *et al.*(2005) suggest positive relationships between the worsening environmental health status of Addis Ababa and the current uncoordinated waste management and refuse disposal practices playing out at all levels in the ten administrative sub-cities of the capital [40, 41]. These levels include the domestic (on single house plot) point of source, neighbourhood (*kebele* or smallest administrative and political unit in Ethiopia) and municipal level. An overview of the stages of waste management in the separation, reuse and recycling of wastes reflect the current modernist approach to the waste management in the city. The first level of source separation is at household level involving street boys, private sector enterprises and waste pickers at municipal landfills. Plastic materials including glass and bottles are considered as valuable and usually sorted out for reuse. Recyclable materials include metal, wood, tyres, electricity products and old shoes. The role of the municipality in recycling is absent as the municipality focuses on the collection, storage, transportation and disposal of waste. Most of the collection of recyclable wastes in the city is done by the informal sector. The recyclable waste materials are used by the local plastic, shoe, and metal factories. The municipality dominates the transportation and disposal of solid waste from garbage containers (secondary collection) to the dumping sites. The role of the private sector in the transportation of solid waste mainly involves the informal sector dominated by the deployment of door-to-door push-cart collection service. The push-carts dump the collected domestic waste, wrapped in sacks, at central collection points from where it is collected and transferred by municipal and private trucks for disposal at the only municipal dumping site known as Reppi or Koshe – 15 kilometres from the city centre. The present method of disposal is crude open dumping, notably, hauling the wastes by truck, spreading and levelling or compacting by bulldozer.

The current uncoordinated approach to waste collection and disposal has been mainly blamed for the high incidence of waterborne pathogens responsible for communicable diseases such as cholera, typhoid, amoebic infections and dysentery that account for four-fifths of all diseases in the country [42]. Kebbede (2004) attributes these endemic diseases to the deteriorating urban environmental conditions of all the country's urban centres. These conditions continue to manifest themselves in the increasingly poor housing shelter, overcrowding in squalid housing neighbourhoods, unsafe drinking water, poor sanitation, water pollution, indoor and air pollution and poor waste management [43]. It has been noted that despite the relatively long history of environmental health practices in the country since the early 1950s, the provision of services in the field still remains below expectations [44]. Kuma and Ali (2005) contend that the progress that has been made in environmental health service coverage so far does not seem to reflect any significant changes since the early 1970s. This is reflected by the current coverage levels of safe drinking water and latrines countrywide of 30% and 13%, respectively [45]. The per capita drinking water meets only half of the minimum requirement of 20 litres per person daily [46]. Arguably, most of the residents living in the slum areas or houses not connected to piped water have to obtain water for domestic use from vendors at high prices well above the tap price. In a study of household access to safe water in Addis Ababa, the UN-Habitat Report (2003) revealed that most (88.5%) of its residents had access to improved water mainly from piped water either into the dwelling or into the yard (67.8%)[47]. The study noted that, however, there are disparities between the ten sub-cities comprising the city. Bole sub-city, where the highest population of high income households lives, has the largest population (98.9%) with access to improved water while Akaki Kaliti and Nefas Silk depend on unprotected wells and springs for their water supply thus exposing them to anthropogenic health risks [48].

Equally, the access to latrines has similar drawbacks. These drawbacks include the limited achievements in environmental health service coverage over recent decades mainly owing to socio-economic factors and poor waste management practices detached from policies. How have these drawbacks given rise to the proliferation of environmental risks endangering the health welfare of most residents in Addis Ababa? The city has inadequate, hygienically deplorable sanitation facilities in a terrain of rivers that have literally turned into open sewers over the years. Kebbede (2004) established that most public and private shared latrines in the city are unventilated, overused, unlined, collapsing and overflowing [49]. The unlit pit latrine superstructures are old and the wooden planks covering latrine pits are unstable and frequently fouled. On average, 34 people share a pit latrine. Over 70% of the latrines are shared by twenty or more users. The densely populated central areas have the highest percentage of households (50% – 60 %) using shared latrine pits. Many of the *kebeles* in these areas have 70% - 75% of their population using these latrines. One of the most poorly serviced, densely populated areas in the city is Addis Merkato. Over 60% of the residents of the Addis Merkato area use shared latrines and 25% have no toilet facilities. The shared latrines, sandwiched in between houses in collapsing superstructures, are overused and overflowing with raw sewage. Cleaning tankers cannot access them. Due to the lack of toilet facilities, roads often overflow with human excreta and garbage. The communal latrines provided are often blocked with all types of garbage and overflow into the streets attracting harmful insects and rodent vectors such as rats.

In the densely populated *kebeles*, the lack of space often forces large numbers of households to share the same latrine. Not surprisingly, residents of the *kebele* slum compounds make do with the inadequate and poorly accessible sanitation facilities and services. The tenants' lack of ownership of such dwellings provides little or no incentive to keep them clean, make improvements in the existing facilities, or build new ones. This also leads to the misuse of the shared latrine facilities, unhygienic and unsanitary conditions.

Private pit latrines are better constructed and maintained than shared ones. Most are ventilated (though some low-income households do not have ventilated pit latrines) and are made of zinc sheets, *chika*, or other scrap materials. Most private pit latrines have fewer than ten users. The proportion of privately owned houses is greater in the outskirts of the city and, thus, private latrines constitute 20% – 25 % of the sanitation used [50]. By contrast, less than 10 % of the sanitation consists of private pit latrines. The central areas are congested and house ownership issues are more likely to restrict construction of private latrines. The use of private pit latrines covers about 15% of the housing areas between the periphery and the centre of the city.

About 11% of the households of greater financial means in the city or more affluent suburbs have private or shared flush toilets. However, since most of these toilets are not connected to the main sewer network, septic tanks, cesspools and open waterways are used instead for discharging sludge [51]. Most of the sludge is washed into the nearby streams during the rainy season or percolates into the underground water table - the main source of borehole water for domestic use in most parts of the city.

There are more than 72 public toilets scattered in the city. However, most of these public toilets are located in central business district and a few are located in the peripheral commercial areas. They are extensively used and only men have access to them as they have no separate sections for women. All are maintained by the municipality and are designed as squatting plates with flush systems. Most of them are dreadfully grubby. Only one of these toilets is connected to a reticulated sewer system. Thirty-eight of them have septic tanks that are connected to storm water drains and streams; the rest have no septic outlet and are emptied about twice weekly by tankers. These toilets are estimated to have 1 800 users/toilet/day. Many of the cisterns are broken and flush continuously. Many are not easy and safe to use, especially for children, the elderly, and disabled people as they are poorly lit. The maintenance and cleaning of public toilets are so poor that many people avoid using them.

In the context of the deteriorating environmental health status of Addis Ababa amplified by recent studies, the case study [52] sought to investigate the question why most urban residents in Addis Ababa live their everyday at the increasing health risks in their living areas. The findings of the case study suggest that an integrated approach to city-wide waste management and refuse disposal interventions incorporating the ecohydrological perspective can serve as a useful point of departure.

3. Aim

The chapter seeks to inform the current waste management policy interventions towards ameliorating the existing environmental health risks in Addis Ababa using an ecohydrological perspective. The major thrust of this approach is problem solving. Such a perspective can help in the understanding of the underlying interactions between local authorities, investors and local residents involved in waste management at the varying socio-economic and spatial scales – the home, neighbourhood through to city level – within a given ecohydrological system.

4. Objectives and research questions

The chapter evaluates the ecohydrological status and sanitation practices of living with environmental health risks in Addis Ababa by drawing on the findings of a recent case study by Mazhindu et al. [53]. The chapter provides an in-depth analysis of the case study findings answering the following key questions:

What are the trends in the spread of water-borne diseases to the urban residents of Ethiopia since 1998 to date?

How are the urban centres managing domestic wastes in the existing environment of both inadequate and poorly managed sanitation services?

What are the existing sanitation facilities and services at domestic and neighbourhood level?

What are the current domestic waste management and refuse disposal practices in the low income neighbourhoods?

How do the members of low income neighbourhoods perceive the waste management practices and performance of the city authorities?

What policy options remain open for sustainable liquid and solid waste management in the city?

5. Description of the study area

With the current estimated 4 million people residing in a built-up area of 290 square kilometres covering ten sub-cities (Fig. 4.). Addis Ababa shares an estimated 30% of the country's urban population signifying a population density of 10 345 persons per square kilometre. The city has arguably one of the highest populations in the world living in dilapidated and poorly serviced slum settlements largely located in the inner-city. Many slum settlements - including extensive very poor informal settlements dominate the flood prone areas in the oldest and overcrowded central parts of the city [54] including Merkato and Kasanchis. Squatters often select land not to be demanded for any other use in order to minimize the possibility of eviction. Such sites are likely to be dangerous and unhealthy. They include hillsides, flood plains, and polluted land sites near solid waste dumps or areas inundated with high levels of noise pollution. The flood prone areas along most of the river banks exhibit the most densely settled parts of the Addis Ababa, thus heightening the propensity of water stagnation and the spread of pathogens in the catchment interface. The population densities exerting pressure on the hydrology and the ecosystem services of the slum areas under study are revealed by the latest demographic transformations taking place in the city.

The most recent census figures published by the Central Statistics Authority of Ethiopia (2007) reveal a total of 52 063 households living in Addis Ketema sub-city covering an area of 85.95 square kilometres, where the most congested slum compounds of Merkato are housed. Addis Ketema sub-city has a population density of an estimated 4 284 households per square kilometer and the population of Merkato is an estimated 31 552 households. Kirkos (Cherkos) sub-city, housing the Kasanchis slum compounds, had a total of 54 398 households occupying a total area of 14.7

square kilometers. The total number of households in the slum compounds of Kasanchis is estimated to be 45 500 households. The least congested but better served slum areas of Meri-Luke feature in Yeka sub-city – home to 90 195 households living in an area of 85.94 square kilometres.

Some recent studies also observe that, elsewhere in the rapidly transforming sprawling landscape of the city, most residents live in poorly constructed sub-standard housing units (Fig. 5 a and b). These housing quarters are inadequately serviced in terms of the existing sanitation facilities due to the fragmented waste management approaches in practice [55]. Abebe (2001) noted that about 60% of the population of Addis Ababa lives below the poverty datum line [56]. According to UN-Habitat (2006) estimates, 80% of Addis Ababa's settlements are considered slum [57]. Khurana (2004) claims that the word "slum", originated from the word "slumber", which meant "a sleepy back alley" or "wet mire" or working class housing built near factories during the British industrial revolution [58]. The Collins English Dictionary (2007) defines a slum as a poor rundown and overpopulated section of a city [59] while the UN-Habitat Report (2003) depicts a slum as "a heavily populated urban area characterized by sub-standard housing and squalor" [60]. Slums happen and can also be perpetuated by a number of phenomena including rapid rural-to-

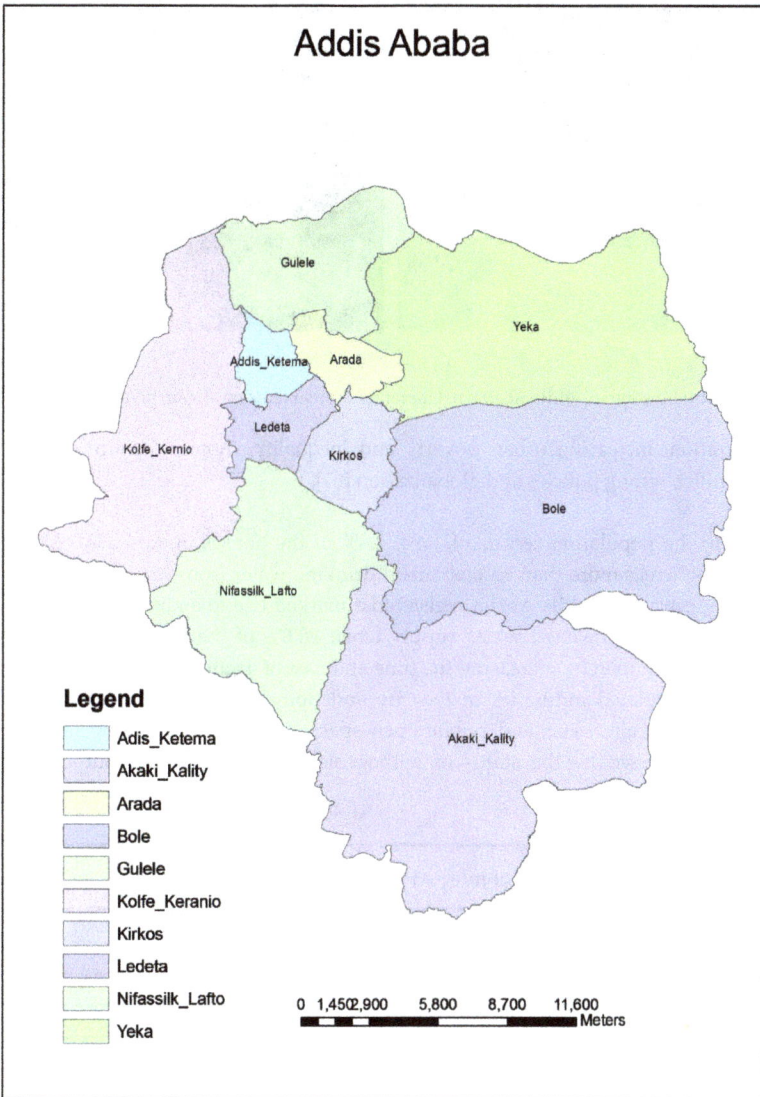

Figure 4. Administrative boundaries of Addis Ababa sub-cities

(a)   (b)

Figure 5. a. Slum housing in Addis Ababa; b. Unplanned settlements on stream banks

urban migration, increasing urban poverty and inequality, insecure tenure on property, political conflict, wrong policies and globalization [61].

According to the population census of 1994, 4.4% of the houses in Addis Ababa had tap water inside whereas more than 45% obtained drinking water from vendors. However, the 2008 housing census in Addis Ababa registered a marked rise to an estimated 88.5% of the housing units with reticulated water supply albeit 28.6% of the households experienced frequent disruption thereby increasing the time and cost of acquiring water [62]. Moreover, the largely unregulated industries in the city and domestic households release harmful pollutants into the air, water and public open spaces, further endangering the health of residents as witnessed by the status of pathogenic infections in Lafto Sub-city in 2008 (Table 2).

Type of disease	Number of infected people	Percent (321 000 total population of sub-city)
Intestinal parasites	18 618	5.8
Common diarrhea	14 445	4.5
Respiratory infection	11 877	3.7
Amoeba	9 309	2.9
Typhoid fever	8 667	2.7
Typhus	8 346	2.6
Dysentery	8 346	2.6
Total	79 608	24.8

Table 2. Sanitation related diseases and infected people in Lafto Sub-city, Addis Ababa (Source: Nifas Silk Lafto Sub-City Health Centre Annual Report 2008)

Communicable diseases attributable to poor sanitation and practices, affecting mainly the underprivileged sections of the population, are considered as the major causes of morbidity, mortality as well as disability in Ethiopia [63, 64]. The high prevalence of communicable diseases in the country has been positively linked with the poorly developed socio-economic and environmental factors that have been inherent for centuries [65]. The rapidly shifting demographic and morphology of the city of Addis Ababa - featuring the rapid population size, widespread unemployment, the unremitting housing shortages, the demand for social and physical infrastructure are worsened by incompatible and unregulated land use activities. Such transformations are manifest by the proliferating squatter settlements in the interstitial spaces of the built environment, the dumping of solid and liquid wastes in open spaces, in the open sewerage drains and streams.

6. Applied methodology

Motivated by the growing body of literature on the environmental health risks to residents in Addis Ababa, Mazhindu et al [66] conducted a case study of the slum areas in Addis Ababa to assess the current domestic sanitation practices and impacts on the ecohydrological status of the rapidly expanding urban centres in Ethiopia. The case study reviewed the findings of earlier studies by Kebbede [67], Van Rooijen et al. [68] and Alamayehu [69] as a starting point.

Using an exploratory design, the study adopted a mixture of purposive and stratified cluster sampling techniques to diagnose the current institutional arrangements for waste management in the Addis Ababa metropolitan area and to identify the distribution, access, usage, quality and maintenance of existing sanitation facilities at local level. The study targeted the slum compounds of Merkato (in Addis Ketema Sub-city), Kasanchis (in Kirkos Sub-city) and Meri Luke in Yeka Sub-city, undoubtedly the most overcrowded experiencing the worst environmental health threats. In comparison to Merkato and Kasanchis, the study considered the slum areas of Meri Luke - in the eastern outskirts of the city – as a mixed density residential area for the largely mixed low and middle income households, as less environmentally threatened. Not surprisingly, however, the findings of the study established that all the slum areas visited exhibit very similar environmental health management threats to both human health and the degraded aquatic systems that increasingly dominate the undulating and thinly forested landscape of the rapidly sprawling city.

Merkato (Amharic for "New Market", popularly just "Mercato", from the Italian for "market") is the local name for the largest open-air marketplace in Africa as well as the neighborhood in which it is located (Fig. 6a.). Merkato is located in Addis Ketema sub-city which is the smallest and most overcrowded of the 9 sub-cities of Addis Ababa. The sub-city has an average population density of 448 persons per hectare. Addis Ketema is viewed as the economic core of the country with transportation access to the rest of the city and the country. At Merkato main bus terminal (Fig. 6b.), about 950 buses serve commuters to all parts of Ethiopia everyday. An estimated 200 000 people visit Addis Ketema sub-city daily either on business or in search of employment [70].

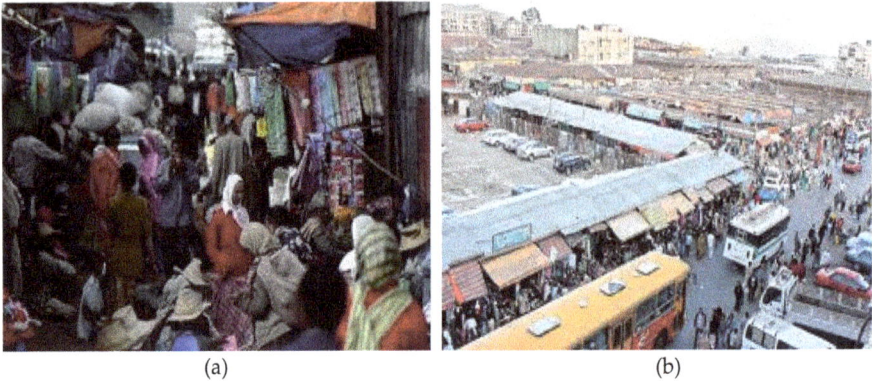

(a)   (b)

Figure 6. a. Merkato Open Air Market Stall; b. Addis Merkato Bus Terminus

The Merkato market covers several square kilometres and employs an estimated 13,000 people in 7100 business units. The open air market has over 120 stores and one massive shopping complex that houses 75 stores. The primary merchandise passing through Merkato comprises local agricultural produce — most notably coffee – cheap synthetic textile and electronic imports from countries in the Middle East and Far East.

The inquiry sought to collect both secondary and primary data on the trends depicting the spread of water-borne diseases in urban Ethiopia since the early 1990's. The study mainly sought to explore the major causes of deteriorating aquatic systems in relation to the worsening environmental health situation playing out in the poorly served and managed sections of the city. This required knowledge about the current waste management policy responsiveness of Addis Ababa concerning access, use, quality and maintenance of existing sanitation facilities and services at all levels. The study adopted the UN-Habitat (2003) definition of an adequate toilet facility as "an easily maintained toilet in each person's home or at a reasonable distance with provision for hand washing and safe removal and disposal of wastes" [71]. In this definitive context, the study assessed the adequacy of toilet facilities on the basis of ownership of toilets, the use patterns of toilet facilities by households, the types of toilets, the physical characteristics of toilets versus number of users, the filling of toilets and emptying periods.

Our exploratory research design deployed staff of three supervisors and four research interviewers, who graduated on the current Urban Management Masters Programme at the University of the Ethiopian Civil Service in Addis Ababa. The questionnaire was initially drafted in English, translated to Amharic, and then pretested in the similar slum housing contexts in the city adjusted to the features of suitability in terms of duration, language appropriateness, content validity, and question comprehensibility. Prior to commencement of the field visits, all study personnel were confident in interviewing skills, content of the semi-structured questionnaire, data quality, and ethical conduct of human research. Each research assistant surveyed an average of seven households per day in Merkato and

Kasanchis slum compounds as well as Meri Luke twice weekly for two months yielding a total sample of 160 people. Using both simple random and convenient sampling designs, a total of fifty households in each of the three study areas was interviewed to gather data on the frequencies, reflecting their opinions on the management of liquid waste at home, the efficiency of the refuse collection practices of the Addis Ababa municipality, the accessibility of sanitation facilities in the locality and the institutional capacity of the municipality for liquid waste management.

The case study initially reviewed the most recent studies (Kebbede 2004; Kuma 2004; Tadesse *et al.* 2004; Bihon 2008; Van Rooijen *et al.* 2009) on the provision of environmental health services in the urban centres of Ethiopia. These studies highlight the current trends of waste management and refuse disposal practices polluting the riverine systems of most urban centres in Ethiopia. To test the significance of the findings of these previous studies, we conducted personal field observation tours of the low income residential neighbourhoods in Addis Ababa. The pilot study helped in drawing the sample frame and units of analysis for the household survey. At the same time, we took pictorial samples of the existing sanitation facilities and domestic waste disposal practices in the study area. The pilot study sought to identify the most densely populated neighbourhoods of Addis Ababa exhibiting high incidences of ecological degradation, inadequate sanitation facilities and the impacts of the current solid and liquid waste management practices.

The study observed that the overwhelming majority (60%) of the houses in the current slums of Addis Ababa were built by feudal landlords of the Emperor Haile Selassie era, ending with the Marxist coup in 1974. The new regime nationalized all land and rental houses in decree number 47/1975. Rental houses were given to *kebeles*, urban dweller associations, for management. The current government, under Meles Zenawi since 1994, has not changed this policy and the state still owns all land, all rented units while the rents remain low and heavily subsidized. But having cut rent by as much as 70%, and passing all revenue to the central government, there is not much left for maintenance and construction. Residents have thus little incentive to move (even if they could afford to) and even less to improve the housing. The impossibility of access to land by the poor has ensured the proliferation of informal structures amid the formal – today a dominant feature of the sprawling cityscape and threat to environmental aesthetics.

7. Results

7.1. Management of liquid waste

UN-Habitat (2003:19) postulates that "a household is considered to have adequate access to sanitation if an excreta disposal system, either in the form of a private toilet or public toilet shared with a reasonable number of people, is available to household members" [72]. To determine the adequacy of access to sanitary facilities, the survey used a number of indicators, namely, the various types of toilet facilities available, the number of households sharing a toilet facility, the location of toilet facilities, structural quality of toilet facilities, maintenance and hygiene practices. Inadequate sanitation was taken to include service or bucket latrines, and latrines with open pits. Similarly, UN-Habitat (2003) considers the sharing of a toilet facility with not more than two households as inadequate [73] augmenting the spread of pathogens through overcrowding.

As reflected in table 3, most 52 (63%) of the existing shared toilets were found in the *kebele* (smallest administrative unit in Ethiopia) slum compounds of Addis Merkato and Kasanchis followed by 28% on the rented housing properties.

Latrine ownership	Number of respondents by status of house ownership							
	Private		*Kebele* house		Rented House		Relatives		Total	
Private	35	68%	2	3%	8	17%	6	12%	51	32%
Shared	4	5%	52	63%	23	28%	3	4%	82	51%
Public	-	-	5	71%	2	29%	-	-	7	4%
No toilet	3	15%	17	85%	-	-	-	-	20	13%
Total	42		76		33		9		160	100%

Table 3. *Latrine ownership by status of house ownership* (Source: Field Survey, September 2009)

An earlier study by UN-Habitat (2003) revealed that the most of the households (61.4%) in the ten sub-cities of Addis Ababa share toilet facilities while most people (51%) share toilets with more than two households [74]. Mazhindu et al (2010) established that only 5% of the shared toilet facilities were privately owned whereas 32% of the privately used toilets comprised mainly the toilets on the privately owned housing properties in Meri Luke, Yeka Sub-city [75]. The majority (85%) of the respondents who had no toilets reside in the Merkato and Kasanchis *kebele* housing compounds. Both Merkato and Kasanchis slum areas are in the most densely populated and overcrowded sub-city of Addis Ketema where the UN-Habitat (2003) study earlier revealed that 85.6% of households shared toilet facilities [76]. Insofar as the number of households sharing a toilet can be considered a critical and defining level of sanitation adequacy, most (68%) of the respondents in Merkato and Kasanchis slum areas indicated that on average seven households share a latrine, albeit in some cases, more than seven households share one latrine. On the whole, most residents (51%) depend on share latrines, followed by 32% sharing private latrines, 4% sharing public latrines while a sizeable population of 13% has no toilet of any kind.

As an important aspect of access, the UN-Habitat (2003) postulates that sanitation facilities should be available without excessive demand in physical effort and time on the user [77]. Our study showed that access to the shared public latrines by households, particularly in the overcrowded Merkato and Kasanchis residential compounds, varied from household to household among other paraphernalia of access variables. The majority of the households in the Kasanchis slum compounds live considerable distances from the nearest shared toilet facilities. For such households, the walking distance to the nearest latrine or toilet facility may stretch from 500 metres to one kilometre – an unrealistic distance to access such facilities especially by the very young, elderly and disabled. Clearly, the location of shared pit latrines is not easily accessible to many members of households living in the congested and marginalized areas without toilet facilities. The problem rests with the layout of facilities indifferent to the interacting linkages between several components in waste management and the conflicting interests among the different stakeholders in the use of space. This is a problem fix that has been scarcely touched in the urban development studies on Ethiopia concerning the spatial linkages between waste management service delivery and utilization in poor neighbourhoods.

Our physical assessments of the existing latrine facilities in all the slum neighbourhoods under study revealed that access was not the only constraint. In fact, most of the shared latrines were not readily usable. The physical characteristics of a latrine indicate its functional adequacy in satisfying the user needs such as hand washing receptacles and its environmental quality in terms of aesthetics and building fabric combine in determining the usability of the facility. The study established only 54% of the respondent households owned latrines complete with housing structures and door, while 27% owned toilets with the housing structure only (without door) and 19% owned open-air latrines.

The survey assessed the internal conditions of the latrine facilities and established that most (68%) of the existing latrines had cement floors whereas 32 % of the respondents indicated that their latrines had either wooden or mud floors. The study revealed that 8.6 % of the shared pit latrines were emptied twice annually but 11.8 % only once (Table 4).

Frequency of latrine emptying	Number of respondents	%
Once per year	18	11.8
Twice per year	13	8.6
More than a year	53	34.9
Never emptied	68	44.7
	152	100.0

Table 4. Frequencies of latrine emptying by annual interval (Source: Field Survey, September 2009).

Almost 45% of the respondent households indicated that they did not need to empty their traditional pit latrines since the pits would be covered with earth when full. Nearly thirty-five percent revealed that they channeled the sewage from the overfilling toilet pits into nearby open drains and waterways allowing it to flow along with domestic waste water ending up in the streams and tributaries that drain Addis Ababa city. Not surprisingly, the quantities of garbage and raw sewage accumulating in the catchments of Big and Little Akaki rivers, which flow into the Aba-Samuel dam, one of the main sources of water to Addis Ababa, have literally turned all the streams into open sewers threatening the city population with pathogens. This situation is made worse by the uncontrolled discharge of toxic liquid and solid industrial waste emitting from a wide range of large and small-scale factories that are clustered within the city that commonly rely on unregulated waste disposal systems [78].

Since efficient liquid waste disposal practices are equally essential for ensuring a safe and livable environment, the study established that there is no commitment on the part of households in observing clean practices to safeguard or enhance the environmental health conditions of their living areas and surroundings. The study observed that most shared toilet facilities in the slum compounds of Merkato and Kasanchis were bereft of water taps and other hygienic ancillaries including hand-washing receptacles, disinfectants and anal cleaning materials. This situation is equally dire in the poorly served slum areas of Meri Luke where the unscheduled and frequent disruptions to water supply diminish the importance of personal hygiene practices to mitigate the increasing burden of communicable diseases and threats to the aquatic systems. Although disruption to water supply was not assessed as an indicator of access by the case study, an earlier study by UN-Habitat (2003:16) revealed that 28.6% of the households in Addis Ababa experienced terminal disruptions in the last two weeks. These disruptions affect mainly households that depend on public tap (34.1%), followed by water piped into yard or plot (31.6%) and water piped into the private dwelling (24.8%) [79].

7.2. Solid waste disposal practices

Our focus group discussions with *kebele* waste management officials reflected that most households were reluctant to pay for the municipality services rendered for solid waste removal. This coincided with the common practice of residents dumping all range of solid waste materials on the river banks, open sewers in the surroundings of their living areas (Fig. 7 a and b.).

(a)   (b)

Figure 7. a. Garbage thrown into nearby stream; b. Scavenging and foraging on dump waste

Table 5 shows that 56% of the 160 respondent households deposited their solid waste in plastic bags while 19% dumped their solid wastes in open spaces, waterways and in the vicinity of their homes. Quite often, waste pickers team up with scavenging domestic animals including stray dogs and cats foraging for food on the dumping sites, tear the plastic bags open causing offensive smells in the surroundings and exposing the unpalatable contents to people in the surroundings.

Primary collection method	Frequency	Percent (%)
Open container	25	16
Closed container	15	9
Plastic bags	90	56
Open space	30	19
Total	160	100
Disposal method		
Door to door	35	22
Homestead yard	30	19
Collection point	15	9
Waterways and open space	70	44
Burning	10	6
Total	160	100

Table 5. *Solid waste collection and disposal methods* (Source: Field Survey, September 2009)

While 22% of the respondent households had their refuse collected by the door to door municipal service in the more affluent Meri Luke area, 19% and 9% of the households in the slum compounds of Merkato and Kasanchis deposited their refuse either in the yards of their homesteads or at central collection points respectively. Disturbingly, 6% of the households burn their solid wastes causing significant air pollution. The study attributed the prevalence of improper solid waste disposal practices to the absence of household garbage bins thus reflecting some of the deficiencies of the municipality in regulating solid waste management.

7.3. Institutional setting for waste management

The study established that the management of waste in the city falls under the Sanitation, Beautification and Park Development Agency in collaboration with the Region 14 Health Bureau. The key players in the waste management sector in Addis Ababa are formal and informal operators in the processes of collection, separation, recycling, reuse and transportation of waste for final disposal at the city dumping site of Koshe. Formal operators are those registered and licensed to work subject to tax and space regulations. These operators include municipal cleaners and private operators authorized by government, whereas informal operators are not registered and have no legal base for the operation of their business. The latter category includes scavengers, unregistered recyclers and re-usable article sellers.

The general tendency in many African cities of associating work in the waste sector with certain ethnic, religious or social groups, has been questioned by Klundert *et al.*(1995:8) who argued that the sector serves as a niche for income generating opportunities and a viable source of livelihoods for the marginalized minorities in urban areas [80]. This argument was refuted by Chekole (2006) whose findings revealed that the informal waste collectors in Addis Ababa constitute mainly the unemployed urban poor and individuals with other reliable income sources [81]. Chekole (2006) revealed the common practice of these informal waste collectors in organizing themselves into informal waste collection enterprises is based on social ties and living in the same neighbourhood.

On recognizing the advantages of participatory involvement in service delivery, the city government of Addis Ababa introduced regulations to promote the involvement of private institutions in the waste management sector. In its recognition of the vital role of the private sector in waste management, the Addis Ababa city government promulgated "Waste Management Collection and Disposal Regulations" No. 13/2004. The regulations stipulate that "service provided by government in the collection, transportation and disposing of solid waste may, through different participatory or transferring methods, be given to private sector investors" [82]. The city government justified this intervention on grounds that the involvement of the private sector – mainly medium and small scale enterprises – would ultimately pay off dividends by regulating citywide solid waste management practices. The city government viewed the pre-collection service offered by the existing informal waste collectors as fragmented and less effective towards achieving an ideally clean city. Moreover, the unregulated practices of informality in the solid waste sector were considered difficult to supervise regarding essential back-up services in the form of equipment and subsidies from government [83].

Our in-depth interviews with *kebele* (neighbourhood) representatives revealed that it is the responsibility of the *kebele* to keep its district clean. The *kebele* administration has the following obligations: to penalise dwellers caught throwing their waste around the containers; to visit the area and observe how dwellers manage to collect their waste; to conduct a campaign and clean the area, and contact the municipality to empty the bin frequently, and to construct latrines for those who do not have such facilities. However, *kebeles* are not able to penalize residents for throwing garbage in ditches and other open spaces, because this is usually done at night or when no one is around to evade prosecution.

7.3.1. Services expected from the Health Bureau

Our focus group discussions with *kebele* officials indicated that the *kebele* administrations responsible for Merkato and Kasanchis were often confronted with residents complaining about the garbage scattered around the containers, the bad smell and the health hazards that they pose in the neighbourhoods. The complainants suggested that solid waste collecting containers have to be emptied frequently albeit a *kebele* administration has no power to put pressure on the municipality to do so. The grieved residents also felt that the door-to-door collection service should comply with the dates and times for the convenience of all households. The *kebele* administration officials proposed regular consultative meetings with the Region 14 Health Bureau for concerted efforts in waste disposal and environmental awareness campaigns in all neighbourhoods.

7.3.2. Waste management malpractices in kebele neighbourhoods

Our focus group discussions with *kebele* officials responsible for waste management highlighted a number of malpractices in the collection and disposal wastes in the surroundings of homes adding to the eutrophication of the Akaki rivers. These practices are most visible both in the inner city of Addis Ababa including Merkato and Kasanchis as well as along the banks of Akaki River and its tributaries serving both as natural sewers and the mainstay of informal irrigation agriculture.

The *kebele* officials argued that since some distant *kebeles* do not have enough or convenient space to place garbage containers, residents are forced to walk long distances to neighbouring *kebeles* – to dump waste. Moreover, the existing garbage containers are not emptied frequently resulting in the accumulation of waste in open spaces. Alternatively, people throw their waste in sewers and ditches. Whenever the areas surrounding waste containers become muddy, people do not go close enough to the containers (skips), rather they just throw the garbage around the containers. The solid waste collection trucks only go where there is an access road. This means only those households that live along or near the road get the service.

8. Discussion

It is now generally agreed that prescriptive measures based on engineering and technological fix are unlikely to restore ecological processes of disturbed ecosystems [84]. The evidence of the degradation of natural ecosystems is overwhelming [85] in the rapidly growing cities of developing nations. Taking into account the concentration of nearly 70% of all industries in Addis Ababa, the uncontrolled discharge of industrial effluents into the Akaki River system is degrading the once pristine aquatic habitats. This has resulted in putting both human health and the absorptive capacity of the existing urban ecosystems under threat.

Ryszkowski (2000) has aptly observed that the integrity of biological and physical or chemical processes is a basic foundation of the modern ecosystem or landscape ecological approaches [86]. He has suggested that the recognition of the functional relationships between waste management practices and ecohydrological systems leads to the conclusion that biodiversity cannot be successfully protected only by isolation from hostile surroundings. Rather, its conservation should rely on the active management of the landscape structures through diversification.

The prevailing liquid and solid waste management practices in the overcrowded slum areas and the scattered industrial clusters in Addis Ababa pose considerable environmental health risks for the populace. This demands the awareness and active involvement of all stakeholders at all levels, whose interests and activities impact on the ecohydrological welfare of the city and its environs.

The daily waste generation in the catchment of Addis Ababa is reported to be 0.252 kg per capita per day and 65% (1482 cubic metres per day) of municipal waste is collected [87]. The balance of 35% adds to the accumulating waste visible in the clogging or poisoning of riverine systems and the uncontrolled disposal of refuse on public open spaces (Fig. 8 a and b.).

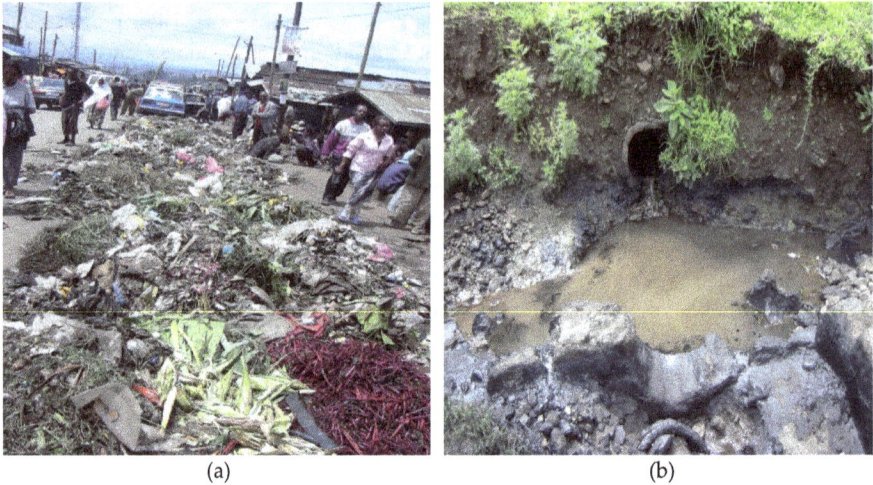

(a)   (b)

Figure 8. a. Refuse dumped on street pavement; b. Photograph 12: Industrial toxic waste water in a nearby stream

The uncollected solid waste has often the common cause of blocked drainages which exacerbates the risk of flooding and vector borne diseases thereby reducing the aesthetic value of green areas and the riverine systems that serve as the sewerage conduits within the densely built up slum areas of Addis Ababa. Climate change is augmenting the occurrence and spatial distribution of waterborne diseases such as malaria and rift valley fever. The incidence of cholera and other pathogens is on the increase due to poor sanitation, flooding of rivers and the extreme droughts [88].

There are increasing public health concerns about the excessive concentration of heavy metal ions (pH, Mn, Cr, Ni), coliform and pathogen pollution in the surface and groundwater of the Akaki River system [89]. Apart from supplying water to the city and serving as a natural sewage reserve, the easily accessible river water is the mainstay of the irrigated agricultural produce on which the city depends for its main fresh food supplies. Prabu (2009) investigated the toxic heavy metal contamination of Akaki River by measuring the concentration levels of seven selected metals Cd, Cr, Cu, Zn, Mn, Fe and Ni in associated with the current morbidity threats to human health and biosystems [90]. His results showed that the concentration levels of these metals in the river exceeded the regulatory limits set by the country's Environmental Protection Authority (EPA 2003). The laboratory tests carried out revealed varying proportions of the presence of all the selected heavy metals in the river – Cd (40%), Cr (64), Cu (15%), Zn (13%), Mn (4%), Fe (2%) and Ni (2%) [91]. However, the total concentration of Fe was two-fold higher than the limit set by the EPA, while Mn concentration was seven fold higher than the limit. The presence of toxic heavy metal concentration in the Akaki River is due to the accumulation of both toxic liquid and solid waste generated by industrial, municipal and domestic activities in the neigbourhoods.

Arguably, as their only accessible source of livelihood, some households draw the polluted waste water from the Akaki River for washing and irrigating their market gardens. Fisseha (2004) observed that 40% of the vegetable supplied to consumers in Addis Ababa city and animal feed comes from the fields directly irrigated by waste water drawn from the polluted rivers during the heavy rainy season or in the dry season when the rivers manifest very low levels of water [92].

Our case study established that the inefficient solid waste management by the municipality has given rise to the accumulation of waste on open lands, in the open drainage system and in the vicinity of many households, causing a nuisance and foul-smelling pools, environmental pollution through leaches from piles (water and soil) and the burning of waste (air pollution), the clogging of drains, and the possible spread of anthropogenic diseases. Kuma (2004) testifies that unattended piles of waste provide a fertile breeding ground for disease carrying insects (mosquitoes, house flies) and rats [93]. The management of solid waste demands a sector-wide approach encompassing the cooperation and integration of government sector agencies, nongovernmental organizations, industry and the community based organizations in addressing the challenges. The importance of inclusive practices of good governance in prioritizing the efficient delivery of waste disposal and management services need to encourage an equitable provision and efficient use of sanitation facilities by all domestic users, industry and public institutions. A leading role in determining policies and projects should be given to local community representatives – especially of the disadvantaged domestic households.

9. Conclusions

Although Addis Ababa is endowed with abundant water resources, surface and ground water, the current health status and levels of safe water supply to sustain basic household hygiene and the public health of most residents, are unsatisfactory by world standards [94].

Arguably, the rapid population growth and expansion of Addis Ababa – mainly due to the in-migration of people coming from all corners of the country in search of better employment opportunities and urban services - pose the city with many challenges of waste management. Some studies have estimated that the rural population migrating to the primate city accounts for 40% of its annual growth [95]. However, coupled with its natural population growth, Addis Ababa happens to be one of the fastest growing cities in Africa. The combined pressures of urban poverty, widespread unemployment and affordable housing shortages are considered responsible for the continued loss of healthy aquatic reserves and the enduring environmental health to nearly 80% of the city' population of 4 million people, mainly housed in the slums and informal settlements. By and large, the emergence and episodic transformation of the city to its present mixed urban-rural zones attributes to the socialization of all land resources by the Derg regime that denied any form of private land ownership. The absence of formal city planning to contain the imploding city population has led to the proliferating informal settlements that now dominate the built environment of today's Addis Ababa. The ecological footprint of informal settlements in the interstitial spaces of the city continues unabated as it eats into the once pristine biological reserves of available open space. The impact of all range of anthropogenic practices including the lack of proper sanitation and ill-disposal of liquid and solid waste on water resources and the health of residents embed themselves in the disturbances wrought by the failure of striking a balance between the consumption of ecosystem services (land, vegetation, water, biota, air) and conserving them in their natural state.

The chapter reviews the empirical findings of recent case studies on the impacts of the inadequacy of sanitation facilities, the related impacts of anthropogenic practices on public health and the degraded ecosystems of the rapidly transforming urban centres of Ethiopia on the basis of our exploratory case study on the slum areas of Addis Ababa. The majority (over 60%) residents in Addis Ababa live in slum settlements featuring poorly constructed houses, poor sanitary conditions, lack of all services (power, running water, and garbage collection), and lack of legal tenure on such residential dwellings. Public services and infrastructure such as running piped water and access roads visibly lack in most of these slum settlements – most of them comprise housing structures called *chekabit* (mixed mud and pole) are virtually overcrowded informal villages exhibiting pitiable living conditions. The inadequate provision and structural weaknesses in the management of sanitation and drainage facilities in the slums and informal settlements literally force residents to get by through resort to urinating and defecating in public open spaces, on river banks and in the interstitial spaces – usually - hidden from public view.

The city residents suffer from the prevalence of pathogens mainly because of the paucity of basic infrastructure and services such as sewers, drains, or services to collect solid and liquid wastes and safely dispose of them. Table 1 is indicative of the high incidences of water related diseases in Addis Ababa attributing to inadequate infrastructure and poor waste management practices steeped in the "throw-away" culture of the citizenry. The diseases include diarrhoea, dysentery, typhoid and intestinal parasites. These pathogens are a cause of many debilitating and endemic diseases that mainly afflict the urban poor households - hidden from public view in the backyard slums of the city. Invariably, the environmental health concerns of the home merge into or are part of the wider problems afflicting the neighbourhoods. The ecological footprint impacting on the riverine ecosystems illustrates the improper siting of the slum structures on dangerous precipices along stream banks. Not surprisingly, squatters often select land that is likely not to be demanded for any other use in order to minimize the possibility of eviction. Such sites are likely to be dangerous or unhealthy. They include hillsides, flood plains, and polluted land sites. The dendritic layout patterns of slum settlements in the city visibly lack in terms of passable access roads thus complicating the regular collection and transportation of domestic garbage. In the rainy season, the walkways turn muddy imposing the environmental health risks of the raw sewage flowing into nearby water bodies and inappropriately sited homesteads.

Our exploratory study confirmed that the deteriorating liquid and solid waste management situation in Addis Ababa imperils the health of the majority poor residents, most of them live in the most densely populated areas of the inner-city. Their uncontrolled domestic waste management practices in turn threaten the ecohydrological sustainability of the catchment basin of Little and Greater Akaki Rivers on which the city solely depends for its current and future growth. The findings of the case study established that there is a small sewerage network of the city, serving only 3% of the population. Most households (about 75%) use pit latrines discharging waste into open drains; about 15% have flush toilets and septic tanks; these likewise often discharging to open drains; a significant minority (about 5%) resorts to open defecation mainly along the stream banks. Such human practices have been construed to be the major culprits responsible for the contamination of water detrimental to the quality of soil and ground water resources through drainage and leaching. The concern about water quality is not out of place. What is needed, perhaps, is a paradigm shift by ecohydrologists that does not only recognize the intertwined ecological processes in urban catchments, but equally, the significance of mutually benefiting waste management practices in the planning and management of the diminishing ecohydrological reserves through predicting the influences of human activities on urban aquatic livelihoods.

The study validates the central argument in recent studies that the anthropogenic threats to groundwater and riverine systems are most prevalent in the overcrowded central areas of Addis Ababa due, in part, to the poorly serviced with basic sanitary facilities [96; 97]. In many cases, refuse collection is restricted to high income residential areas. There are no regular collections of solid wastes in the slum compounds. The uncollected refuse soon attracts rodents, flies and other vermin. The attendant threats have largely been blamed on the fragmented approach to sanitation and liquid management in Addis Ababa. This scenario has resulted in the accumulation of waste on open lands, in open drains, the living area of many people and the eutrophication of the city's traditional water sources and spread of water borne diseases (Tables 1 and 2). The study contends that the present system of solid and liquid waste management in Addis Ababa relies entirely on the municipality for the provision of the full range of waste collection and disposal services. This is proving to be a formidable task, and except for the privileged areas and institutions, the services offered have been recognized – at best - as largely inadequate. Even in the privileged or affluent areas of the city, refuse accumulates for weeks unattended. The current top-down approach arguably neglects the many constituent activities and actors of waste management in addressing the array of problems on a sector-wide basis.

Throughout this chapter, there have been frequent references to the need to direct multi-disciplinary energies in addressing the perennial challenges of waste management by examining the interactions between poor anthropogenic practices and the threats to both residents and the ecosystem services on which their livelihoods depend. Excessive preoccupation within individual disciplines, espoused in fragmented approaches to managing waste, lead to the neglect of the very challenges that desperately need attention as they unfold - in time and space. Clearly, the new approach will need new kinds of policy conceptualizations require the processes of collaboration and active involvement of all stakeholders that must be developed and improved. At present, while waste management in municipalities is the sole responsibility of government agencies, if liberty of private action is to be retained, constructive collaboration can safeguard against excessive public control of development especially where the institutional capacity is limited. The roles and perceptions of both institutions and individuals in managing the accumulating waste quantities on open spaces and aquatic habitats of the city need to change in order to move out of the "waste management problem fix" towards adopting and implementing strategies that seek to protect and conserve the deteriorating urban biosphere reserves. These biosphere reserves include a variety of environmental, biological, economic and cultural situations ranging from largely undisturbed regions to cities [98].

Although the implementation of urban biosphere reserves (UBRs) is not widespread towards fulfilling sustainable development in practice, they are still spatially influential, open to innovative technologies and remedies for the outputs of the existing unsustainable urban living experiences [99]. There is a greater consensus on the importance of integration of bio-diversity with urban planning [100]. Apart from protecting biological diversity in and around urban areas, Azime Tezer (2005) advocates that the conservation of biodiversity will add more value to urban life through public public awareness of the economic benefits to be gained. Such an initiative primarily envisages a multi-disciplinary framework of watershed management comprising metropolitan administration, local governments, non-governmental organizations, research institutions and community representatives. An UBR can be part of a city, or surround a city, but not necessarily designated as a whole city. The reality urges greater attention to the reconciliation of development pressures with ecological units and biodiversity in urban areas.

Such strategies may need to consider the problem of multiple-deprivation, for, despite the existing range of social services and facilities, there appear large sections of urban residents whose relative standard of environmental health is deteriorating rapidly. An illustrative example, though not typical, is the inadequate father fails to meet the needs of a large family, who cannot a paying job and cannot afford any kind of descent home, who is unregistered by the municipality and ultimately degenerates into squatting on the precipice of a stream bank in the centre of the city. Hence the approach to many of society's problems need necessarily be synoptic in discipline, scale and agency. Our environment consists not only of density, and the accumulating waste in aquatic habitats but also of education, religion, social services, police and the planning process itself. Eversley (1978) rightly argued that "there is no such thing as an impersonal threat to the ecosystem: only our ignorant mishandling of the situation, and a lack of belief in the capacity of human dignity to cope with the new dangers" [101] constitutes the major threat.

Author details

Elias Mazhindu

Ethiopian Civil Service University, Addis Ababa, Ethiopia

Trynos Gumbo
African Doctoral Academy, Stellenbosch University, South Africa

Tendayi Gondo
Department of Urban and Regional Planning, University of Venda, Thohoyandou, South Africa

Acknowledgement

The authors are indebted to the Institute of the Polish Academy of Sciences, European Regional Centre for Ecohydrology under the auspices of UNESCO, the Department of Applied Ecology, University of Lodz, Poland, and the Division of Water Resources, Ethiopia – for organising the International Symposium "Ecohydrology for Water Ecosystems and Society in Ethiopia" in Addis Ababa - at which the findings on which this chapter is based were presented. We are grateful to Getawa Desta Tegegne and Semaegzer Aligaz Ligassu, both masters students at the Ethiopian Civil Service University, Addis Ababa, for producing the photographs that depict the current waste disposal practices and aquatic disturbances in Addis Ababa.

10. References

[1] Adesina, A. 2007. Socio-spatial Transformation and the Urban Fringe Landscape in Developing Countries, Paper presented at the United Nations Institute for Environment and Human Society, Munich, Germany.

[2] Satterthwaite, D. 2007. The transition to a predominantly urban world and its underpinnings, Human Settlements Discussion Paper Series, International Institute for Environment and Development, London.

[3] UN-Habitat 1989. Urbanization and Sustainable Development in the Third World: An Unrecognized Global Issue. Nairobi.

[4] Kebedde, G. 2004. Living With Environmental Health Risks: The Case of Ethiopia, Aldershot, England

[5] Ibid.

[6] R. M. K.Silitshena. 1989. Urban Environmental Management and Issues in Africa South of the Sahara.

[7] Mosha , A.C. 1990. "A Review of Sub-national Planning Experieinces in Tanzania" in Helmsing & Wekwete K.H. (eds) *Subnational Planning in Southern and Eastern Africa, Approaches, Finance and Education,* Biling & Sons Worcester, U.K.

[8] Mwafongo, W. K.1991. Rapid urban growth: Implications for urban management in Malawi. Paper presented at the RUPSEA Conference on Urban Management in Southern and Eastern Africa, Lilongwe, Malawi.

[9] Hill , H. 1992. Concrete and clay: Angola's parallel city. Africa South of the Sahara, June.

[10] CSA (Central Statistics Authority) 1999. Urban Inequities Survey. In: *Paper on Housing for the Poor in Addis Ababa.* Addis   Ababa Administration Housing Agency, Addis Ababa.

[11] UN-Habitat. 1989. In R.M.K. Silitshena, Urban environmental management and issues in Africa south of the Sahara. http://archive.unu.edu/unupress/unupbooks/ Acessed 4 April 2012.

[12] Ibid., page 11.

[13] Ibid., page 11 – 12.

[14] Tadesse, T. 2004. Solid Waste Management, University of Gondar in collaboration with Ministry of Health, Ethiopia.

[15] Schuebeler, P. 1996 Conceptual Framework for Municipal Solid Waste Management, *Urban Management and Infrastructure*, Collaborative Programme on Municipal Solid Waste Management in Low-Income Countries, UNDP/UNCHS (Habitat) / World Bank, SDC

[16] Bartone, C., Janis Bernstein, Josef Leitmann, Jochen Eigen and UNCHS-Habitat (1994). Toward Environmental Strategies for Cities: Policy Considerations for Urban Environmental Management in Developing Countries, Washington DC: World Bank.

[17] Zalweski, M., Janauer, G.A., Jolankai, G. (eds). 1997. *Ecohydrology. A new paradigm for the sustainable use of aquatic resources.* UNESCO, Paris, IHP-V Technical Documents in Hydrology 7.

[18] Zalewski, M., McCain, M.E. (eds). 1998. *Ecohydrology. A list of scientific activitiesof IHP-V Projects 2.3/2.4.* UNESCO, Paris, IHP-V Technical Documents in Hydrology 21.

[19] Zalewski, M., Yohannes Zerihun Negussie, Magdalena Urbaniak. 2010. *Ecohydrology for Ethiopia – regulation of water biota interactions for sustainable water resources and ecosystem services for societies.* International Journal of Ecohydrology & Hydrobiology, Vol. 10, No. 2 – 4:102.

[20] Ibid., page 102.

[21] Zalewski M. (ed). 2000. Ecohydrology. *Ecohydrological Engineering.* Special Issue 16: 1 – 197).

[22] Zalewski, M., Yohannes Zerihun Negussie, Magdalena Urbaniak. 2010. *Ecohydrology for Ethiopia – regulation of water biota interactions for sustainable water resources and ecosystem services for societies.* International Journal of Ecohydrology & Hydrobiology, Vol. 10, No. 2 – 4:102.

[23] Ibid., page 102.

[24] Ibid., page 102.

[25] Bartone, C.L and Bernstein, J.D (1993). Improving Municipal Solid Waste Management in Third World Countries. Resources, Conservation and Recycling; 8; 43-45.

[26] Dierig, S. 1999. *Urban Environmental Management in Addis Ababa: Problems, Policies, Perspectives and the Role of NGOs.* Hamburg African Studies, Technical University of Berlin, Berlin.

[27] Amiga A. 2002. *Households' Willingness to Pay for Improved Solid Waste Management,* unpublished thesis, Addis Ababa University.

[28] Kebedde, G. 2004. Living With Environmental Health Risks: The Case of Ethiopia, Aldershot, England

[29] Kuma, A. and Ahmed Ali. 2005. An overview of environmental health status in Ethiopia with particular emphasis on its organization, drinking water and sanitation: A literature survey, *Ethiopian Journal of Health Development*, 19(2): 89 – 103.

[30] Tadesse, T. 2004. Solid Waste Management, University of Gondar in collaboration with Ministry of Health, Ethiopia.

[31] Bihon, A.K. 2008. Urban Inequities Survey. In: *Paper on Housing for the Poor in Addis Ababa*. Addis Ababa Administration Housing Agency, Addis Ababa, pp. 10–12.

[32] Van Rooijen, D. and Taddesse, G. 2009. *Urban sanitation and wastewater treatment in Addis Ababa in the Awash Basin, Ethiopia*. 34th WEDC International Conference, Addis Ababa,Ethiopia,.http://www.wedcknowledge.org/wedcopac/opacreq.dll/fullnf?Search_l ink=AAAA:1744:83419421

[33] Abebe, Z. 2001. Urban renewal in Addis Ababa: a case study of Sheraton and Casanchis Projects, Ethiopian Civil Service College, Addis Ababa, Ethiopia.

[34] Kuma, A. and Ahmed Ali. 2005. An overview of environmental health status in Ethiopia with particular emphasis on its organization, drinking water and sanitation: A literature survey, *Ethiopian Journal of Health Development*, 19(2): 89 – 103.

[35] Tessema, F. 2010. *Overview of Addis Ababa City Solid Waste Management*, Presentation at Workshop on Solid Waste Management in Addis Ababa, Ethiopia.

[36] Alebachew, Z., Legesse, W., Haddis, A., Deboch, B., Biruk, W. 2004. Determination of BOD for liquid waste generated from student cafeteria of Jimma University: A tool for the development of scientific criteria to protect aquatic health in the region. *Ethiopian Journal of Health Science* 14(2): 101 – 110.

[37] Lidgi, E. E. and Nigussie, A. 2007. *Ecohydrology as an important tool for integrated water resources management, IRWD in the Nile Basin.* Paper presented to the First Eco-hydrology Component Workshop, Entebbe, Uganda.

[38] Gondo, T., Trynos Gumbo, Elias Mazhindu, Emaculate Ingwani, Raymond Makhanda. 2010 *Spatial Analysis of solid waste induced ecological hot spots in Ethiopia: where should ecohydrology begin?* International Journal of Ecohydrology and Hydrology, Vol. 10 No. 2 – 4:287 – 295.

[39] Amiga, A. 2002. Households' Willingness to Pay for Improved Solid Waste Management – The Case of Addis Ababa, Masters Thesis, University of Addis Ababa.

[40] Kebedde, G. 2004. Living With Environmental Health Risks: The Case of Ethiopia, Aldershot, England.

[41] Kuma, A. and Ahmed Ali. 2005. An overview of environmental health status in Ethiopia with particular emphasis on its organization, drinking water and sanitation: A literature survey, *Ethiopian Journal of Health Development*, 19(2): 89 – 103.

[42] Kebbede, G. 2004. Living With Environmental Health Risks: The Case of Ethiopia, Aldershot, England.

[43] Kebbede, G. Ibid.

[44] Kuma, A. and Ahmed Ali. 2005. An overview of environmental health status in Ethiopia with particular emphasis on its organization, drinking water and sanitation: A literature survey, *Ethiopian Journal of Health Development*, 19(2): 89 – 103.

[45] Kuma, A. and Ahmed Ali. Ibid.

[46] Ibid.

[47] UN-Habitat. 2003. Urban Inequities Report: Addis Ababa, Nairobi, Kenya.

[48] Ibid.

[49] Kebbede, G. 2004. Living With Environmental Health Risks: The Case of Ethiopia, Aldershot, England.

[50] UN-Habitat. 1989. In R.M.K. Silitshena, Urban environmental management and issues in Africa south of the Sahara. http://archive.unu.edu/unupress/unupbooks/ Acessed 4 April 2012.

[51] Alemayehu, T. 2001 *The impact of uncontrolled waste disposal on water quality in Addis Ababa*, SINET: Ethiopian Journal of Science 24(1): 93 – 104).

[52] Mazhindu, E., Gumbo T. and Gondo, T. 2010. Living with environmental risks – the case of Addis Ababa in *Journal of Ecohydrology & Hydrobiology*, Volume 10, Number 2 – 4, 2010, pages 281 - 286, Versita, Warsaw. Accessed: 11 July, 2011.

[53] Ibid. pages 281 – 286.

[54] Ibid. pages 281 – 286.

[55] Van Rooijen, D. and Taddesse, G. 2009. *Urban sanitation and wastewater treatment in Addis Ababa in the Awash Basin, Ethiopia*. 34th WEDC International Conference, Addis Ababa, Ethiopia,.http://www.wedcknowledge.org/wedcopac/opacreq.dll/fullnf?Search_link=A AAA:1744:83419421

[56] Abebe, Z. 2001. Urban renewal in Addis Ababa: a case study of Sheraton and Casanchis Projects, Ethiopian Civil Service College, Addis Ababa, Ethiopia.

[57] UN-Habitat (2006) The State of the World's Cities Report 2006/2007. Earthscan.

[58] Khurana M.L. 2004 In: Shelter Development Through Cooperatives: A Strategy for Poverty Alleviation and Slum Improvement for Asia and the Pacific Region, Chapter 3.

[59] Collins English Dictionary (2007) Harper Collins Publishers, Great Britain.

[60] UN-Habitat 2003. *Urban Inequities Report, Cities and Citizens Series 2*. UN-Habitat. Addis Ababa

[61] Ibid.

[62] Bihon, A.K. 2008. Urban Inequities Survey. In: *Paper on Housing for the Poor in Addis Ababa*. Addis Ababa Administration Housing Agency, Addis Ababa, pp. 10–12.

[63] Helmut K. and Zein A.Z (eds) 1993. The ecology of health and diseases in Ethiopia. Westview Press, Boulder: USA.

[64] Federal Democratic Republic of Ethiopia, Ministry of Health, Federal Democratic Republic of Ethiopia. 1998. Programme Plan of Action for Health Sector Development Programme, Addis Ababa, Ethiopia. Federal Democratic Republic of Ethiopia. Ministry of Health,2001/02. Planning and Programming Department. Health and health-related indicators. Addis Ababa: Ethiopia Federal Democratic Republic of Ethiopia. Ministry of Health, 2002/03. Planning and Programming Department. Health and health-related indicators. Addis Ababa: Ethiopia

[65] Kumie A. and Ahmed Ali 2005. An overview of the environmental health status in Ethiopia with particular emphasis on its organization, drinking water and sanitation. *A literature survey in Review Article. Ethiopian Journal of Health Development. page 89.* Addis Ababa University Press: Addis Ababa.

[66] Mazhindu E. *et al.*, Ibid.

[67] Kebbede, G 2004. Ibid.

[68] Van Rooijen, D. and Taddesse, G. 2009. *Urban sanitation and wastewater treatment in Addis Ababa in the Awash Basin, Ethiopia.* 34th WEDC International Conference, Addis Ababa,Ethiopia,.http://www.wedcknowledge.org/wedcopac/opacreq.dll/fullnf?Search_link=AAAA:1744:83419421

[69] Alemayehu, T. (2001). The impact of uncontrolled waste disposal on surface water quality in Addis Ababa. SINET: Ethiopian Journal of Science 24(1):93-104

[70] Alem, S. 2007. New Partners for Local Government in Service Delivery: Solid Waste Management in Addis Ababa, Master of Science Thesis in Urban Management, University of Technology, Berlin.

[71] UN-Habitat (2003) 'the Challenge of Slums: Global Report on Human Settlement', London: Earthscan Publishing.

[72] UN-Habitat 2003 Urban Inequities Report: Addis Ababa.

[73] Ibid.

[74] Ibid.

[75] Mazhindu, E., Trynos Gumbo and Tendayi Gondo 2010. Living with environmental health risks – a case study of Addis Ababa" in *Ecohydrology & Hydrology* Vol. 10, No. 2 – 4, 2010, "Ecohydrology for Water Ecosystems and Society in Ethiopia.

[76] UN-Habitat 2003 Urban Inequities Report: Addis Ababa.

[77] Ibid. page 14.

[78] Alemayehu, T., Waltenigus, S., Tadesse, Y. 2003. *Surface and ground water pollution status in Addis Ababa, Ethiopia.* http//:www.un.urbanwater.net/cities/addababa.html.

[79] UN-Habitat 2003 Urban Inequities Report: Addis Ababa, page 16.

[80] Klundert van de, A. and Lardinois, I., 1995. Community and Private formal and informal) Sector Involvement in Municipal Solid Waste Management in Developing Countries, A Paper Presented at a Workshop organized by the Swiss Development Cooperation (SDC) and Urban Management Programme in Ittingen, Switzerland.

[81] Chekole, F. Z. 2006. Controlling the Informal Sector: Solid Waste Collection and the Addis Administration. 2003 – 2005, MPhl. Thesis, Norwegian University of Science and Technology, Trondheim, Norway.

[82] AACG (Addis Ababa City Government). 2004. Waste Management Collection and Disposal Regulations of the Addis Ababa City Government, Negarit Gazeta of the City Government of Addis Ababa. Berhanena Selam Printing, Addis Ababa.

[83] Chekole, F. Z. 2006. Controlling the Informal Sector: Solid Waste Collection and the Addis Administration. 2003 – 2005, MPhl. Thesis, Norwegian University of Science and Technology, Trondheim, Norway. Page 26.

[84] Zalewski, M., Janauer, G.A., Jolankaj, G. 1997. *Ecohydrology: a new paradigm for the sustainable use of aquatic resources.* Technical Documents in Hydrology No. 7, UNESCO, Paris.

[85] Zaitsev, Y. P. 1992. *Recent changes in the trophic structure of the Black Sea.* Fisheries Oceanography, Vol. 1:180 – 189.

[86] Ryszkowski, L. 2000. The coming change in the environmental protection paradigm. In: Crabbe, P., Holland, A. L., Ryszkowski, L., Westra, L.(eds). *Implementing ecological integrity.* Kluwer Acad. Publishers, Dordrecht, pp. 37 – 56.

[87] Tadesse, T. (2004) Solid Waste Management, University of Gondar in collaboration with Ministry of Health, Ethiopia.

[88] Shongwe S.V. 2009. *The impact of climate change on health in the East, Central and Southern African (ECSA) Region.* Commonwealth Health Ministers' Update.

[89] Prabu, P.C. 2009. *Impact of Heavy Metal Contamination of Akaki River on Soil and Metal Toxicity on Cultivated Vegetable Crops,* Journal of Environmental, Agricultural and Food Chemistry, Vol. 8(9): 818 -827.

[90] Ibid.

[91] Ibid. page 822.

[92] Fisseha, I. 2004. Metals in leafy vegetables grown in Addis Ababa and toxicological implications, *Ethiopian Journal of Health Development* 16 (3), 295–302

[93] Kuma, T. 2004. Accounting for Urban Environment. www.ictp.trieste.it/~eee/Workshops/smr1597/Kuma_1.doc

[94] Ibid. 2004

[95] Ibid. 2004

[96] UN-HABITAT, United Nations Human Settlements Programme. 2008. Addis Ababa Urban Profile. UN-Habitat, Nairobi..

[97] Tadesse, T. 2004. Solid Waste Management, University of Gondar in collaboration with Ministry of Health, Ethiopia.

[98] Tezer, A. 2005. *The Urban Biosphere Reserve (UBR) concept for sustainable use and protection of urban aquatic habitats: case of the Omerli Watershed, Instanbul,* International Journal of Ecohydrology & Hydrobiology, Vol. 5(4): 312.

[99] Ibid., page 313.

[100] Alfsen-Norodom, C. 2004 Urban Biosphere and Society: Partnership of Cities In: Alfsen-Norodom, C., Lane, B.D., Corry, M. (eds) *Urban Biospehere and Society: Partnerships of Cities, Annals of the New York Academy of Sciences, 1023, 1 – 9.*

[101] Eversley, D. E. C. 1978. *"The special case – managing human population growth",* in L.R. Taylor (ed), *The optimum Population for Britain,* London, 1970, Chapter 8.

Chapter 2

Solid Waste Management in African Cities – East Africa

James Okot-Okumu

Additional information is available at the end of the chapter

1. Introduction

This chapter analyses solid waste management trends in East African cities from the colonial time to the present, where the cities have moved from the purely centrally controlled systems monopolised by the urban authorities to the current mixture of both public and private systems in varying combinations, that involve many actors(service providers) serving the different urban communities. The main challenges associated with this transition in solid waste management systems are described and compared among the major cities.

In most developing countries it is the urban authorities that are responsible for waste management. Waste management is one of the most visible urban services whose effectiveness and sustainability serves as an indicator for good local governance, sound municipal management and successful urban reforms. Waste management therefore is a very good indicator of performance of a municipality.

Information for the preparation of this chapter came entirely from publications and reports on waste management in Urban Councils of the East African Community (EAC) Countries of Kenya, Tanzania and Uganda. The chapter will examine the management of solid waste from the source to final disposal and will describe and compare waste management in East African cities. This chapter will also examine E-wastes which is becoming a significant management issue in East African urban centres.

2. Trends in solid waste management in East African cities

Waste management in urban centres of East Africa has for a long time been centralised (Liyala 2011), with the use of imported refuse truck (Rotich *et al.*, 2006; Okot-Okumu & Nyenje 2011) that collect wastes from sources or transfer point and deliver to designated waste dumps. Municipal solid waste management (MSWM) system in East Africa has changed from the colonial days in the 40s, 50s and early 60s when it was efficient because of the lower urban population and adequate resources (Okot-Okumu &Nyenje 2011) to the current status that displays inefficiencies. The centralised waste management system has evolved into the current management mixtures that include decentralised as well as the involvement of the private sector.

The storage, collection, transportation and final treatment/disposal of wastes are reported to have become a major problem in urban centres (ADB 2002; Kaseva & Mbuligwe 2005; Okot-Okumu & Nyenje 2011; Rotich *et al.*, 2006). The composition of wastes generated by the East African urban centres is mainly decomposable organic materials (Table 1) based on the urban community consumption that generates much kitchen wastes, compound wastes and floor sweepings (Oberlin, 2011; Okot-Okumu & Nyenje 2011; Scheinberg, 2011; Simon, 2008; Rotich *et al.*, 2006). This calls for efficient collection system to avoid health, aesthetics and environmental impacts. The global trend of increased use of electrical and electronic goods is also evident in EAC where E-waste is becoming a significant threat to the environment and human health in EAC urban centres (Blaser & Schluep, 2012; NEMA 2010 & UNEP, 2010; Wasswa & Schluep, 2008).

Waste composition (%)	Dar es Salaam	Moshi*	Kampala[#+]	Jinja	Lira	Nairobi*
Biowaste	71	65	77.2	78.6	68.7	65
Paper	9	9	8.3	8	5.5	6
Plastic	9	9	9.5	7.9	6.8	12
Glass	4	3	1.3	0.7	1.9	2
Metal	3	2	0.3	0.5	2.2	1
Others	4	12	3.4	4.3	14.9	14
kg/cap/day	0.4	0.9	0.59	0.55	0.5	0.6
Percent collection	40	61	60	55	43	65
Population	3,070,060	183,520	1,700,850	91,153	107,809	4,000,000
Population paying for collection (% of total population)		35	ND	ND	ND	45

ND= Not determined
+ KCC, 2006; [#]NEMA 2007;*Scheinberg et al., 2010

Table 1. Composition of solid wastes generated in East African urban centres

2.1. Solid waste characteristics, generation rate and household management

Waste generation rates of the urban centres in this study are shown in Table 1. Waste sources are households(residential), commercial premises, markets, institutions, industries and health care facilities (Table 2) as illustrated by the case of Uganda, which is similar to other EAC urban centres (Kaseva & Mbuligwe 2005; Liyala 2011; Oberlin 2011; Rotich *et al.*, 2006; Scheinberg 2011). Residential areas or households are the major contributor of wastes followed by markets and commercial areas respectively (Kaseva & Mbuligwe 2005; Kibwage

Solid waste streams	Contribution in weight %	Waste characteristics Comments	Comments
Domestic (Residential)	52 -80	Major: food wastes. Minor: paper; plastic; textiles; glass; ceramics; ashes; leather; compound wastes	Waste quantity increasing with population increase -E-waste is emerging as significant - Wastes collection by: urban councils; private companies, NGOs and CBOs
Markets[a]	4 -20	Major: vegetable wastes (leaves, stalks), spoilt fruits Minor: damaged packaging materials (e.g., sacks, bags, paper, timber)	Markets in all municipalities - Number increasing - Waste collection: urban councils and private collectors
Commercial (excluding markets)	3.7-8	Major: packaging materials; food wastes; scrap metals Minor: glass, hazardous wastes (e.g. contaminated containers, batteries and cleaning textiles)	Shops, hotels, restaurants, offices, open pavement trading - Mobile open air traders - Increasing business - Increasing waste volumes - E-wastes has become significant Waste collection: urban council and private collectors
Institutional (e.g. Government and private-Ministries, Educational establishments, sports facilities, clubs)	5	Major: food wastes, stationery Minor: packaging (e.g., cardboard, paper, plastics)	- Expanding in numbers with population increase - E-wastes has become significant Waste collection: mainly by private companies.
Industrial (manufacturing)	3	Various types depending on industry (e.g., decomposable wastes from food industries, non-degradable such as broken bottles and plastic containers	Production wastes: by-products and damaged items - Broken bottles: recycled or dumped - E-wastes has become significant - Plastic: recovered, re-used, recycled or dumped - Scrap metals: recycled or dumped - Recycling plants available in the EAC

Healthcare(hospitals, clinics, drug shops)	1	Major: domestic type of wastes Minor: hazardous(e.g., anatomical, contaminated materials, sharps)	Major hospitals treats own hazardous wastes. - Clinics dump with other wastes - Domestic: collected by private companies. E-waste is becoming significant
Others	11-11.4	Examples: street sweepings, public park wastes, construction wastes	collected by: Urban council and private companies

[a] Merchandise for urban markets comes along with enormous amounts of wastes (e.g. leaves, stalks, grass, sacks, and branches) from the countryside. (Source: Okot-Okumu & Nyenje 2011; Oberlin 2011).

Table 2. Solid waste streams and the estimated contribution to the urban waste load

2002; Oberlin 2011; Okot-Okumu & Nyenje 2011). Densely populated urban zones (e.g. slums) have low income households with waste generation estimated between 0.22 and 0.3 kg/cap/day. Solid waste generation by the higher income households is estimated between 0.66 and 0.9 kg/cap/day on average (Kaseva & Mbuligwe 2005; Kibwage 2002; Oberlin 2011; Okot-Okumu & Nyenje 2011; Oyoo et al., 2011; Scheinberg 2011).

Overall waste generation rate for EAC urban centres vary on average between 0.26 (low income) and 0.78 (high income) kg/cap/day (Kaseva & Mbuligwe 2005; Kibwage 2002; Oberlin 2011; Okot-Okumu & Nyenje 2011; Oyoo et al., 2011; Rotich et al., 2006; Scheinberg 2011). Similar waste generation rates have been reported for developing countries of other regions of the world (Achankeng, 2003; Supriyadi et al., 2000; Vidanaarachchi et al., 2006). Low income urban communities generate lower waste volumes because they buy little and are less wasteful in consumption. In contrast the higher income groups have higher disposable income and purchase larger volumes of consumable goods, that have high waste portions and also practice a more wasteful consumption pattern (EWAG, 2008; Okot-Okumu 2008; Scheinberg et al., 2011).This observation is consistent with what has been reported by other authors (Hina Zia & Devadas V, 2007; Scheinberg, 2011; Passarini et al., 2011; Supriyadi et al., 2000). Low income urban community spend most of their disposable income for purchase of food items most part of which are consumed and little disposed, while the higher income groups purchase a variety of goods some with associated wastes in form of non-consumables (e.g. packaging, containers, etc.).Total waste generation by the urban councils generation rate is associated with national GDP per capita as illustrated in Scheinberg, 2011. Therefore developing economies such as countries in Africa and Asia have lower waste generation rates (≤ 1.0 kg/cap/day) compared to developed economies (> 1.5 kg/cap/day) as reported by IPCC 2006 and Scheinberg, 2011.

The quality of the urban council wastes can be illustrated by a study done in Uganda in preparation for the composting project (NEMA 2007) that indicated pH (5.7 – 6.9); moisture content (50–75%); Relative humidity(75-155%);volatile solids (66-79%); decomposable organic carbon (DOC 74-86%). Study (NEMA 2007) done in Uganda indicate methane emission potential from such urban wastes to vary between 0.9 and 4.12 Gg/yr. The high decomposable biowaste contents and the optimal moisture content (for aerobic decomposition) of the solid waste make it suitable for composting (Chakrabatrti *et al.*, 2009; NEMA, 2007; Kumar 2006). Composting is being practiced in more than 11 urban councils of Uganda under the Clean Development Mechanism (CDM) pilot project promoted by the World Bank (NEMA 2007; Kumar 2006). In Dar es Salaam composting was initiated by women CBO (KIWODET) operating in Kinondoni (Oberlin & Sza'nto' 2011). The KIWODET composting project was suspended because of land use pressure and negative consumer attitude. Oberlin & Sza'nto' (2011) argue that even though successful composting can arise from local community capacity, lack of municipal integration and support leaves such technically viable initiatives vulnerable to external factors. Aerobic composting is apart from economic benefits that may accrue are environmentally important because it eliminates GHG emission that would occur during waste decomposition at dumpsites or landfill (NEMA, 2007; Kumar 2006). EAC countries should consider composting as an option for the implementation of an integrated approach to solid waste management.

Bulk density of the waste varies between 180 and 310 kg/m^3 comparable to wastes from other African countries (Palczynski, 2002) that are typical of low income countries. Little is known of waste from urban agriculture that has emerged and this together with poor sanitation in the peri-urban areas (Asomani- Boateng & Haight, 1999; Okot-Okumu, 2008) pose high risk to human health.

Household wastes are stored in bins by the affluent and in sacks, plastic bags, cut jerry cans, cardboard boxes by the low-income households, and a large percentage of domestic waste storage containers (e.g. sacks, polythene bags and boxes) used by the poorer urban community are dumped with the wastes(Figure1). There is no sorting as such, but households separate components of wastes considered of value such as vegetables and food leftover (*for animal feeds – used at source or sold, sometimes given free*), plastic bags (*reuse*), bottles- plastic/glass (*reuse and sale*), tins (*reuse and sale*) and scrap metals (*for sale*) are separated by some households from waste that is usually stored mixed. Sorted/separated wastes are either reutilised at source or sold to itinerant buyers who afterwards sell them to middlemen who supply recycling industries. Waste separation also takes place at transfer stations (e.g. bunkers, skips, road verges,), on transit to the landfill and at the landfill or dump sites.

        (a)   (b)   (c)

Figure 1. Waste containers disposed with wastes (A)- plastic jerry cans in in Soroti, Uganda (B) – sacks, plastic bags and cardboard boxes disposed with waste in Kampala, Uganda *(A&B are photos by author, 2010)* (C)- wastes in plastic bags in Dar es Salaam Tanzania *(source: Simon , 2007).*

2.2. Waste transfer stations

The generated wastes are transported to transfer points (Skips, bunkers, standby trailers, open lots see Figs 2 and 3 mostly by the waste generators (e.g. households, commercial premises, market traders) themselves or hired (informal) labour, before collection by urban council workers or private operators. Industries, large institutions (e.g. educational, hospitals), shopping malls, large markets have their own transfer stations served by skips , bunkers, trailers and other waste containment facilities.

(a)   (b)   (c)

Figure 2. (A) Skip amidst wastes in Lira, Uganda(2006) (B) community managed skip at a collection point in Kamwokya, Kampala Uganda (2005)(*A&B photos by author,*) (C) A skip in Mwanza, Tanzania *(source: Ishengoma 2007)*

(a)   (b)

Figure 3. Waste transfer points/methods: (A) -A bunker in Makerere, Kampala Uganda (*(photos by author, 2009); (B)* - a tractor trailer in Dar es Salaam, Tanzania (source: *Simon, 2007*)

2.3. Waste collection and transportation

Three main methods of wastes collection can be identified as the informal primary or pre-collection phase mainly from households to community collection points (e.g. skips, bunkers or open roadside) mostly by households, hired labour. The secondary phase collection is from community transfer points to final disposal sites or landfills and is mostly by formal institutions like urban councils and private operators. Private operators mostly collect wastes directly from generating sources (door to door). Typical waste management scheme in East African urban centres is illustrated by Figure 4. Private operators collect waste at negotiated fees with the individual clients. Industries and shopping malls in most cases contract private waste collectors to pick wastes from their premises, while community markets and hospitals still rely mainly on urban council collection. Other collection modes take the form of a "summon to bring" system, where a truck is parked at a location and a horn (hooting) summons people to deliver wastes to the truck.

The frequencies of household waste collection vary between low-income and high income groups. The high-income groups dispose waste often 3 times a week that is determined by the frequency of collection by most contractors of 2-3 times a week similar (Kaseva & Mbuligwe 2005; Okot-Okumu & Nyenje 2011).

Much time is spent collection as on waste is manually loaded onto trucks by urban council workers. Percent of waste collected vary between 35 and 68, which is comparable to other urban councils in developing countries (Vidanaarachchi et al., 2006; Palcznki, 2002; Supriyadi et al., 2000; Scheinberg, 2011). The introduction of private operators has increased solid waste collection levels compared when it was dependent entirely on the urban councils (Kaseva & Mbuligwe 2005; Oberlin 2011; Okot-Okumu & Nyanje 2011). However most of these reported collection efforts only apply to wastes that have reached community collection points (Transfer points).

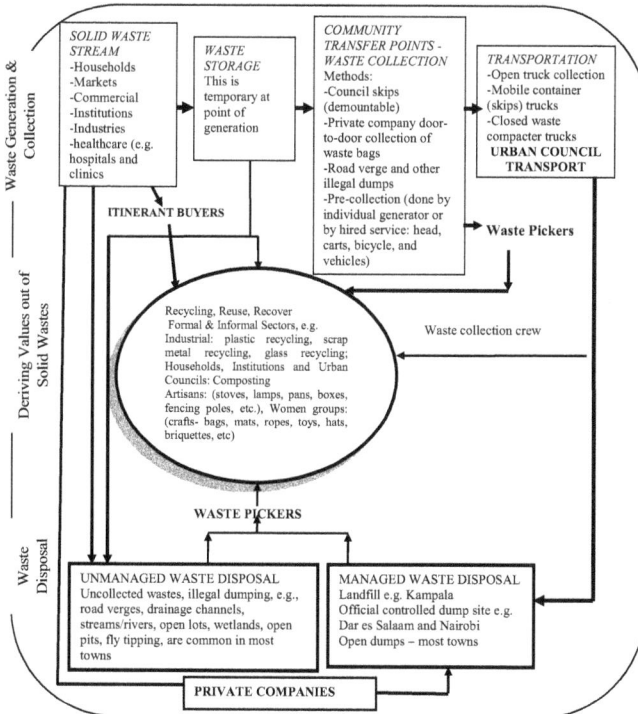

Figure 4. Typical waste management scheme in EAC urban centres (*source: Okot-Okumu & Nyenje 2011 with modifications*)

This means a higher percentage of urban solid waste do not reach the legal disposal points but end up in the environment. Open dumping is the most common waste disposal methods in urban areas (Oberlin 2011; Okot-Okumu & Nyenje 2011). Where skips and waste bunkers are too far the communities dump wastes indiscriminately and some disposal points are often overflowing with uncollected wastes (Figure 2). The use of skips has been terminated in many parts of Kampala since 2002. Skips were found to be linked to lack of cleanliness and most urban residents were dissatisfied with its use (Katusiimeh *pers comm*). The current operating systems in Kampala are open ground disposal and in the remaining skips. Communities without access to transfer stations resort to open disposal methods which include burning, burying, using of wastes as animal feeds and indiscriminate disposal. There is rampant littering caused by the indiscriminate disposal of wastes in storm drainage channels, road verges and open lots.   The carelessly disposed wastes block storm water drains causing floods and also cause health hazards and poor aesthetic.

Institutions like universities, schools, hospitals and business complexes are often served by the private companies, while those not served transport their wastes individually to community collection points. The urban poor receive very low to no waste collection services due to inaccessible roads, unplanned facilities and neglect by the urban councils. Waste collection in East African urban centres is not based on the total amount of waste generated but rather on the level of income of the service area (Kaseva & Mbuligwe 2005; Okot-Okumu & Nyenje 2011). Satisfaction level for waste collection is higher for private operators compared with the urban councils. This can be attributed to the low waste collection frequency causing nuisance to communities. In most urban areas only a small fraction of the wastes generated daily is collected and safely disposed. For example in the cities 45% (Rotich *et al.*, 2006), 43% (Okot-Okumu 2008; Okot-Okumu & Nyenje 2011) and 30% (Oberlin, 2011) are collected for Nairobi, Kampala and Dar es Salaam respectively. Collection of solid wastes is usually concentrated in the city centres and high income neighbourhoods and even then these are irregular.   Common collection and transport modes are covered compressor trucks, open trucks and trailers. Wastes on transit are often uncovered causing littering, odour and aesthetics problems (Fig 5).

(a)   (b)

Figure 5. Waste transport: (A) - overloaded open truck (the sacks contain sorted plastics) in Kampala, Uganda *(photos by author, 2010)* and (B) - a skip conveying truck in Kampala Uganda *(photos by author, 2005)*

NGOs and CBOs are also participating in waste collection, recycling and disposal in the urban councils (Tukahirwa 2011; Oberlin 2011, Okot-Okumu & Nyenje 2011; Liyala 2011). NGOs and CBOs focus mainly on the less privileged urban communities where they serve more than half the population compared to urban councils and private companies combined. CBOs are much more well established in Dar es Salaam and Nairobi compared to Kampala where they focus mainly on the urban poor solid waste management (Tukahirwa 2011).There is also the informal waste collection by waste picker (also known as 'scavengers') who salvage from public bins, disposal sites and along streets; itinerant waste buyers who buy or exchange (barter) items from households such as bottles, plastics and old newsprints for mostly fruits and vegetables (in Kampala). There are also waste buying kiosks for scrap metals and plastics. The informal waste collectors operating in urban councils deal directly with households, markets and other establishments. It is reported (Wang, et al., 2008) that even though these informal practices have positive contribution to urban waste management, they also bring about social problems[1]. Barter systems could be explored by urban councils for the poorer zones of the municipalities to reduce the negative impacts solid wastes and eliminate cheating while at the same time improving on community nutrition (Okot-Okumu & Nyenje 2011). The public-, private- and informal sectors can all work together with the community to improve solid waste management (Chakrabrati 2009; Liyala 2011; Oberlin 2011; Okot-Okumu & Nyenje 2011; Tukahirwa 2011). The community can be involved more effectively through CBOs that are already working with them in waste management projects.

Wastes of value such as plastics, cardboards and scrap metals separated stating at source, at transfer points, on transit and at disposal sites. Some of the separated wastes are sold to artisans and women groups who convert them into goods such as hats, bags, necklaces, baskets, door rugs, mats and seedling cups that are sold to the community as crafts (Fig 6).

Urban councils in many African countries concentrate on waste collection with other aspects of the waste stream having little attention (Achankeng, 2003; Kasenga et al., 2003). Solid waste collection in itself is very much below satisfaction as reported by many authors (Kaseva & Mbuligwe 2005; Okot-Okumu & Nyenje 2011). The collected waste is often disposed indiscriminately without concern for human health impacts and environmental degradation that includes soil, surface and ground water pollution. Indiscriminately dumped wastes in places such as open lots, roadsides and drainage channels is known to cost about 2-3 times per ton the cost of communal collection as noted by Anjum and Deshazo (1996) quoted by Kaseva & Mbuligwe (2005).

2.4. Final disposal of waste

Almost all waste disposal sites in EAC urban councils are in what are considered wasteland like old quarry sites or valleys close to wetlands that are not prioritised for other uses as is

---

[1] Problems of waste pickers are nuisance and social disruption. Itinerant waste buyers sometimes cheat on price, quality and weight or recyclables and in some cases stealing. Informal waste collectors compete for zones allocated to formal collectors causing financial losses to contracted collectors.

Figure 6. Recycled wastes (A)-bags; (B) - necklaces; (C) –tree seedling in plastic cups, (D) - hats and baskets, all from plastic wastes and (E) - bicycle carrier from scrap metals.

often the case in developing countries (Johannessen & Boyer 1999). The disposal sites are therefore in most cases located in environmentally sensitive areas in lowlands areas such as wetlands, forest edge or adjacent to water bodies. They often do not have liners, fences, soil covers and compactors as is in most developing countries (Johannessen & Boyer 1999). In Uganda, Kampala city has upgraded its waste dump at Kitezi to a sanitary landfill (Fig 7). The landfill was funded by the World Bank and has been managed since 1999 by private companies. Though built to standard with a leachate treatment plant, there is some leachate leakage before the treatment plant and this is polluting the surrounding environment with heavy metal (Skoog 2004).

Waste dump sites receive mixed wastes of various origins that include domestic, industrial, medical and commercial wastes. The waste dumps pose real hazard to workers, waste pickers and stray animals that visit the sites. Most of the waste workers do not wear proper protective gears. Major problems from solid waste disposal sites include pollution from leachate, odour, vermin, attraction of scavengers and poor accessibility. Wastes are often dumped at easily accessible points on the way or close to the dumping sites causing serious pollution and aesthetic impacts.

The openness of landfill sites provides free access to waste pickers to sort valuable items for sales. Waste pickers work under no clear control and do not follow any safety and health regulations. Waste collection workers also pick out wastes of value en route to the landfills and sell them to middlemen. The major wastes picked are plastics (e.g. jerry cans, bottles, plates, and basins) and paper and cardboard. Formalisation of waste pickers groups can make them more effective, make them follow health and safety regulations and protect them against exploitation. Achankeng (2003) argues that as the waste pickers get more organised

(a)   (b)

Figure 7. Waste disposal: (A) - The Kitezi landfill in Kampala where waste pickers can be seen at work (B) – the impact of indiscriminate disposal on Nakivubo channel, Kampala (wastes washed from the catchment floating on storm water).

through formalisation of operation conflicts with formal collectors ensues especially regarding areas of operation. Achankeng (2003) reports such conflicts in Cameroon and quotes Kamel (2001) for such conflicts in Cairo. Such conflicts however could be avoided if urban councils formalise all waste operation activities and set clear rules of operation whereby zones of operation and all other requirements are strictly adhered to. This is only possible for waste pickers if organised in formal groups that can be legally registered, monitored and supervised as reported in studies by Mbeng *et al.*, (2009) in Cameroon and Nzeadibe (2009) in Nigeria.

The private sector is to some extent involved in waste recycling of items such as plastics, paper and cardboard. This apart from providing source of livelihood to the waste pickers is helping in removing wastes that could have ended in the environment causing pollution and aesthetics impairment.

2.5. E-wastes in East Africa urban centres

There is no comprehensive data on electronic wastes in EAC urban councils even though it forms a significant component of disposed waste as regards risks to the environment and the urban community. A study by Wasswa & Schluep 2008 indicates that 2,000 tons of e-wastes were generated mainly from computers in Uganda in 2007. There is no formal e-waste management in Uganda and formal recyclers do not exist. For Kenya e-wastes generated were 11,400 tonnes from refrigerators, 2,800 tonnes from TVs, 2,500 tonnes from personal computers, 500 tonnes from printers and 150 tonnes from mobile phones (NEMA & UNEP010). Although Kenya has developed guidelines for e-waste management (2010), e-waste management is still largely handled by the informal sector (*Jua Kali*). Tanzania produces between 2,000 and 3,000 tons of e-waste annually and there is no formal recycling. Feasibility studies were carried out for piloting e-waste management in Tanzania and Uganda (Blaser & Schluep 2012; Wasswa & Schluep 2008). Results indicate difficulties in achieving any profitable venture because of the low quantities of e-wastes generated currently. In all the three countries there is no clear disposal mechanism and large stocks of e-wastes are being held in storage by consumers. However recycling options for parts from e-wastes such as plastics, metals (ferrous, aluminium and copper) do exist and should be explored. Recycling options for printed wiring boards, CRT tubes and hazardous fractions (e.g. PCBs, mercury, batteries) do not exist

3. Solid waste management challenges

All the EAC countries have policy, legal and institutional framework for waste management where urban councils are charged with the tasks to manage urban wastes (Liyala 2011; Oberlin 2011; Okot-Okumu & Nyenje, 2011; Tukahirwa, 2011). The duties and responsibilities are spelt out in a number of pieces of national legislations mainly in the area of public health, environmental management, urban planning and local governance (Kaseva & Mbuligwe 2005; Liyala 2011; Oberlin 2011; Okot-Okumu & Nyenje 2011). The urban councils are responsible for the implementation of these instruments including ordinances and bye-laws. In the (EAC), environmental policymaking remains largely a function of the central government, but implementation of policies and legislation is devolved to the Local Governments (Liyala 2011; Oberlin2011; Okot-Okumu & Nyenje 2011; Tukahirwa 2011). The existing laws on waste management are not being effectively enforced (Liyala 2011; Oberlin 2011; Okot-Okumu & Nyenje 2011; Simon 2007), which may be attributed to inherent weaknesses of the laws themselves. The informal sector and the community therefore operate with little or no regulation at all.

Waste management is poorly financed because it is not a prioritised activity in all urban councils. Funds for the operation of the urban councils are mainly from external sources (over 50 %) like the central government and donors in the form of grants (Liyala, 2011). This means fiscal autonomy has not been realised by the EAC urban councils as observed by Okot-Okumu & Nyenje (2011). The central governments do not adequately cost-evaluate the decentralised environmental management functions implemented by the urban council (Okot-Okumu & Nyenje 2011). National priorities usually differ from environmental management activities causing low remissions in these sectors by the central governments. The study by Liyala (2011), of Kisumu (Kenya), Mwanza (Tanzania) and Jinja (Uganda) clearly illustrates the solid waste management financing dilemma. It is difficult to solve the dilemma because urban council local revenue sources are limited and locally raised revenues are in some cases as low as 3% of the total annual local authority budgets. In Kenya and Tanzania the community pay for waste management, while in Uganda there is unwillingness to pay although in all these countries, Local Governments are by law (Local Government Acts) are given the powers to charge fees for waste services (Liyala 2011). The problem is compounded by the inability by a large percentage or urban community to pay for waste collection services due to low income levels in the EAC region. Therefore households not served by waste collection have developed their own waste management systems. The most common household waste management methods identified are waste burning and backyard burying or indiscriminate open dumping (Liyala 2011; Oberlin, 2011; Okot-Okumu & Nyenje 2011; Simon, 2007). Waste composting is still small-scale and insignificant, often by households and mostly for individual household gardens, while anaerobic biogas production is limited and in the homes or farms of high-income people in peri- urban or rural areas. These are informal setups that are not easy to assess or control. Composting and anaerobic biogas production are therefore often on individual basis and insignificant. Uganda is however piloting composting in eleven urban councils that have scored different levels of success (NEMA 2007; Okot-Okumu & Nyenje, 2011). In Tanzania KIWODET composting activities demonstrated that successful composting initiatives can arise from local capacity (Oberlin & Sza'nto', 2011). The main problems of the composting project are sorting which is not done at source but on delivery of wastes, mixed wastes of all categories increasing health risk to workers and lack of market for the compost. Some authors (Matete & Trois, 2008; Mbeng et al., 2009; Mbuligwe & Kasenga, 2004; Rotich et al., 2006;Wang et al., 2008) have identified recycling, composting and biogas production as feasible options with social, economic and environmental benefits by reducing amounts of waste disposed, saving the environment and generating income for communities the lack of municipal integration and support leaves composting, which is technically viable, to be strongly vulnerable to external factors (Oberlin & Sza'nto', 2011).In East Africa these innovative methods for waste management remain un-researched denying interested individuals among the urban communities information on such projects.

An integrated approach to solid waste management involving a mix of centralised urban council-controlled conventional methods with decentralised innovative decentralised alternatives such as 3Rs; composting and anaerobic biogas production can help to attain sustainability in waste management. This is the preferred approach in the region after the realisation by the EAC governments that the solid waste management monopoly by the urban councils is not succeeding to meet expected results, and the outcome is the acceptance of a holistic approach to solid waste management in the EAC that involves the community, private collectors, CBOs, NGOs and the informal sector working together under a decentralised arrangement (Okot-Okumu & Nyenje, 2011; Liyala, 2011; Tukahirwa et al., 2010; Simon, 2008). Mbeng et al., (2009) identified information as critical for the success of 3Rs, composting, waste prevention and waste minimisation in urban councils since most communities lack vital knowledge for effective implementation of these methods of waste management.

Some authors (Liyala 2011; Oberlin 2011; Tukahirwa 2011) identify common causes for poor waste management services as the inadequate policy and legislation, lack of political will, lack of public commitment, lack of technical capacity, poor financing. A different group of authors think it is seldom technical (Scheinberg 2011) but rather politics, economics or institution (Wilson et al., 2010)

Little investment has been made in MSWM research, resources and human capacity development. The only coordinated major research that has been done on waste management in EA cities is the one under PROVIDE where some of the MSc and PhD research results have already been published (Liyala 2011; Oberlin 2011; Simon 2007) .

The efforts by the Central and Local Governments, international organisations such as International Labour Organisation (ILO), Non-Governmental Organisations (NGO) and

CBOs to train for awareness and capacity building for MSWM are not coordinated and are also discontinuous causing duplication of efforts, therefore have insignificant impacts on target communities. Even though urban councils contract private operators to collect wastes, the urban councils themselves are still the main waste collectors and the combined efforts of the urban councils with the private sector have not yielded the levels of success expected. This is evident by the common scenes of uncollected wastes on roadsides and in drainage channels, streams and wetlands in urban and peri-urban areas.

The problem of MSWM in East Africa is compounded by the rapid urban population growth caused by rural to urban migration overstretching resources. The rising urban population and increasing industrial activities means larger volumes of wastes that pose threat to public health and the environment since they are predominantly decomposable organic (Table 1) and E-wastes are also increasing in the waste stream. Zurbrugg (1999) noted that the problems of MSWM are of immediate importance in many urban areas of the developing world and waste management is known as one of the key issues in urban management aside from water and sanitation. Municipal wastes therefore constitute one of the most crucial health and environmental problem of African urban councils (Achankeng 2003; Adebilu & Okekunle 1989; Asomani-Boateng & Haight 1999; Kaseva & Mbuligwe 2005). It is evident from Kaseva & Mbuligwe (2005) for Tanzania, Rotich et al., (2006) for Kenya and Okot-Okumu & Nyenje (2011) for Uganda that urban areas in East Africa have been experiencing serious solid waste management failures.

The prevailing attitude of the public towards waste collection and disposal or treatment is poor (Liyala 2011; Oberlin 2011). The urban communities generally do not participate in waste management responsibly and this is not helped by the inability of the urban councils to enforce existing waste management laws (Liyala 2011). Political interference caused by personal interests has in some cases obstructed opportunities to implement ordinances or bye-laws. Political interference weakens environmental management institutions and creates a community that is difficult to work with for environmental management (Okot-Okumu & Nyenje 2011).

There are also the negative factors of attitude and culture that have prevented in some cases the very important element of public participation as noted by some authors (Kaseva & Mbuligwe 2005; Palczynski 2002; Rotich et al., 2006; Yhedgo 1995). The low standard of living (poor pay), education (high illiteracy levels) and the economy (low GDP per capita) are influencing factors that cause low levels of willingness to participate in public management matters. The combination of all these factors together with the urban council weaknesses that cause management failures have led to the accumulation of wastes in neighbourhoods leading to environmental degradation and threat of disease epidemics such as cholera, diarrhoea and parasites. Socio-cultural and attitude problems in waste management may be addressed gradually through public education to sensitise the communities, while economic issues can be addressed by providing livelihood opportunities (employment) within the waste management activities.

There is need to explore the opportunities for the 3Rs and composting in urban waste management among urban communities to minimise waste while at the same time providing social (e.g. clean and healthy neighbourhood) and economic (e.g. sale of recycled or recyclable materials) benefits. To successfully adopt sustainable methods of waste management by the communities, Mbeng *et al.*, (2009) suggested making awareness programmes simple and accessible to change the mindset of urban residents to perceive waste as resources (goods) rather than something without value. To address community level waste problems pre-collection/ primary collection needs better organisation and strengthening by communities working together with urban councils and CBOs to chart the most suitable waste minimisation and collection methods. Integrated waste management approach that employs decentralised community based systems involving NGOs/CBOs targeting the peri-urban poor and the more centralised urban council and private operator systems that target the central business areas and the rich and middle class estates should be explored by the urban councils. Such systems can be promoted through community participation and education involving CBOs and the informal sector. There is need for political support for such initiatives of waste management strategies to succeed.

4. Conclusions

The demand for solid waste collection has steadily increased in the East African urban councils as urban population increase with the accompanying expansion of settlements mostly occupied by the peri-urban poor (in informal settlements) that receive little or no waste services at all. The waste collection and disposal levels are low in all urban councils in East Africa resulting in waste piles that cause environmental degradation and health hazard. Waste management is a decentralised function of urban councils but its funding is predominantly external and the urban councils do not prioritise waste management in their annual plans. These have combined to cause poor allocation of resources and ineffective solid waste management by urban councils.

The predominantly conventional waste management methods have failed because they do not effectively address local conditions such as culture, financing system, institutional framework, technical and human capacities, socio- political situation and waste characteristics. There is therefore need for urban councils to explore opportunities for innovative integrated approach for sustainable waste management such as the 3Rs, composting, anaerobic biogas production that involve all stakeholders including the community and the informal sector. The process from planning to implementation should be all inclusive to ensure consensus building for success. The role of the private sector, NGOs, CBOs and the informal sector should be strengthened to minimise waste in the environment while at the same time providing social and economic benefits to communities especially the urban poor. This requires long-term planning by the urban councils that involve all the stakeholders.

Author details

James Okot-Okumu
Department of Environmental Management; School of Forestry Environmental and Geographical Sciences; College of Agricultural and Environmental Sciences; Makerere University Kampala Uganda

Acknowledgement

I acknowledge the urban councils and individual authors whose publications I obtained information from when writing this chapter. I also thank all individuals that gave time for me to discuss with them issues of solid waste management.

5. References

ADB., 2002. Report Prepared for Sustainable Development and Poverty Reduction Unit: Solid Waste Management Options For Africa. Côte dÍvoire

Achankeng, E., 2003. Globalization, Urbanisation and Municipal Solid Waste Management in Africa. African Studies Association of Australia and the Pacific. Conference Proceedings – Africa on a Global Stage.

Adebilu, A.A., & Okenkule, A.A., 1989. Issues on Environmental Sanitation of Lagos mainland Nigeria. The Environmentalist 9(2), 91-100.

Asomani-Boateng, R., & Haight, M., 1999. Reusing organic solid waste in urban agriculture in African Cities: A challenge for urban planners. Urban Agriculture in West Africa. Contributing to Food Security and Urban Sanitation, In Smith. O. (ed.),International Development Research Centre, Ottawa, Canada and Technical Centre for Agricultural and Rural Cooperation ACP-EU, Wageningen, Netherlands, pp. 138-154.

Blaser, F., & Schluep M., 2012. E-waste. Economic Feasibility of e-Waste Treatment in Tanzania Final Version, March 2012. EMPA Switzerland & UNIDO.

Chakrabrati S., Majumder A., & Chakrabrati., S., 2009. Public-community participation in household management in India: An operational approach. *Habitat International*. 33 pp. 125-130.

Hina Zia., & Devadas V., 2007. Municipal solid waste management in Kanpur, India: obstacles and prospects. *Management of Environmental quality: An International Journal*. 18 (1) pp. 89-108

Johannessen, L.M., & Boyer, G., 1999. Observations of Solid Waste Landfills in Developing Countries: Africa, Asia, and Latin America. The International Bank for Reconstruction and Development. The World Bank.Washington D.C.

Kaseva, M.E., & Mbuligwe, S.E., 2005. Appraisal of solid waste collection following private sector involvement in Dar es Salaam. Habitat International 29, 353-366.

Kibwage, J.K., 2002. Integrating Informal Recycling Sector into Solid Waste Management Planning in Nairobi City. PhD Thesis. Maseno University.

Kumar, S.N, 2006. Report on Setting up compost Plants for Municipal Solid wastes in Uganda. EMCBP- II World Bank & National Environment Management Authority, Kampala Uganda.

Liyala C.M., 2011. Modernising Solid Waste Management at Municipal Level: Institutional arrangements in urban centres of East Africa. PhD Thesis. Environmental Policy Series. Wageningen University. The Netherlands.

Matete N., & Trois, C., 2008. Towards Zero waste in Emerging countries – A South African experience. Waste Management. 28, 1480-1492.

Mbeng, L.O., Phillips, P.S., & Fairweather. R., 2009. Developing Sustainable Waste Management Practice: Application of Q Methodology to construct new Strategy Component in Limbe- Cameroon. The Open Waste Management Journal 2: 27-36.

NEMA -National Environment Management Authority., 2007. Clean Development mechanism (CDM)–Uganda solid waste composting project. Analysis Report. - 2006. *State of Environment Report for Uganda 2006/7*. NEMA, Kampala.357pp.

NEMA & UNEP., 2010. Guidelines for E-Waste Management in Kenya. Ministry of Environment and Mineral Resources, Kenya.

Nzeadibe, C., 2009. Solid Waste Reforms and Recycling in Enugu Area, Nigeria. *Habitat International*: 33 pp. 93-99

Oberlin, A.S., 2011. The Role of Households in Solid Waste Management in East Africa Capital Cities. PhD Thesis. Environmental Policy Series. Wageningen University. The Netherlands.

Oberlin, A.S., & Sza´nto´ G. L., 2011. Community level composting in a developing country: case study of KIWODET, Tanzania. *Waste Management & Research*. 29(10) 1071–1077

Okot-Okumu, J., 2008. Solid waste Management in Uganda: Issues Challenges and Opportunities. POVIDE programme Workshop. The Netherlands

Okot-Okumu, J., & Nyenje. R., 2011. "Municipal solid waste management under decentralisation in Uganda." Habitat International 35, pp. 537 543.

Palczynski, J. R., 2002. Study on Solid Waste Management Options for Africa. African Development Bank.

Passarini, F.; Vassura, I.; Monti, F; Morselli, L; & Villani, B., 2011. Indicators of waste management efficiency related to different territorial Conditions. *Waste management*: 32 pp. 785-792.

Rotich, H. K.; Yongsheng, Z; & Jun, D., 2006. Municipal solid waste management challenges in developing countries: Kenyan case study. Waste Management 26 (1), 92-100

Scheinberg A., 2011. Value Added: Modes of Sustainable Recycling in the Modernisation of waste Management Systems. PhD Thesis. Wageningen University. The Netherlands.

Scheinberg A *et al.*, 2011. Assessing Recycling in Low- and middle-income Countries: building on modernised mixtures. *Habitat International* 35 (2) pp. 100198.

Simon A.M., 2008. Analysis of Activities of Community Based Organizations Involved in Solid waste Management, Investigation Modernized Mixtures Approach. The Case of Kinondoni Municipality, Dar es Salaam. MSc Thesis. Wageningen University.

Skoog, K., 2004. Waste management in Kampala, Uganda and the impact of Mpererwe Landfill. Minor field study No.274. Swedish University of Agricultural science.

Supriyadi, S., Kriwoken, L.K., & Birley, I., 2000. Solid waste management solutions for Semarang, Indonesia. Waste Manage Res 18: 557-566

Tukahirwa, J.T., 2011. Civil Society in Urban sanitation and Solid waste Management. PhD Thesis. Wageningen University. The Netherlands.

Vidanaarachchi, C.K., Yuen, S.T.S., & Pilapitiya, S., 2006. Municipal solid waste management in the southern province of Sri Lanka: Problems, issues and challenges. Waste Management 26: 920-930.

Wang, J., Han, L., & Li, S., 2008. The collection system for residential recyclables in communities in Haidian District, Beijing: A possible approach for China recycling. Waste Management 28, 1672-1680.

Wasswa, J., & Schluep, M., 2008. E-Waste Assessment in Uganda. A situational analysis of e-waste management and generation with special emphasis on personal computers. Uganda Cleaner Production Centre, Kampala Uganda and EMPA Switzerland, UNIDO, Microsoft.

Wilson D.C *et al.*, 2010. Comparative Analysis of Solid Waste Management. In: Cities Around the World. Paper Delivered at the UK Solid waste Association, Nov.2010.

Zurbrugg, C., 1999. The challenges of solid waste disposal in developing countries, 2003. Available from: <www.sandec.edu>.

Chapter 3

Solid Waste Management in Malaysia – A Move Towards Sustainability

Jayashree Sreenivasan, Marthandan Govindan,
Malarvizhi Chinnasami and Indrakaran Kadiresu

Additional information is available at the end of the chapter

"Make less, buy less, use less, throw away less"
Akkiko Busch

1. Introduction

Waste management is a crucial area related to the economic status of a country and the lifestyle of its population. Solid waste management can be defined as a discipline associated with the control of generation, storage, collection, transfer and transport, processing and disposal of solid wastes (Tchobanoglous 1993) and in spite of the aggressive economic development in Malaysia, the solid waste management is relatively poor(MMHLG 1988; Nesadurai 1999).The main objective is to improve waste minimization strategy and control. Modern waste management is shifted to a more flexible waste hierarchy concept, also called as 3R (reduce, reuse, recycle) policies (Tanaka 1999; Wilson 2007). The developing Asia counts as the fastest and largest waste generator globally and a closer inspection reveals a mix of general and specific elements of policy dynamics in the evolution and adoption of waste management policies (UNCRD et al. 2009).

Global 3R Initiative aims to promote the "3Rs" (reduce, reuse and recycle) globally in order to build a sound material-cycle society through the effective use of resources and materials and it was agreed upon at the G8 Sea Island Summit as a new G8 initiative and the UN Millennium Development Goal (MDG) aims to ensure environmental sustainability because of the prevalence of unsustainable production and rapid consumption of virgin raw material/ natural resources. It is achievable through effective and efficient 3R programmes which are vital to reverse the trends of environmental unsustainability. 3R initiatives in Asian regions were officially launched at the 3R Ministerial Conference hosted by the Government of Japan in April 2005 (Visvanathan, Adhikari, & Ananth, 2007).

Today, waste and waste management has given rise to many pressing issues (Björklund, 1998; Japan International Cooperation Agency, 2006) such as expensive land prices, strict environmental regulations (Fullerton & Kinnaman, 1995), health and safety issues, improper management of waste disposal sites (Ministry of Housing and Local Government Malaysia, 2005), landfill spaces becoming limited (Bartelings & Sterner, 1999), policy problems (Choe & Fraser, 1999), and the unwillingness of local communities to accept new technologies and facilities in 'their own back yards' (Petts, 1995). Failing in managing solid waste leads to increased operation cost and damaging the environment (Agamuthu, 2001; United Nations Development Programme Malaysia, 2008; Weitz, Thorneloe, Nishtala, Yarkosky, & Zannes, 2002). In Malaysia, waste management and waste minimization is not the sole responsibility of Local authorithies but most government agencies like the Ministry of Housing and Local Government, Ministry of Environment, Ministry Of Health, the various academic institutions and NGOs should work together to achieve this.

2. Integrated solid waste management –Problems and issues

An integrated solid waste management involves a combination of techniques and programs to suit their local needs specifically.In Malaysia, until the late 1960s, city streets were cleaned by the local district health office and the Local Government Act 1976 and the Street, Drainage and Building Act 1974 were passed for public cleansing services and sanitary disposal. Malaysian laws were too general and were far from satisfactory due to lack of resources and faced municipal budget constraint. The budget for waste collection was ranging from 20% to 70%, according to the size of the municipality (Hassan et al. 2000). Dumping of wastes in open fields and rivers are common even until today and a study of waste disposal behaviour in Kuala Lumpur indicated that 31.9% of waste were disposed by open burning, while 6.5% were dumped into the river system (Murad & Siwar 2007).Hence the environmental safety concern in Malaysia was secondary and most municipalities had a tough time in finding new disposal sites as, the existing disposal sites were nearly exhausted (Hassan et al. 2000).

Kuala Lumpur, is on dire need to reduce its dependence on landfills due to its population density and an alternative solution such as incinerator is difficult to implement. Hence managing solid waste in Malaysia is still a big challenge. Malaysia is looking towards innovative solutions to the problems of inadequate and inefficient services provided by local authorities as far as waste management practices are concerned. A waste audit is a formal process which quantifies the amount and types of waste generated. Audits can be performed on office waste, municipal waste, commercial and industrial waste and construction waste through different means like visual waste audits, waste characterization, and desktop audits. Waste audits are a key to establish waste and source reduction programs and should be understood by the Government. Table 1 shows solid waste composition of selected locations in peninsular Malaysia (Wahid 1996). Table 2 shows the predicted results of total solid waste generated (per day and per year) (Nasir 2004).Table 3 shows the prediction of SWG of various sectors (Fauziah 2003).

Waste Composition	Kuala Lumpur	Saha Alam	Petaling Jaya
Garbage	45.7	47.8	36.5
Plastic	9.0	14.0	16.4
Bottles/Glass	3.9	4.3	3.1
Paper/Cardboard	29.9	20.6	27.0
Metals	5.1	6.9	3.9
Fabric	2.1	2.4	3.1
Miscellaneous	4.3	4.0	10.0

Table 1. Solid Waste Composition of Selected Locations in Peninsular Malaysia(Wahid 1996)

Year	Population of K.L. city Millions	MSWG Kg/Cap./day	MSWG Tons/day	MSWG Tons/year
2009	2.43	1.66	4029.85	1470895.25
2011	2.63	1.72	4534.78	1655194.70
2013	2.85	1.79	5102.97	1862584.05
2015	3.08	1.87	5742.35	2095957.75
2017	3.33	1.94	6461.85	2358575.25
2019	3.60	2.02	7271.50	2654097.50
2021	3.90	2.10	8182.59	2986645.35
2023	4.21	2.19	9207.84	3360861.60

Table 2. Prediction of Total MSWG of Kuala Lumpur (Nasir,2004)

Year	Residential (48%)	Street Cleansing (11%)	Commercial (24%)	Institutional (6%)	Construction & Industry (4%)	Landscape (7%)
2009	1934.33	443.28	1025.97	241.79	161.19	282.09
2011	2176.69	498.83	1088.35	272.09	181.39	317.43
2013	2449.42	561.33	1224.71	306.18	204.12	357.21
2015	2756.33	631.66	1378.16	344.54	299.69	401.96
2017	3101.69	710.80	1550.84	387.71	258.47	452.33
2019	3490.32	799.86	1745.16	436.29	290.86	509.00
2021	3927.64	900.09	1963.82	490.96	327.30	572.78
2023	4419.77	1012.86	2209.88	552.47	368.31	644.55

Table 3. Prediction of Sectoral SWG of Kuala Lumpur (Tons/day)(Fauziah 2003)

3. Government initiatives and milestones

Over the past 20 years, a wave of decentralisation has swept the globe, as national governments have handed responsibilities to lower levels of government and if state and provincial governments are given correct incentives, decentralisation can stimulate greater competition and efficiency(Francis 2010). Federal Cabinet as early as 6 September 1995 had decided to privatise responsibilities of the Local Authorities (LA) in 1998. Since 1 January 1997, the solid management responsibility of 48 LA has been privatized to 2 concession companies i.e. Alam Flora for the Central Region and Southern Waste for the Southern Region while the North was under interim regime for a year. Legislation to streamline the strategies and measures in the Strategic Plan were to be enacted. Solid Waste Management was under Local Government Act, 1976; Street, Drainage and Building Act 1974 and now it is under National Solid Waste Management Department and Solid Waste and Public Cleansing Management Corporation Act 2007(www.kpkt.gov.my). 3Rs in Malaysia was first launched in late 1980s and the campaigns were focused mainly on the recycling activities but unfortunately it failed to improve the existing waste management practice. Policy for Integrated Solid Waste Management in Malaysia – 2001, National Strategic Plan for Solid Waste Management in Malaysia – 2005 and Master Plan on National Waste Minimization - 2006 were introduced. According to the report of the Government for the UN Conference on Human Environment "Solid waste collection is satisfactory but the disposal of solid waste is a problem like those in any countries and an organized programme in this direction is needed. The local authorities in many cases are hampered by lack of trained personnel, financial resources, and knowledge."

3.1. Action plan for a beautiful and clean Malaysia

Ministry of Housing and Local Government(MHLG) produced the Action Plan for a Beautiful and Clean Malaysia (ABC) document in 1988 which had outlined the following:

- Local authorities should be strengthened to be able to establish efficient and effective systems of MSWM in their areas.
- A regional approach for MSWM should be encouraged, to improve the economic and technical level.
- All urban centres should prepare and implement MSWM plans extending into the future including periodical revisions.
- All MSW generated in urban and semi-urban areas should be collected and disposed of adequately in such a manner that would not create public health, workers' health and environmental problems and would be technically and financially viable.
- The generator of waste who is supported by the Rural Environmental Programme of the MHLG should dispose of all municipal solid wastes generated in rural areas adequately.
- Reduction of solid waste generation especially that of packaging wastes and household chemical wastes should be encouraged involving the producers and distributors of consumer goods as well as consumers themselves.
- MSW should be treated as a resource and all efforts must be made to recycle and recover most of the materials that are presently burnt and buried.
- MSWM services should be self-financing and an appropriate user charge or any other methods to attain the self-financing objective should be imposed on beneficiaries of the service.

- The private sector should be encouraged to be contractors for MSW collection and disposal services. In addition, the national automobile industries and other related industries should be encouraged to produce locally all the vehicles and the equipment necessary for MSWM.
- The public should be continuously educated on cleanliness and resources recovery through health and environmental education, cleanliness campaigns and strict enforcement of the anti-litter by-laws.
- Land for MSWM disposal should be identified and reserved for the purpose.
- Research and development about MSWM should be strengthened to cope with the ever changing environment(Zaini 2002)

3.2. Solid Waste and Public Cleansing Management Corporation Act 2007

SWMPC Act 2007 was approved by Parliament on 17 July 2007 and gazetted on 30th August 2007 by vesting executive power to the Federal Government to implement solid waste management and public cleansing. The corporation viewed the issue on the overall basis and not merely collection of garbage and construction of dumps and is responsible to monitor, supervise and enforce solid waste management and public cleansing in the country. It also inculcates public awareness for sustainable management of public waste and cleansing and is also responsible for recycling technology.

Department of National Solid Waste Management was created to propose policies, plans, and strategies along with setting standards, specifications and codes of practices and to enforce the law and regulations, set guidelines, monitor and give approval. The Act defines Solid Waste as, any scrap material or other unwanted surplus substance or rejected products arising from the application of any process; any substance required to be disposed of as being broken, worn out, contaminated or otherwise spoiled. The Act focuses on recycling and has a special allocation for separation of wastes at the source. With the staff count of 900 at 52 district and state offices nationwide, the corporation is optimistic towards making Malaysia a clean country in line with its vision with the support of the citizens.

Improper disposal of household hazardous wastes like pouring down the drain, on the ground, into storm sewers, or putting them out with the trash can pollute the environment and pose a severe threat to mankind.Services of SWM are separation, storage, collection, transportation, transfer, processing, recycling, treatment and disposal of controlled solid waste which are classified into 8 categories namely commercial, construction, household, industrial, institutional, imported, public and others which can be prescribed from time to time. The act provides power for Federal Government to enter into agreement with any person to undertake, manage, operate and carry out solid waste management services or public cleansing and to establish PSP Tribunal. The key aspects are:

- Local Authority will not be responsible on SWMPC.
- Local Authority staff to be given options to join concession companies.
- Integrated system of solid waste management – concessionaires vs others
- Priorities to 3R

3.2.1. Reasons for federal take over

- Lack of human and financial resources to manage solid waste and public cleansing
- Integrated system and holistic approach for solid waste management
- Interim Privatisation Period Too long – difficult to secure loans
- Environmental Degradation

3.3. 3rd outline perspective plan (2001-2010)

The government considered the adoption of a comprehensive waste management policy including the installation of incinerators for efficient disposal of waste and to formulate strategies for waste reduction, reuse and recycling. 3Rs were re-launched in 2001 by Ministry of Housing and Local Government (MHLG) and the current recycling rate is 5%.

3.4. 8th Malaysian plan (2001-2005)

- "The adoption of a comprehensive waste management policy to address the issues of waste reduction, reuse and recycling;"
- The conduct of "relevant studies and demonstration projects to ascertain the viability and the acceptability of a waste recycling industry"; the introduction by local authorities of "various initiatives and appropriate economic approaches such as incentives and collection charges to reduce the amount of household waste;" and
- "A clearing house mechanism be established to facilitate industrial symbiosis, whereby one industry's waste could be another's resource." (8thMP:550)

3.5. 9th Malaysia plan

National Strategic Plan for Solid Waste Management was implemented and it upgraded the unsanitary landfills and constructed the construction of new sanitary landfills and transfer stations with integrated material recovery facilities. It also aimed at establishing a comprehensive, integrated, cost-effective, sustainable and socially acceptable SWM based on waste management hierarchy that give priority to waste reduction through 3R, intermediate treatment and final disposal by providing comprehensive, standardized and efficient quality services. They also aimed at establishing legal, regulation and institutional bodies and adopt a much environmentally friendly, cost-effective, proven SWM technology (9th MP).

3.6. 10thMalaysia plan

The Ministry of Natural Resources and Environment is to develop an environmental performance index(EPI) to gauge the environmental management performance of every state in collaboration with Universiti Teknologi Malaysia (UTM) under the 10th Malaysia Plan (2011-2015). The cabinet has agreed to the ministry's proposal which also had the support of the various federal agencies. Malaysia is at 54th position among 163 countries worldwide under the Global EPI 2010 based on quantitative data obtained from the World

Health Oganisation, United Nations Global Environmental Monitoring System, government agencies, NGOs and academia. Solid waste is one of the three major environmental problems in Malaysia. It plays a significant role in the ability of Nature to sustain life within its capacity. Currently, over 23,000 tonnes of waste is produced each day in Malaysia. However, this amount is expected to rise to 30,000 tonnes by the year 2020. The amount of waste generated continues to increase due to the increasing population and development, and only less than 5% of the waste is being recycled.

Despite the massive amount and complexity of waste produced, the standards of waste management in Malaysia are still poor. These include outdated and poor documentation of waste generation rates and its composition, inefficient storage and collection systems, disposal of municipal wastes with toxic and hazardous waste, indiscriminate disposal or dumping of wastes and inefficient utilization of disposal site space. Rivers represent the lease of life which pulses through the earth. It is a finite and only source of water.

In Malaysia, there are almost 1800 rivers. Sadly, more than half of these rivers have been polluted and destroyed. Improper solid waste management contributes greatly to river pollution. Improper solid waste management (SWM) also contributes to climate change – decomposing waste produces methane and production of new products to meet demand emits greenhouse gases and utilizes natural resources(10th MP).

4. Management strategies

The strategies were formulated for immediate Safe closure of 16 landfills in critical areas; and to upgrade non sanitary landfills and to build new sanitary landfills and incinerators. It also concentrates on enhancing quality of services through Key Performance Indicators, monitoring and quality control, courses and training and to improve delivery system – inventories and database, clear guidelines and regulations. They have given more emphasis to bring public awareness and information dissemination through dialogue, seminar and mass media.

The climatic changes have made garbage disposal dumps as the only method for efficient garbage disposal in Malaysia. Incinerator is unable to meet the disposal and government has upgraded 30 of the 175 existing waste disposal dumps into sanitary facilities by 2010. The implementation of solid waste management strategy based on 'waste hierarchy' is practised by emphasising reuse and improving the quality of products that can be recycled. Government promotes the private sector to invest in green technology in order to boost the efficiency of environmental-friendly products.

At the United Nations Organisation Conference On Climate Changes in Copenhagen, Denmark (COP 15), Prime Minister Datuk Seri Najib Tun Razak stated Malaysia's commitment to cut the percentage of carbon dioxide emissions by 40 per cent by the year 2020 with the help of developed nations. The government has started a pilot project on waste separation in Putrajaya in order to create public awareness on recycling and the joint-venture effort by Solid Waste Management Department, Putrajaya Corporation, Alam Flora and Konsortium SSI-Schaefer is aiming at reducing 40 per cent of the volume of garbage sent for disposal. Every household was provided with two garbage bins, one for organic waste and the other for non-organic waste that can be recycled and this facilitates the respondents to recycle and reduce the amount of rubbish sent to disposal sites. The organic waste can be turned into compost and can be used for other purposes and lengthening garbage disposal lifespan (Syed, 2009).

5. Waste minimization in Malaysia (1995 – present)

The ever increasing per capita waste generation gave rise to a doubt on whether disposal is a sustainable solution and hence an alternative thinking based on the principles of waste hierarchy 3R, became more popular as a policy goal. As a policy objective, the waste minimization goal requires socialization of the 3R idea on a larger scale and this urged the government to focus on waste hierarchy. Information campaigns were staged to promote 3R aiming to increase awareness and to change attitude and behaviour. Malaysia prefers a State-led approach to waste management. Waste minimisation usually requires knowledge about the production process, cradle-to-grave analysis (the tracking of materials from their extraction to their return to earth) and detailed knowledge of the composition of the waste. The two effective ways for waste minimization are through firms' production system and technical changes and through a regulatory system that may finance the modification of internal organization .

5.1. Waste minimization, resource recovery and climate benefits

In Malaysia, waste minimization programs cannot be carried out effectively without a reliable data on waste composition and generation (Hassan et al. 2000)1. The amount of Malaysian solid waste being separated at source for recycling purposes was less than 2% in the year 1992 but the senior government officials believe that the actual rate could be as high as 15%. The 'National Recycling Program' was initiated in 2000 and in 2005, Malaysia released the 'National Strategic Plan for Solid Waste Management (2000-2020)' and waste minimization is recognized as one of the priorities. Article 102 of the Act stipulates that the government can place responsibility for the collection of products on the manufacturer, assembler, importer, or dealer (Pedersen 2008).

5.2. Waste minimization hierarchy

It is everyone's legal and moral responsibility to minimize the amount of waste produced and to dispose waste in a fashion that has the least impact on the environment. The aim of a Waste Management Hierarchy is to minimize the amount of waste from entering the landfill/dump sites. Three top initiatives in the waste management hierarchy is the 3Rs initiative, i.e. Reduce, Reuse and Recycle. To cultivate a 3R culture in a society, it is important to train groups of people by creating an awareness programme towards implementing 3Rs initiative (Hashim,2011). The 3Rs principle helps to improve waste management system and to reduce human ecological footprint. It paves way to improve the economic activities, tend to reduce environmental impacts from waste disposal and prevents the loss of resources and lengthen the operating lifespan of landfills.3Rs is more successful in developed countries than among the developing nations (Agamuthu *et al.*, 2001).

The hierarchy of MSWM is an internationally accepted and practised concept in many countries throughout the world especially in developed countries. For example, study by Cooper (1996) and Clarke (1993) indicates that the concept is used as a guideline for planning modern MSWM facilities. Under full privatisation or concession period, contractors will roughly try to match the hierarchy of MSWM starting with waste minimisation, waste separation and recycling, waste processing such as incineration and composting and finally disposal to the landfill. This integrated strategy requires participation at all levels: government, industries, public and the waste management concessionaires (Zain 2002).

'Waste hierarchy' is being established to help the government manage their waste according to a sustainable agenda. Waste management hierarchy is 'a concept that promotes a cyclical approach to waste management' (Challenger, 2007). The main objective of the waste management hierarchy is to minimize the environmental effects of waste disposal (Rasmussen et al., 2005; Wolf, 1988). This hierarchy is used as a main framework to develop waste management policies. Waste hierarchy which has been developed in the 1970s (Challenger, 2007; Rasmussen et al., 2005), was placed in the following order (Challenger, 2007; Kirkpatrick, 1993; Rasmussen et al., 2005): waste minimization/prevention/reduction, reusing, recycling, composting, incineration and disposal. Barr (2007) defines waste hierarchy order as a *waste management behaviour* which relates to recycling, reusing and reduction. Refer figure 1 for solid waste minimization. Table 4 depicts the Goal-attitude-outcomes of waste minimization hierarchy.

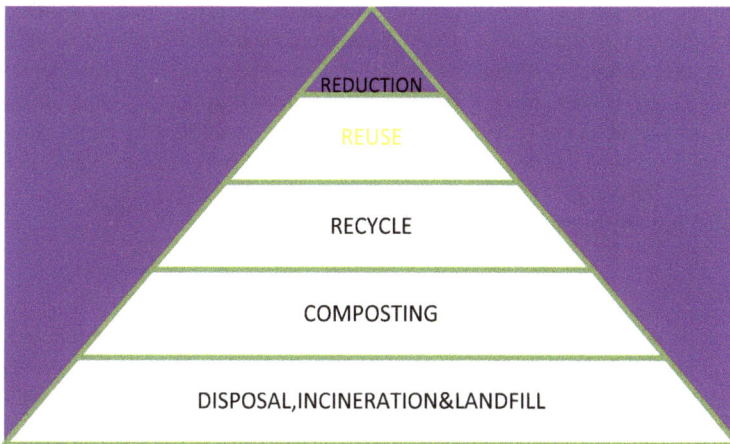

Figure 1. Waste minimisation hierarchy

Goal	Attribute	Outcome
Reduce	Preventive	Most Desirable
Reuse	Predominantly ameliorative, part preventive	
Recycle	Predominantly ameliorative, part preventive	
Treatment	Predominantly assimilative, partially ameliorative	
Disposal	Assimilative	Least Desirable

Table 4. Waste Minimization Hierarchy (Gertsakis & Lewis, 2003)

Waste management hierarchy is undergoing several changes. Incineration, which has been in the hierarchy in the first stage of the evolution, has being criticized due to the cost (Rasmussen et al., 2005) and impact to the environment (Connett & Sheehan, 2001). Therefore, in the recent hierarchy, incineration has been pulled out from the hierarchy and replaced by treatment (Gertsakis & Lewis, 2003); or thermal treatment (Sarifah Yaacob, 2009); or recovery (Pongrácz, Phillips, & Keiski, 2004); or waste to energy (Ministry of Housing and Local Government Malaysia, 2005).It is essential to start by educating people with knowledge. It reduces the amount of wastes through the following steps:

5.2.1. Source reduction/Waste reduction

It is also known as waste prevention, means reducing waste at the source. It can take many different forms, including reusing or donating items, buying in bulk, reducing packaging, redesigning products, and reducing toxicity. Source reduction is also important in manufacturing and can save natural resources, conserve energy, reduce pollution, reduce the toxicity of our waste; and save money for consumers and businesses alike. It includes any activity that reduces or eliminates the generation of waste. Waste reduction helps to create less waste in the first place - before recycling. Since it avoids recycling, composting, landfilling, and combustion it helps to reduce waste disposal and handling costs.Waste reduction can be achieved at several levels, such as reduction of per capita waste generation through public education and government policy initiatives. Separation of recyclable materials can be very useful.

5.2.2. Reuse

It is defined as re-employment of materials to be used in the same application or to be used in lower grade application. To **reuse** is to use an item more than once. This includes conventional reuse where the item is used again for the same function, and new-life reuse where it is used for a different function. In broader economic terms, reuse offers quality products to people and organizations with limited means, while generating jobs and business activity that contribute to the economy (wikipedia).

5.2.3. Recycling

It includes using a waste material for another purpose, treating and reusing it in the same process. Recycling is a series of activities that includes the collection of used, reused, or unused items that would otherwise be considered waste; sorting and processing the recyclable products into raw materials; and remanufacturing the recycled raw materials into new products. Consumers provide the last link in recycling by purchasing products made from recycled content. Recycling also can include composting of food scraps, yard trimmings, and other organic materials. Recycling prevents the emission of many greenhouse gases and water pollutants, saves energy, supplies valuable raw materials to industry, creates jobs, stimulates the development of greener technologies, conserves resources for our children's future, and reduces the need for new landfills and combustors.

Recycling includes the reuse or recovery of in-process materials or materials generated as by-products that can be processed further on. It also improve production efficiency, profits, good neighbor image, product quality and environmental performance. At the moment there is no organised programme for recycling in Malaysia. Efforts are made to come up with their own programme and objective with a single recycling programme with both short-term and long-term perspectives. The short-term measures will mobilise the stakeholders towards active recyclable generators and enhance their participation whereas long-term measures will aim towards an increased diversion of waste for recycling, and a collection system.

Annual allocation for awareness creation among public reached RM70 million (US$18 million). Poster, pamphlets, bulletin, and electronic medium such as television, radio, websites, school busses, Light Rail Transit(LRT) billboards were used and exhibition were conducted and carnivals and seminar were held. Awareness among the public was high but only few were practicing. Recycling facilities were insufficient and inappropriately located. The available facilities were recycling bins, recycling centers, silver boxes, centers, recycling lorry and mobile collection unit (van) and charity recycling boxes. Improvement in recycling practices was possible by bringing awareness, but the task involves huge cost. It can create many job opportunities and calls for a policy that can effectively handle the issue.

5.2.4. Composting

It includes elementary neutralization and composting achieves the microbiological degradation of organic matter to produce an organic product for use in agriculture, etc. Even though the technology of composting MSW is well established, only a few of the refuse composting plants around the world are economically successful. The drawbacks commonly experienced with composting are its high cost and low value of the compost products. Subsequently, composting in Malaysia is not pursued as a solution to MSW disposal problems because the quality of product depends on the waste and hence waste separation is very important. Composting can play a key role in diverting organic waste away from disposal facilities. A compost plant also requires more area. Lack of suitable markets for compost and lack of economies of scale for quantities for the recyclable market is also a major problem. Apart from the above, a landfill disposal will still be required for component of waste that is not suitable for composting.

5.2.5. Disposal/Landfill operation/Incineration

The final product that cannot be recycled, reduced, reused or energy recovered goes to the land field for disposal. In Malaysia landfill is more preferred way of disposal than incinerator because again of the demographic behaviour. Landfilling manage the waste that cannot be reduced or recycled. Disposal decisions depend on the cost, land availability, population characteristics, and proximity to waterbodies. They include a large disposal area which contains numerous smaller cells and the solid waste is deposited in these cells daily, using specially designed bulldozers, and covered with a thin layer of soil or some alternative cover. There is abundant unused land and ex-mining ponds which needs to be refilled for development purposes. Hence Malaysian govt prefers landfills to inceneration. It is cheaper to operate and maintain compared to incenerators. It may be feasible where landfill is scarce and located in a very remote area from the actual MSW generation centre. Modern incineration and flue gas cleaning technologies make waste incineration an environmentally viable method.

5.2.6. Implications of waste minimization

It helps to cut down the operational cost of waste management and act as an indicator to show that the country is a developed nation. It increases the lifespan of landfills and gives greener environment and improves the climatic changes (green technologies) by preventing the pollution.

6. Barriers to 3R implementation

The 2007 law had provided for the 'federalization' of waste management, and this is a devolution of authorities to the lowest possible level, which is highly essential for waste management. 3R implementation in Malaysia was privatized and this had created more problems than a effective solutions (Milne 1992; Sun & Tong 2002). The awareness of public on 3Rs is low, in spite of the Malaysian government's funding for public information campaigns. There is a lack of policy to promote 3Rs and a low public participation. In 1988, the Action Plan for a Beautiful and Clean (ABC) Malaysia was introduced but had only minimal responses from the general public (Hassan et al 2000). In April 2009, the Ministry of Energy, Green Technology and Water was established to handle green technology development in Malaysia, and the government has encouraged the private sector to invest in green technology to promote the usage of more environmentally sound waste management towards facing the changes in the global environment.

7. Tips for effective practice of 3R's

- Avoid purchasing items that are over packed
- Reduce the amount of waste created by the household by shopping smartly

- Reuse items around your home
- Recycle all paper, cardboard, rigid plastic, aluminum, steel cans and glass bottles and jars
- Compost the household's green and organic waste

8. Conclusion

Policy makers in the developing countries must develop the institutional capacity to respond in long-term policy development by integrating the goals and objectives of the state, community, and business. Developing countries may want to combine basic strategy of developing waste management capacity along with the promotion of international cooperation to upgrade it into a effective 3R-based policies. The demographic pattern, the life-style and behaviour of the citizens as well as the foreign workers with different culture and life-styles should be taken into consideration before forming the blue print or the master plan for waste minimization. Since the waste management system is privatised, it will be able to improve the quality of service and it's efficiency. Currently, the privatisation of the waste management system in Malaysia has not reached full privatisation. The system is still in an interim period, and is not running as expected due to some problems arising from the lack of funds, the length of the interim period, and the unavailability of financial resources. Problems faced by consortia have led to the inefficient operation of the waste management system. These problems affect the future planning for waste management in Malaysia, and frustrate the implementation of privatisation. Government incentives, the NGO's contributions through their social community obligations, assistance from Danida from Netherlands, Jica from Japan, UNICCEF, UNEP, WHO and other world bodies should contribute to help developing nations like Malaysia to keep up with their level of waste minimization plans and programmes. These Earth-loving actions will certainly allow us to reduce the negative impact on the earth by adding deeper meaning and joy to our lives.

Author details

Jayashree Sreenivasan, Marthandan Govindan,
Malarvizhi Chinnasami and Indrakaran Kadiresu
Multimedia University, Malaysia

9. References

Agamuthu, P. (2001). *Solid waste: principles and management: with Malaysian case studies*. Insitute of Biological Sciences, University of Malaya, Kuala Lumpur.

Barr, S. (2007). Factors influencing environmental attitudes and behaviors: A UK case study of household waste management. *Environment and behavior, 39*(4), 435.

Bartelings, H., & Sterner, T. (1999). Household waste management in a Swedish municipality: Determinants of waste disposal, recycling and composting,*Environmental and resource economics, 13*(4), 473-491.

Björklund, A. (1998). Environmental systems analysis waste management. *Licentiate thesis, KTH.*

Cashore, B., and M. Howlett. 2007. Punctuating Which Equilibrium? Understanding Thermostatic Policy Dynamics in Pacific Northwest Forestry. American Journal of Political Science 51:532-551.

Challenger, I. (2007). Can we fix it? Lets hope so! Turning the waste management hierarchy the right way up. *WasteMINZ Annual Conference.*

Che Mamat, R., T.L. Chong, Public's Role in Solid Waste Management: In IMPAK, quarterly DOE update on environment, development and sustainability. Issue (4), 2007, pp. 5-7.

Choe, C., & Fraser, I. (1999). An economic analysis of household waste management. *Journal of Environmental Economics and Management, 38*(2), 234-246.

Cooper. J. 1996. Integrated waste management in Vienna. Journal Waste Management., 16-17.

Clarke. M. J. 1993. Integrated municipal solid waste planning and decision-making in New York City: The citizen's alternative plan. Journal Air and Waste Management. 43 (4) : 453-462.

Fauziah, S.H. and P. Agamuthu, A Comparative Study on Selected Landfills in Selangor, Kuala Lumpur Municipal Solid Waste Management, Institute of Biological Sciences, Faculty of Science, University Malaya, 2003.

Francis, H.(2010).Malaysia's federalism-a good thing going to waste,Penang Economic Monthly, November.

Fullerton, D., & Kinnaman, T. C. (1995). Garbage, Recycling, and Illicit Burning or Dumping. *Journal of Environmental Economics and Management, 29*(1), 78-91. doi:10.1006/jeem.1995.1032

Gertsakis,J.,&Lewis,H.(2003).Sustainability and Waste Management Hierarchy.

Hashim, K. S.M, Abdul, H and Mohamed S.R. , Haneesa ,Z. (2011) *Developing conceptual waste minimization awareness model through community based movement: a case study of Green Team, International Islamic University Malaysia.* In: Persidangan Kebangsaan Masyarakat, Ruang dan Alam Sekitar (MATRA 2011) , 16-17 November 2011, Pulau Pinang.

Hassan, M. N., R. Abdul Rahman, L. C. Theng, Z. Zakaria, and M. Awang. 2000. Waste recycling in Malaysia: problems and prospects. Waste Management & Research 18:320-328.

Hezri, A. A., and M. N. Hasan. 2006. Towards sustainable development? The evolution of environmental policy in Malaysia. Natural Resources Forum 30:37-50.

Japan International Cooperation Agency.(2006,July).The study on National Waste Minimization in Malaysia.

Kirkpatrick, N. (1993). Selecting a waste management option using a life-cycle analysis approach. *Packaging Technology and Science, 6*, 159–159.

Kraft, M. E., and N. J. Vig. 1994. Environmental policy from the 1970s to the 1990s: continuity and change in N. J. Vig, and M. E. Kraft, editors. Environmental policy in the 1990s: toward a new agenda. CQ Press, Washington, D.C.

Malaysia, (1971), Government of , Report for the United Nations Conference on Human Environment. April 1. 22p.

Malaysia, Statement of, (1972), UN Conference on Human Environment, Stockholm, 1-5 June. 4p.

Malaysia, (2001). Eighth Malaysia Plan (2001-2005) (8MP). Kuala Lumpur: Percetakan Nasional Berhad. p.550.

Malaysian Ministry of Housing and Local Government (1988), Action plan for the beautiful and Clean Malaysia (ABC Plan) Kuala Lumpur.

Malaysia, (2001). Eighth Malaysia Plan (2001-2005) (8MP). Kuala Lumpur: Percetakan Nasional Berhad. p.550.

Ministry of Housing and Local Government Malaysia. (2005). National Strategic Plan for Solid Waste Management.

Milne, R. S. 1992. Privatization in the ASEAN States: Who Gets What, Why, and With What Effect? Pacific Affairs 65:7-29.

Murad, M.W. and C.Siwar, Waste Management and Recycling Practices of the Urban Poor:A Case Study in KaulaLumpur city,Malaysia.Online Journal of Waste Management Research,ISWA,UK,2006.

Murad, W., and C. Siwar. 2007. Waste management and recycling practices of the urban poor: a case study in Kuala Lumpur city, Malaysia. Waste Management & Research 25:3-13.

Nasir, M.H., Lecture notes. Faculty of Environmental studies, UPM, 2004.

Nesadurai, N., 1999, The 5R Approach to Environmentally Sound Solid Waste, Paper presented in Seminar on "Local Communication and the Environment" organized by EPSM, 24-25th Oct., 1998 Shah's Village Hotel, 1999.

Pedersen, A. 2008. Exploring the clean development mechanism: Malaysian case study. Waste Management & Research 26:111-114.

Petts.J.(1995).Waste management strategy development :a case study of community involvement and consensus-building in Hampshire. Journal of Environmental Planning and Management,38(4),519-536.

Pongracz, E., Phillips, P.S., & Keiski, R.L.(2004).Evolving the Theory of Waste Management-Implications to waste minimization. Proceedings of the Waste minimization and Resources Use Optimization Conference, June 10th(pp.61-7)

Rasmussen, C., Vigso, D., Ackerman, F., Porter, R., Pearce, D., Dijkgraaf, E., &

Sarifah Y.(2009).Solid waste management Hierarchy-application towards the concept of green technology. Green technology on waste management: current knowledge and practices.Presented at Green Technology on waste management:current knowledge and practices,Kaula Lumpur.

Syed. A(2009) Bernama http://www.epa.gov/waste/nonhaz/municipal/hierarchy.htm

Sun, Q., and W. H. S. Tong. 2002. Malaysia Privatization: A Comprehensive Study. Financial Management 31:79-105.

Tanaka, M. 1999. Recent trends in recycling activities and waste management in Japan. Journal of Material Cycles and Waste Management 1:10-16.

Tchobanoglous, G., H. Theisen and S. Vigil, Integrated Solid Waste Management-Engineering Principles and Management Issues, published by Mc Graw- Hill, New York, 1993.

United Nations Development Programme Malaysia. (2008). Malaysia developing a solid waste management: model for penang.

UNCRD, UNEP-RRCAP, and IGES. 2009. National 3R Strategy Development: a progress report on seven countries in Asia from 2005 to 2009. United Nations Centre for Regional Development, Institute for Global Environmental Strategies, Nagoya.

Visvanathan, C., Adhikari, R., & Ananth, A.P. (2007).3R Practices for municipal solid waste management in Asia.

Vollebergh, H. (2005). *Rethinking the Waste Hierarchy*. Environmental Assessment Institute.

Wahid, A.G., Mohd Nasir Hassan, Muda, Domestic and Commercial waste: Present and Future Trends. CAP-SAM National Conference on the State of Malaysian Environment, Penang: RECDAM,1996.

Weitz, K. A., Thorneloe, S. A., Nishtala, S. R., Yarkosky, S., & Zannes, M. (2002). The impact of municipal solid waste management on greenhouse gas emissions in the United States. *Journal of the Air & Waste Management Association*, 52(9),1000-1011.

Wilson, D. C. 2007. Development drivers for waste management. Waste Management & Research 25:198-207.

Wolf,K.(1988).Source reduction and waste management hierarchy. J.AIR POLLUT. CONTROL ASSOC., 38(5),681-686.

Zaini, S., Gerrard, S., A, Jones, P., and Kadaruddin, A. 2002. Policy, challenges and future prospect of solid waste management in Malaysia. Proceeding on International Sustainable Development Research Conference. University of Manchester. 8-9 April 2002. 391-398.

http://irep.iium.edu.my/id/eprint/8655

http://www.fayette.k12.il.us/isbe/science/pdf/sci_8.pdf

http://www.epa.gov/tribalcompliance/wmanagement/wmwastedrill.html

http://www.gecnet.info/index.cfm?&menuid=83

http://www.eria.org/pdf/research/y2009/no10/Ch11_3R.pdf

ttp://www.epa.gov/wastes/nonhaz/municipal/hierarchy.htm

http://www.gecnet.info/index.cfm?&menuid=83

http://www.gecnet.info/index.cfm?&menuid=83

http://irep.iium.edu.my/id/eprint/8655

http://irep.iium.edu.my/id/eprint/8655

http://www.fayette.k12.il.us/isbe/science/pdf/sci_8.pdf

http://www.epa.gov/tribalcompliance/wmanagement/wmwastedrill.html

Household Solid Waste Management in Jakarta, Indonesia: A Socio-Economic Evaluation

Aretha Aprilia, Tetsuo Tezuka and Gert Spaargaren

Additional information is available at the end of the chapter

1. Introduction

Rapid population growth in Jakarta has posed serious challenges. The urban population is expected to increase by 65% by 2030 compared to its level in 2006 (ADB, 2006). This condition presents a serious challenge for the management of waste in urban areas. The major urban centres in Indonesia produce nearly 10 million tonnes of waste annually, and this amount increases by 2 to 4% annually (Ministry of Environment, 2008). Jakarta uses a major landfill located at Bantar Gebang in the suburban town of Bekasi, and the landfill only absorbs approximately 6,000 tonnes per day. As the capacity of the landfill decreases over time, the waste service providers – in particular, the government – are confronted with the need to reorganise the present system for the treatment and management of solid waste. However, the issue of proper waste management is not just a government task but is a shared responsibility that includes the citizens and households of Jakarta, who are the main end-users of waste management facilities and services. When reorganising solid waste management systems, understanding the role of households, their attitudes, their waste handling practices and their interactions with other actors in the waste system is therefore essential (Oosterveer et al, 2010; Oberlin, 2011).

The largest stream of municipal solid waste in Indonesia flows from households followed by traditional markets (Aye and Widjaya, 2006). Solid waste management (SWM) usually relates to both formal and informal sectors. In Indonesia, the formal sector includes municipal agencies and formal businesses, whereas the informal sector consists of individuals, groups and small businesses engaging in activities that are not registered and are not formally regulated. In solid waste activities, the informal sector refers to recycling activities that are conducted by scavengers (itinerant waste pickers) and waste buyers. (Sembiring and Nitivatta, 2010).

Engineers and other decision-makers in the public domain have often found that their technical suggestions have been met with scepticism and even resistance by the public (Corotis, 2009). One of the solutions to dealing with this challenge is to conduct a quick scan, which is a first step toward collecting information about a particular issue in a specific context (Merkx, van der Weijden, Oostveen, van den Besselaar, and Spaapen, 2007). Quick scans may precede or run parallel to economic cost-benefit analyses, thereby making the inputs into the technical design-phase based on real-life conditions much more significant. Quick scans provide information regarding social (non)acceptance rates, and they can be used to determine expected levels of public acceptance. A social quick scan could thereby highlight aspects and dynamics that govern the so-called 'primary phase' of the solid waste management system in which households and informal waste pickers play an important role. Actors in the primary phase are responsible for the generation, collection, storage, and transportation of domestic solid waste. The behaviours and opinions of these actors are key variables that explain the success or failure of MSW policies. These variables, referring to the social dynamics of waste management, have not been discussed in-depth in the solid waste management literature, which is dominated by technical science and supply-side thinking. Therefore, studies focusing on interactions between the real-life conditions of householders (on the one hand) and the providers and regulators of solid waste management services (on the other hand) are crucial for developing and designing future waste management policies.

Prior studies (e.g., Bohma, Folzb, Kinnamanc, and Podolskyd, 2010; Aye and Widjaya, 2006; Sonneson, Bjorklund, Carlsson, and Dalemo, 2000; Reich, 2005) have discussed and estimated the impact of economic factors in domestic solid waste management. These studies have linked household participation and behaviour to economic assessments with the concept of willingness to pay (e.g., Purcell et al, 2010; Bruvoll et al, 2002; and Berglund, 2006), and the studies have discussed the role of economic factors in the feasibility of various socio-technological options and scenarios to be realised. The economic analysis of our study was performed against the background of five predetermined MSW management scenarios. In addition to the baseline scenario involving the use of a landfill, the scenarios proposed for this study include 20% recycling and 25% landfill usage combined with either communal composting (scenario 2), anaerobic digestion (scenario 3), centralised composting (scenario 4), or landfill gas for energy generation (scenario 5). This study also aims to estimate potential revenues from sorted recyclable materials. Moreover, householders' willingness to pay for other people to sort their waste is analysed under the assumption that the government authorities demand at-source waste sorting.

The first objective is to identify the existing situation, both for households' actual waste behaviours and for their perceptions regarding the present situation. The second objective is to understand the perception of households' roles and willingness to pay with respect to the possible future organisation of solid waste management, which is in line with the scenarios constructed in this study.

2. Scenarios for household solid waste management

Waste management options that would lower CH_4 and N_2O emissions would be regarded favourably (McDougall et al., 2001). Landfill gas consists primarily of methane and carbon dioxide, both of which are 'greenhouse gases', and landfill gas has therefore become significant in the debate over global warming and climate change. Methane is considered to be responsible for approximately 20% of the recent increase in global warming (Lashof and Ahuja, 1990), and landfills are thought to be a major source of methane. The Clean Development Mechanism (CDM) scheme allows a country with an emission-reduction or emission-limitation commitment under the Kyoto Protocol to implement emission-reduction projects in developing countries. Such projects can earn saleable certified emission reduction (CER) credits, each of which is equivalent to one tonne of CO_2, which can be counted toward meeting Kyoto targets (UNFCCC, 2011). A CDM project might involve, for example, landfill gas to energy (waste-to-energy) and anaerobic digestion, from which revenues are generated along with the greenhouse gas reduction.

One objective of this study is to evaluate the economy of each of the waste management scenarios. The scenarios were defined based on both existing and feasible treatment methods for household waste (e.g., IPCC (2006), Oosterveer and Spaargaren (2010), and Aye and Widjaya (2006)), whereas the fraction of waste treated per scenario – both the organic and inorganic fractions – was established using figures found in the literature, such as Japan Bank for International Cooperation (2008) and Yi, Kurisu, and Hanaki (2011).

The majority of biowaste (75%) is treated with the waste treatment method in each scenario, and the rest of biowaste that cannot be treated is disposed of in the landfill. In terms of recycling, 20% of the inorganic waste is assumed to be recycled, considering the portion of inorganic waste that is recyclable and can be sorted by householders. The non-recyclable fraction of inorganic waste is disposed of in the landfill. Prior to defining the scenarios, field observations were conducted. The following flow chart for the waste management system in Jakarta is based on these observations:

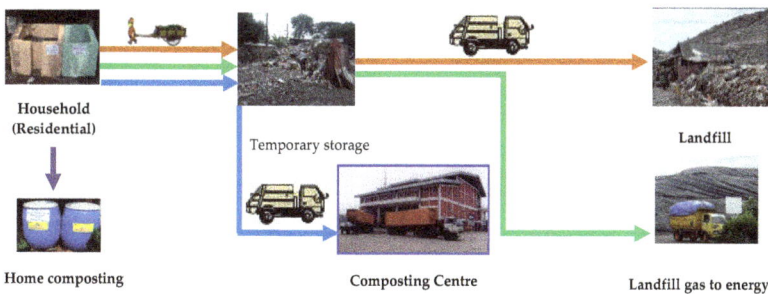

Figure 1. Flow chart of the household solid waste management system in Jakarta.

At few parts of Jakarta, the residents already employ source-separation for composting and recycling purposes. However most of Jakarta residents do not conduct at-source waste separation. Temporary storage sites are established to reduce hauling distances for the collection trucks, thereby lowering transportation costs. These sites are categorised as depots, and hand carts to transfer the waste to the garbage trucks are stored there. Depots also include a base for the handcarts, which is usually located on the side of the road, a trans-ship (shipping/transfer) site, and a waste collection point made of concrete. There are 1,478 temporary storage sites available in Jakarta (Cleansing Department, 2010). At the temporary storage sites, waste is transferred to waste trucks by either manual labour or shovel loader. The waste is subsequently transported to either a composting centre or a landfill. There is no intermediate treatment at these temporary storage sites; however, the efficiency of transfer to disposal and composting sites is increasing. According to the JETRO report (2002), the temporary storage sites increase the effectiveness of collection vehicles from 1.7 to 3 trips per day. (Pasang, 2007). This efficiency is due to the fact that the waste is pooled at the temporary storage sites and is easily collected and transported to the disposal site. By contrast, collecting the waste from various points would reduce the efficiency of collection.

The system boundaries and scenarios proposed in this study are as follows:

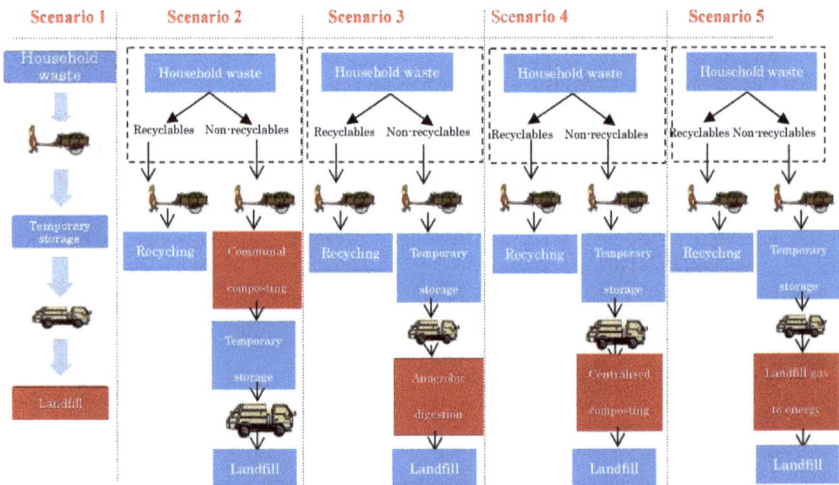

Figure 2. System boundaries and scenarios for waste management

This study compares five scenarios (see Fig. 2) for handling waste from households in Jakarta. The current operation of landfill use (open dumping) was included in the baseline business-as-usual (BAU) scenario for comparison. For the communal and centralised composting scenarios (scenario 2 and 4), the remaining waste or scrap that are not be composted would be delivered to the final disposal. As the incineration of waste is largely not feasible in non-OECD countries, due to cost and frequently unsuitable waste composition (UNEP, 2010), incineration is not included in the scenarios in this study.

Table 1 shows the fraction of each type of waste treated using the modes of treatment specified in each scenario. The fraction of waste treated per scenario was established with the help of figures found in such publications as the Japan Bank for International Cooperation / JBIC (2008); Yi, Kurisu, and Hanaki (2011); and Oberlin (2011). The majority of the biowaste (75%) is treated with the waste treatment method or technology characteristic of the particular scenario, and the rest of the biowaste that cannot be treated is disposed of in the landfill. In terms of recycling, 20% of the inorganic waste is assumed to be recycled, considering the portion of inorganic waste that is recyclable and can be sorted by householders. The non-recyclable fraction of inorganic waste is disposed of in the landfill. Waste pickers, part of the informal sector, play a role in sorting waste and extracting usable materials, such as metal and paper.

Waste types	Scenario 1	Scenario 2		Scenario 3		Scenario 4		Scenario 5	
	L	CC	L	AD	L	CE	L	LFE	L
Biowaste (%)	100	75	25	75	25	75	25	75	25
Inorganic waste (%)		0	100	0	100	0	100	0	100

L: landfill. CC: communal composting. AD: anaerobic digestion. CE: composting center. LF: landfill gas to energy.

Table 1. Fraction of waste treated per scenario[1]

The following table explains the characterisation of the terms 'biowaste' and 'inorganic waste'.

Waste category	Sub-categories	Type of waste	Composting and recycling potentials [a]
Food scraps (kitchen waste)	N/A	Biowaste	CO
Garden waste	N/A	Biowaste	CO
Paper & cardboard	Newspapers	Inorganic	RE
	Magazine	Inorganic	RE
	Other paper	Inorganic	NRE
	Card packaging	Inorganic	NRE
	Other card	Inorganic	NRE
Wood	N/A	Inorganic	NCO
Textile	N/A	Inorganic	NRE
Disposable diapers	N/A	Inorganic	NRE
Rubber & leather	N/A	Inorganic	NRE
Plastic	Refuse sacks	Inorganic	RE
	Other plastic film	Inorganic	NRE
	Clear plastic beverage bottles	Inorganic	RE
	Other plastic bottles	Inorganic	RE
	Food packaging	Inorganic	RE

[1] The fraction of "inorganic" waste to be recycled is assumed at 20%

	Other dense plastic	Inorganic	NRE
Metal	Steel beverage cans	Inorganic	RE
	Steel food cans	Inorganic	RE
	Batteries	Inorganic	NRE
	Other steel cans	Inorganic	RE
	Other ferrous metal	Inorganic	RE
	Aluminum beverage cans	Inorganic	RE
	Aluminum foil	Inorganic	RE
	Other non-ferrous metal	Inorganic	RE
Glass (pottery & ceramics)	Brown glass bottles	Inorganic	RE
	Green glass bottles	Inorganic	RE
	Clear glass bottles	Inorganic	RE
	Clear glass jars	Inorganic	RE
	Other glass	Inorganic	NRE
Other (ash, dirt, dust, soil, e-waste	N/A	Inorganic	NRE

ª Based on Thanh, N.P., et al. (2010). CO: compostable. NCO: non-compostable. RE: recyclable. NRE: non-recyclable.

Table 2. Characterisation of waste based on types and the potentials for composting and recycling

3. Methodology

3.1. Social analysis materials and methods

For the purposes of this study, surveys were conducted in order to investigate the perceptions and behaviour of householders in terms of waste management. Householders were also observed to assess their willingness to sort waste, their willingness to pay, and their perceptions of their own role and that of waste service providers in order to improve performance in the future. Parts of the questionnaire were constructed with reference to previous studies (Bruvoll et al., 2002; Berglund, 2006), particularly the questions regarding personal motives and willingness to pay. These questions were complemented with issues beyond personal motives, such as likeliness to sort if benefits were provided, difficulties encountered in sorting, and participation in home composting and communal composting activities.

The questionnaire included both open and closed questions. The closed questions were designed for ease of answering by the respondents with the aim of collecting the maximum appropriate responses, whereas the open questions were intended to encourage respondents to provide further elaboration on certain questions.

The social and economic analyses of our study were performed through household surveys. The following areas were covered by the questionnaire:

- Part I: Demographic information concerning the respondents' educational background, family income, occupation, age, and household size.
- Part II: Questions concerning waste-related costs and revenues. Responses to these questions generate information on the cost of waste services, potential revenues from

the sale of recyclable waste, and costs of and revenues derived from composting activities.

- Part III: Questions concerning issues regarding waste sorting. Responses to these questions reveal willingness to sort, perceptions of sorting, and willingness to pay others to sort waste.
- Part IV: Questions concerning the solid waste management practices of households. Responses to these questions create a better understanding of waste storage, the scheduling and frequency of waste collection, and the perceptions of households in the primary phase of waste management systems.
- Part V: Questions regarding possible future roles in the waste management system. Responses to these questions reveal how respondents wish to participate in waste management in the future and future improvements to waste service provision.

3.2. Methods for the economic analysis

The financial and economic analysis refers to a prior study by Aye and Widjaya (2006). The costs and benefits of each of the waste management scenarios are estimated by processing information obtained from surveys of the landfill administrator, communal composting officers, the Cleansing Department, and householders. The study makes use of secondary data provided by the government and by the landfill gas-to-energy-generation administrator. These sources provided (sometimes confidential) information, such as landfill operation cost breakdowns and financial aspects of the certified emission reduction rights from the methane gas flaring project.

3.3. Sampling of respondents

A stratified random sample was used to select respondents. Stratified random sampling is a technique that attempts to ensure that all parts of the population are represented in the sample to increase the efficiency and decrease the estimation error (Prasad, N., s.a.). The sample used in this study was therefore based on population demographics and represented all families in Jakarta.

The survey was designed to consider the features of waste collection and of the disposal systems and flows. It was conducted in Central Jakarta, North Jakarta, West Jakarta, South Jakarta, and East Jakarta, the five municipalities of Jakarta city. Fig.3 shows a map of Jakarta City and the locations of the target areas corresponding to the five municipalities.

According to BPS Statistics Indonesia (2009), the percentages of the population of Jakarta with low, middle, and high incomes are 60%, 30%, and 10%, respectively. The annual average income of the low-income group is USD 2,284 or IDR 20.6 million per annum. The annual average income of the middle-income group is USD 5,356 or IDR 48.2 million, and the annual average income of the high-income group is greater than USD 14,198 or IDR 127.8 million.[2]

---

[2] 1 USD = 9,000 IDR

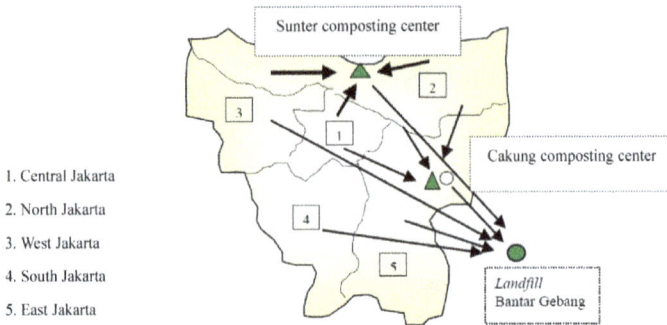

Figure 3. Waste Collection and Disposal Flow (JBIC, 2008)

To obtain a cluster sample, households were selected based on a zoning plan for the regions of the city. In addition, proportionate stratified random sampling was used. The household samples were divided according to the economic or income levels, and samples were taken from each income level within each region. The economic status of the respondents was determined from the responses to the questionnaires (Rahmawati et al., 2010). The questionnaires included demographic characteristics, such as family size. This information was used to estimate the amount of waste generated per capita.

The method of cluster sampling is applied, of which the selection of household sample is divided based upon the zoning of city region. Additionally, proportionate stratified random sampling where the household samples are divided upon the economic or income level and the samples were taken from each income level within each region. The economic statuses of respondents were determined by the responses of the questionnaires (Rahmawati et al, 2010). The questionnaires also cover the demographic characteristics such as the size of family to determine the amount of waste generated per capita.

The size of the sample was determined with the following statistical formula for estimating proportions in a large population (Dennison et al., 1996 and Mc. Call, C.H. Jr., 1982):

$$n = \pi (1 - \pi) Z^2 / \epsilon^2 \qquad (1)$$

where n is the estimated number of individuals required in the sample, π is the proportion to be estimated in the population, Z is the desired level of confidence, and ϵ is the acceptable level of error.

This study used a maximum error level of 0.05, with an associated 95% confidence level, as the desired reliability. A value of 0.50 was assumed for π. Substitution of these values in the equation above gave the required sample size of 384.2. The sampling interval (k) was determined as

$$k = \frac{N}{n} \qquad (2)$$

where N is the population size and n is the sample size.

The population numbers that were previously divided according to the income level distribution were further divided by the number of sub-districts per region. Based on the sample size calculation for the Jakarta survey and the total number of 2,030,341 households in the city, the sample size was rounded to 100 respondents for each combination of sub-district and income level according to the regional and income level distribution.

4. Social and behavioural aspects of the scenarios

4.1. Characteristics of household respondents

Based on the sample of 100, 58% respondents were female, and 42% respondents were male. The ages of the respondents ranged from 15 to more than 55 years with the majority (29%) between 25 and 34 years. Twenty-three percent were between 35 and 44 years, 18% were over 55 years, 17% were between 45 and 54 years, and the remaining 13% were between 15 and 24 years. In terms of education level, 37% had tertiary education, 22% had secondary education, 17% had undergraduate education, 12% had a diploma, 9% had a primary school education, 2% had a postgraduate degree, and 1% had no education. The occupation for the majority was private employee (37%), whereas 34% were housewives, 10% did not specify their occupation, 7% were retirees, 5% were maids, 4% were students, and 3% were civil servants.

Regarding income level, 38% earned between IDR 651,000 and 1,290,000 (ca. USD 76.6 to 152) per month, 26% earned between IDR 1,290,000 and 5,000,000 (ca. USD 152 to 588) per month, 17% earned between IDR 5,001,000 and 10,000,000 (ca. USD 588 to 1,176) per month, 8% earned between IDR 10,001,000 and 15,000,000 (ca. USD 1,176 to 1,764) per month, 7% earned IDR 0 – 650,000 (ca. USD 0 to 76.6) per month, and 4% earned more than IDR 15,001,000 (ca. USD 1,764) per month.

4.2. People's behaviours concerning the waste management system

The majority of people surveyed (67%) store waste that is to be collected from the household for disposal in a plastic waste bin in front of their house; 14% store it in brick garbage bins, and 12% store it in plastic bags. The various types of waste storage containers located in front of houses in Jakarta are depicted in Fig. 4: plastic waste bins, brick garbage bins, and plastic bags.

Figure 4. Various devices for waste storage in front of houses in Jakarta (plastic waste bin, brick waste bin, and plastic bag)

Regarding the location of waste bins within the household, most of the people interviewed gather the waste in a container located in one main room of their residence (66%), rather than locating waste bins in every room (24%). The bins are normally are served by a daily schedule of waste collection (52%).

	Percentage of total respondents
Waste bins within the house	
Waste bins located in each room	24%
Waste bins is pooled in one main room	66%
Other	10%
Waste bins outside the house	
Brick garbage bin outside the house	14%
Plastic waste bin outside the house	67%
Plastic waste bags to be given to waste transporters for disposal	12%
Other	7%

Table 3. Location and storage of household waste

The time of waste collection varies widely, depending on the area of residence, but waste is primarily collected before 11 a.m. The waste collectors do not usually give any particular notification prior to collection, but instead they directly collect the waste in front of people's houses (55%), rather than providing notification by a loud call-out to the household.

The waste collectors who transport waste from households to the temporary storage site are informal workers hired by neighbourhood associations or private companies. These waste collectors all use hand carts with an average capacity of up to 100 kg. The average use period of a hand cart is 7 years, and their frequency of breakdowns is two or three times per year. When a hand cart breaks down, its repair is the responsibility of either the waste collector or the hirer, depending on the degree of damage and prior consent.

4.3. Communal composting

There are communal composting facilities for composting biowaste in several areas of the municipality. There are usually 10 neighbourhood units (*Rukun Tetangga*) within 1 neighbourhood cluster (*Rukun Warga*) in which approximately 680 households reside and are involved in the communal composting initiative (Waste Management Task Force, 2008). Each communal composting facility is usually equipped with a composter 2 x 3 m² in size that is used for composting biowaste. Composter is used as instrument to make decomposition of biowaste as fertilizer, which can be used for organic farming. A shredding machine is usually also available at the facility. Biowaste is collected by manual labourers who transport it to the composting facility.

Figure 5. Communal composter

Of all of the respondents surveyed in this study, 88% claimed that there are no communal composters in their area of residence. Among the respondents who indicated that communal composters are available, only 7% claimed to be actively involved in communal composting activities. These respondents were mostly housewives and retirees. All respondents who were actively involved in communal composting claimed that they do not receive any financial incentive whatsoever to participate in communal composting. All of the respondents who were actively involved in communal composting are users of compost produced by communal composters. As users of the product, these responders perceive the product as being of high quality (86%). The compost products are mainly purchased by householders and small to medium enterprises.

No	Statement	Strongly disagree	Partly disagree	Disagree	Agree	Partly agree	Strongly agree	Don't know
a.	I know that I can purchase / have access to compost produced from waste in the communal composter	0%	29%	0%	43%	14%	14%	0%
b.	I am a consumer of the compost produced from the communal composter	0%	0%	0%	57%	0%	43%	0%
c.	The compost produced by the communal composter is high quality.	0%	0%	0%	43%	14%	43%	0%

Table 4. Perceptions of respondents who are active in communal composting

Regarding home composting, of all of the respondents surveyed, 8% own and use a home composter. All of the respondents who conduct home composting use the product for personal purposes. The composters are purchased by householders who compost their organic household waste at home. The average cost of a home composter is 121 thousand IDR (approximately 14 USD). With an average production of 5.3 kg per month and taking into account the average price of regular compost of 7.7 thousand IDR per kg (ca. 0.9 USD per kg), these respondents have potential revenue of 41 thousand IDR per month (approximately 4.8 USD per month). The primary difference between home and communal composting is the instrument and location of composting.

Figure 6. Typical home composters

4.4. Landfill gas to energy

There are currently several private companies investing in and operating landfill gas to energy generation systems. In these waste-to-energy schemes, MSW is utilised as feedstock to generate energy. There are positive impacts from the implementation of waste-to-energy projects, such as green house gas (GHG) emission reduction, improved air quality in landfills, reduction of methane emissions through methane capture, leachate management, disease vector control (less disease contagion from rats, flies, and vermin to people in urban centres), reduced passive emissions of landfill gases (LFG), and reduced air pollution from landfill fires and open burning of household waste (UNFCCC, 2009).

Figure 7 shows the practice of using a geometrix membrane cell cover to provide anaerobic conditions for the waste, and gas collection pipes are used to harvest methane gas contained in the waste. This technology requires minimal initial capital investment.

Figure 7. Landfill gas to energy generation

4.5. Perceptions of roles within the waste management system

Apart from the waste collection and transportation fees that are charged by waste service providers, the perceptions of respondents regarding waste fees were studied. The questionnaire responses revealed that people generally perceive that they have paid a fee to the government. However, in reality, the government does not charge any fee for waste services. Waste fees vary in accordance with the agreement with the neighbourhood association, and they are collected from households to pay for the services of waste collectors at the average amount of USD 2.4 per month.

In terms of provision of services, the majority of respondents (44%) agreed that commercial services should be involved in managing waste, despite the consequences of increased fees. Forty-seven percent of the respondents strongly agree that waste management is a shared responsibility to which they should be held responsible as citizens. By contrast, almost 49% of the respondents strongly agree that government and waste providers are fully responsible and must provide better service.

Regarding the performance of service providers, 50% of the respondents agree that there is currently a lack of regular service for waste collection, 39% agree that there is pollution from litter that is not properly managed in their respective residential area and that there is scattered waste resulting from careless waste collection. Despite these shortcomings, 40% of the respondents still trust that their waste is properly managed, treated, and disposed of by their waste service provider.

No	Statement	Strongly disagree	Partly disagree	Disagree	Agree	Partly agree	Strongly agree	Don't know
a.	Municipal government is responsible for waste management as I pay a waste levy/fee to them.	10%	4%	16%	25%	16%	28%	1%
b.	Commercial services should be involved to manage the waste properly, even if increased market-rate fees are a consequence.	5%	1%	6%	44%	17%	25%	2%
c.	Waste management is a shared responsibility to which I am, as a citizen, also held responsible.	0%	1%	1%	37%	14%	47%	0%
d.	Government and waste providers are fully responsible and must provide better waste management service.	1%	1%	3%	33%	13%	49%	0%
e.	If household waste sorting is required, women should be the ones who conduct it.	25%	9%	34%	13%	13%	6%	0%
f.	Maids are the ones responsible for managing my household waste	29%	5%	36%	13%	9%	6%	2%
g.	There is a lack of regular waste collection services	3%	5%	12%	50%	11%	17%	2%

h.	There is pollution/littering that is not properly managed in my residential area	1%	16%	17%	39%	17%	9%	1%
i.	There is waste that is scattered as a result of careless collection	10%	10%	16%	33%	11%	20%	0%
j.	I trust that the waste is managed, treated, and disposed of properly by waste providers.	6%	12%	20%	38%	9%	14%	1%

Table 5. Perceptions regarding the current performance of waste management

4.6. People's willingness to sort and willingness to accept waste sorting practices

Regarding waste sorting, most of the people (81%) do not usually conduct waste sorting at home (e.g., sorting organic from inorganic waste). However their responses regarding agreement to consider waste sorting were quite high, with 73% indicating that they would consider sorting their waste at home. The respondents agree (34%) and strongly agree (25%) that if required, both sexes should be responsible for conducting sorting within the household.

Of all the respondents who have already incorporated waste sorting into their daily activities, 44% have been conducting waste sorting for less than a year, and 26% have been doing it for 1 – 5 years. The actors who motivate them to sort their waste include early adopting family members and neighbours (31%) and community leaders (25%).

	Percentage of total respondents
Respondent already conducts waste sorting at home	
Yes	19%
No	81%
Period for which waste has been sorted	
Less than 1 year	44%
1 – 5 years	26%
5 – 10 years	18%
more than 10 years	12%
Respondents who do not yet sort their waste but would consider it	
Yes	59%
No	22%
Do not know	81%

Table 6. Waste sorting activities of householders

The respondents who already conduct waste sorting mainly agreed that the reasons for them to sort waste are shown in table 7.

No	Statement	Strongly disagree	Partly disagree	Disagree	Agree	Partly agree	Strongly agree	Don't know
a.	It is recommended by my community group	6%	0%	18%	49%	12%	12%	3%
b.	To get additional income by selling recyclable/reusable materials to scrap dealers	6%	0%	18%	46%	12%	15%	3%
c.	To contribute to a better environment	0%	0%	6%	37%	21%	33%	3%
d.	To compost biowaste with my home composter	3%	0%	15%	43%	15%	18%	6%
e.	To compost biowaste at the communal composter	0%	0%	21%	43%	21%	9%	6%
e.	It is a pleasant activity in itself that brings me satisfaction	0%	9%	15%	43%	9%	24%	0%

Table 7. Reasons for sorting waste

Following the preceding section discussing the actual behaviours, in this section, we move on to address future behaviour, including the willingness to incorporate behavioural changes in the future. For this purpose, the respondents were given the following question: *"If the following benefits were provided, how willing would you be to sort waste?"*

Benefits/assistance provided	Willingness to sort waste						
	Least willing	Partly willing	Un- willing	Willing	Partly willing	Very willing	Don't know
Financial incentive	3%	5%	7%	35%	22%	22%	3%
Provision of knowledge on how to sort	3%	5%	6%	40%	16%	28%	2%
Free waste sorting bins	0%	1%	2%	32%	15%	49%	1%
Information about benefits of sorting for the environment	0%	3%	7%	34%	29%	26%	1%
Information about benefits of sorting for public health	0%	2%	6%	33%	26%	32%	1%
Information about how the sorted waste will be treated	0%	3%	6%	45%	27%	18%	1%
Free home composter for composting biowaste	4%	4%	5%	40%	28%	17%	2%

Table 8. Willingness to sort waste if the following benefits/assistance were provided

Most of the respondents were found to be willing to consider waste sorting if information on how the sorted waste will be treated is provided (45%), if knowledge on how to sort the waste properly is provided (40%), and if a free home composter for composting household biowaste is provided (40%). Financial incentives would also increase respondents' willingness to consider waste sorting (35%), as would free waste sorting bins (32%) and the provision of information on the benefits of sorting to the environment (34%) and public health (33%).

Forty-three percent of the respondents who already sort their waste agree that the unavailability of sufficient incentives/benefits to sort waste makes it difficult for them to sort waste, even though a mechanism for the treatment of sorted waste is already established. The respondents who have not yet practiced waste sorting agree (44%) that they know how to sort waste properly, but there is still a lack of information on the advantages of sorting (48%), and there is no assurance that the waste transporters will not mix the sorted waste at the temporary storage site (41%).

No	Statement	Strongly disagree	Partly disagree	Dis-agree	Agree	Partly agree	Strongly agree	Don't know
a.	Sufficient incentives/benefits for sorting waste are not provided	11%	4%	14%	43%	10%	17%	1%
b.	A mechanism for the treatment of sorted waste is already established	6%	11%	18%	37%	15%	8%	5%
c.	I know how to properly sort waste	6%	8%	19%	44%	9%	13%	1%
d.	There is a lack of information on the advantages of sorting	1%	1%	14%	48%	18%	16%	2%
e.	There is no assurance that the waste transporter will not mix the sorted waste at the transfer station.	2%	0%	10%	41%	11%	33%	3%

Table 9. Considerations that make it difficult or easy for respondents to sort waste

4.7. Willingness to pay others to conduct waste sorting

Willingness to pay (WTP) provides an indication of the extent to which sorting at the source is perceived as a cost for the household and of the size of this cost in monetary terms (Bruvoll et al, 2002). Debate on the best method for estimating WTP continues, whether open-ended or closed-ended questions should be included in the questionnaire. Sterner (1999) conducted studies on WTP aiming to ascertain how much people would be willing to pay in cash for environmentally sound waste management, and open-ended questions were used. Similarly, the study by Berglund (2006) used the open-ended question approach to prevent response bias.

Although some cost data on waste handling processes are relatively easy to extract from the literature and surveys, other data, such as the time devoted by households to sorting waste, are more difficult to obtain (e.g., Bruvoll, 1998; Reich, 2005). The value placed on the time households spend on sorting waste constitutes a substantial share of the total cost of recovery. One line of thought is that households' time devoted to sorting waste on a daily basis should be seen as a cost to society, due to the opportunity cost of the time in terms of foregone leisure (Berglund, 2006). If government authorities were to require at-source waste sorting, respondents' willingness to pay is shown in table 10.

	Percentage of total respondents
If the government requires waste sorting, the respondent is willing to pay someone to sort their waste	
Yes	42%
No, respondent will sort their own waste	57%
Do not know	1%

Table 10. Willingness to pay

The respondents who agreed to pay others to sort their waste are willing to pay an average of 16.5 thousand IDR (approximately USD 1.87) per month. Another means of determining WTP is to estimate the labour cost per hour of sorting.[3] The minimum regional wage in Jakarta is 1.1 million IDR (approximately USD 124) per month, as per Jakarta Provincial Governor Regulation No. 167/2009. This wage corresponds to USS$ 0.78 per hour, assuming a 20-day work month and an 8-hour workday.

4.8. People's perceptions of future roles in the waste management system

According to the responses to the questionnaires, if appropriate mechanisms, incentives, and technical information are provided, the majority of respondents agree to play future roles, such as

- Being involved in communal composting (37%) and home composting (31%)
- Learning to sort waste properly (50%).

Despite agreeing to adopt more roles in the future, most of the respondents do not wish to be involved in monitoring and evaluation of the overall waste management system in their community.

No	Statement	Strongly disagree	Partly disagree	Dis-agree	Agree	Partly agree	Strongly agree	Don't know
a.	I wish to be involved in community composting to produce compost from my household biowaste.	3%	4%	21%	37%	19%	14%	2%
b.	I wish to be able to produce compost from my household waste by using a home composter.	3%	9%	27%	31%	16%	12%	2%
c.	I wish to be able to sort waste properly.	3%	0%	4%	50%	12%	29%	2%
d.	I wish to be involved in the monitoring and evaluation of the overall waste management system in my community.	6%	11%	34%	25%	13%	9%	2%

Table 11. Roles respondents wish to play in the future if appropriate mechanisms, incentives, and technical information are provided

4.9. People's perceptions regarding future roles of other waste management actors

The majority of respondents strongly agree that there are several improvements and roles that the government and other waste management actors should make in the future, such as:

---

[3] Prior research by Sterner (1999) reported the average time spent on sorting is half an hour per week.

- Providing more regular waste collection (54%)
- Proper handling, treatment, and disposal of waste to reduce pollution (53%)
- Providing information to citizens regarding the methods of waste treatment and disposal and providing overviews on the waste management system (45%).

Forty-three percent of the respondents also agree that waste management actors should actively involve citizens in waste management decision-making processes.

No	Statement	Strongly disagree	Partly disagree	Dis-agree	Agree	Partly agree	Strongly agree	Don't know
a.	Provide more regular waste collection	0%	0%	1%	38%	7%	54%	0%
b.	Provide proper handling, treatment, and disposal of waste to reduce pollution.	0%	0%	0%	36%	11%	53%	0%
c.	Inform citizens concerning how the waste is treated and disposed of and the overall waste management system.	0%	0%	0%	36%	19%	45%	0%
d.	Actively involve citizens in decision-making regarding waste management issues.	1%	1%	6%	43%	16%	33%	0%

Table 12. Respondents' aspirations regarding future improvements and future roles of waste-system actors

5. Economic aspects of the scenarios

The economic assessments of the five scenarios distinguished here consist of cost-benefit analyses with two main components: an economic cost-benefit estimate and an ecological cost-benefit estimate. The first section focuses on the financial costs and benefits from an economic point of view, and the potential revenues from recycling sorted waste are estimated. The second section focuses on the benefits from greenhouse gas (CO_2) emission reduction and co-products, such as compost and electricity, the economic value of which is estimated.

5.1. Financial cost-benefit analysis of the waste management scenarios

The costs were estimated as follows:

$$C_{ET} = C_L + C_C + C_E + C_P + C_{OM} + C_T, \tag{3}$$

where C_{ET} = estimated total cost, C_L = cost of land acquisition, C_C = construction cost, C_E = cost of equipment provision and installation, C_P = cost of planning, design, and engineering, C_{OM} = cost of operation and maintenance, C_T = cost of transportation.

The revenues were estimated as

$$R_{compost} = (S_{compost} \times P_{compost}) \tag{4}$$

$$R_{electricity} = (S_{electricity} \times P_{electricity}) \tag{5}$$

$$R_{product} = R_{compost} + R_{electricity} \tag{6}$$

where

$R_{compost}$ = Revenue from compost (USD per annum)

$S_{compost}$ = Selling price of compost (per tonne)

$P_{compost}$ = Production of compost (tonnes per annum)

$R_{electricity}$ = Revenue from electricity (USD per annum)

$S_{electricity}$ = Selling price of electricity (USD per kWh)

$P_{electricity}$ = Production of electricity (USD per annum)

$R_{product}$ = Revenue from products

Because some of the values on which the estimates of this study were based are from documents that were published in different years (e.g., 2008 and 2009), the values of these parameters in the year 2011 were estimated from the existing values with the following formula:

$$p = \frac{y}{(1+r)^2} \tag{7}$$

where

p = Value for the present year (2011)

y = Value for year y (existing value based on the year for which the value was available in a published document)

r = Interest rate (annual) at 6.5%

t = Time disparity between the present year and the year for which the information was published

5.1.1. Scenario 1

The information on the quantity of waste disposed in the landfill is taken from a reference document, and the investment costs for Scenario 1 are based on the data obtained from the landfill operator PT Godang Tua (2011). There is no revenue from the products generated in the baseline scenario.

5.1.2. Scenario 2

Information on the quantity of waste composted by communal composting and land acquisition were estimated from the reference document JBIC (2008). Information regarding other investment costs and revenues was based on the survey of communal composting officers. The cost of labour is the labour cost at the communal composting site, which is IDR

200,000 per month per person or USD 847 per annum for a total of 3 labourers. The operation and maintenance (O&M) costs also include the cost of fuel for the waste shredders (USD 127 per annum), the costs of fermentation chemicals (USD 28 per annum), the purchase of additives, such as bran and molasses (USD 14 per annum), packaging costs (USD 11,294 per tonne per annum), and maintenance of the facility (USD 85 per annum). The average production of compost is 706 tonnes per annum with an average revenue of USD 118 per tonne.

5.1.3. Scenario 3

The costs and benefits of Scenario 3 are estimated based on the data from a prior study by JBIC (2008). The estimates include revenue from selling electricity to the grid with an estimated average production of 20 GW per annum and a selling price of USD 0.10 per kWh.

5.1.4. Scenario 4

The costs and benefits were estimated based on the data from JBIC (2008). The centralised composting in Scenario 4 is on a larger scale compared to communal composting, as the facility usually serves several areas of the municipality. The estimated production cost of compost at the centralised composting site is USD 47,000 per tonne per annum with an average selling price of USD 39 per tonne of compost.

5.1.5. Scenario 5

The cost-benefit estimate for Scenario 5 is based on UNFCCC (2009). Revenue derives from the sale of electricity with an estimated average production of 17.8 GWh per annum.

5.1.6. Transportation

Fuel consumption costs are added into the cost estimate for each scenario. The total fuel cost is assessed for transport from the temporary storage site of each municipality in Jakarta to the landfill, anaerobic digestion site, or communal composting facility. The total fuel consumption is determined from fuel efficiency (L/km) data, the distance from each area of the city to the solid waste disposal or treatment site, the waste load (tonnes per vehicle) based on JBIC (2008), the total waste transported per annum, and the total number of trips per annum. The price of diesel fuel in Indonesia at the time of study was USD 0.53 per litre. The field observations conducted in this study indicated that household waste that is placed in storage units in the front of houses is subsequently taken to a nearby temporary storage facility by waste transport operators using handcarts. The household waste is subsequently taken by waste trucks from the temporary storage facility to the landfill or to a composting centre.

The estimation also takes into account waste transportation-related costs, such as the wages for transporting waste from households to temporary storage and those for transporting waste from temporary storage to the waste treatment or disposal facility (USD per annum).

The data were obtained from surveys with the waste transporters. The revenues from the recycling of recyclables in each scenario except for the baseline scenario were estimated from the potential revenues (USD per annum).

The total transportation cost is estimated as

$$C_T = \Sigma \, (F_{con} . F_i) + (W_H . H_T / H_S) + (W_T . T_T), \qquad (8)$$

where

C_T = Cost of transportation

F_{con} = Fuel consumption (litres per annum)

F_i = Cost of fuel i (USD per litre)

W_H = Wage for transporting waste from households to temporary storage (USD per person per annum)

T_H = Number of household to temporary storage waste transporters

W_T = Wage for transporting waste from temporary storage to a waste treatment / disposal facility (USD per person per annum)

T_T = Number of temporary storage to waste treatment / disposal facility waste transporters

H_T = Total number of households

H_S = Total number of households served per waste transporter

The wages are estimated from the survey of waste transporters. The average wage for transporting waste from households to temporary storage is USD 1,115 per person per annum, whereas the average wage for transporting waste from temporary storage to waste treatment or disposal facilities is USD 1,501. This difference in wages is due to different wage systems. Those transporting waste from temporary storage to the waste treatment or disposal facilities have official contracts from the Cleansing Department. Those transporting waste from households to temporary storage typically have informal contracts with the neighbourhood associations, and their wages are lower than those of the official contract holders.

Subsequent to all values being estimated for the year 2011, the total cost per tonne of waste is estimated as follows:

$$C_T = C_i / Q_i \qquad (9)$$

The total revenue per tonne of waste is estimated as

$$R_T = R_i / Q_i \qquad (10)$$

where

C_T = Total cost per tonne of waste (USD per annum)

C_i = Total cost per tonne of waste treated per scenario i (USD per annum)

R_T = Total revenue per tonne of waste (USD per annum)

R_i = Total revenue from scenario i (USD per annum)

Q_i = Quantity of waste treated per scenario i (tonne per annum)

Note that the quantity of waste treated differs in each scenario due to the capacity of the waste treatment plant. Therefore, the estimation assesses the cost-benefit ratio per tonne of waste treated.

The cost-benefit estimates for each scenario are presented in Table 13.

	Scenario 1 (Landfill)	Scenario 2 (Communal composting)	Scenario 3 (Anaerobic digestion)	Scenario 4 (Central composting)	Scenario 5 Landfill gas to Energy
Quantity of waste (tonne per day)	6,000	200	250	1,000	298
Quantity of waste (tonne per annum)	2,190,000	73,000	91,250	365,000	108,919
Annual rate			6.5%		
Investment cost (in thousand USD per annum):					
Land acquisition	92.6	0.13	134	258	65
Construction	3,145	1.2	740	359	0.141
Equipment	15	0.0	643	463	67
Planning, design and engineering	453	0.0	422	166	4,764
Total investment cost	3,706	0.14	1,939	1,246	138
Operation and maintenance cost	318	12.4	6,767	6,557	357
Transportation cost	1,920	655	1,920	696	1,920
Total cost (in thousand USD per annum)	5,943	669	10,626	8,500	2,414
Revenue:					
Compost production (tonnes per annum)		706	0	46,976	0
Selling price (USD/tonne)		118	0	40	0
Electricity production per annum (in thousand kWh)		0	20,071	0	17,849
Selling price (USD/kWh)		0	0.11	0	0.11
Total revenue and tipping fee savings (in thousand USD per annum)		959	2,303	1,873	2,048
Revenue:cost ratio	0	1.4	0.217	0.220	0.8

Table 13. Comparison of cost-benefit results for the five scenarios

The revenue:cost ratio is estimated as

$$RCR = R/C \qquad\qquad (11)$$

where

RCR = Revenue:cost ratio

R = Total revenue (USD per tonne per year)

C = Total cost (USD per tonne per year)

The total cost and total revenue were estimated per annum, for which the assumed project life is 20 years. According to the estimates, the communal composting of Scenario 2 has the highest potential in terms of the benefit:cost ratio. The second-best option is the landfill gas to energy of Scenario 5. The third-best option is central composting (Scenario 4) followed closely by anaerobic digestion (Scenario 3). The baseline scenario of landfill use (Scenario 1) has the worst potential, as it does not yield any revenue from products.

5.1.7. Potential revenue from recycling of sorted recyclable waste

In addition to the economic evaluation for each of the scenarios, this study also estimates the potential revenue from sorted recyclable waste based on primary data on the quantity of recyclable waste from households and selling prices of recyclable materials obtained from field surveys. The potential revenue from these waste products is shown in Table 14.

Waste category	Sub-category	Average selling price (USD per kg)	Average quantity sold per household (kg per month)	Revenue potential (USD per annum)
Paper and cardboard				
	Newspapers	0.17	3.57	14,684,065
	Magazine	0.21	1.75	8,869,442
	Carton boxes	0.25	4.43	27,130,412
Plastic				
	Refuse plastic sacks	0.33	1.00	8,121,364
	Plastic bottles	0.27	1.75	11,617,372
Metal		0.45	1.04	11,529,765
Glass		0.23	1.36	7,668,986
Textiles	Used clothes and fabrics	1.04	1.00	25,319,547
Total			15.90	114,940,952

Table 14. Potential revenue from recycling of recyclable waste in Jakarta

5.2. Benefits from greenhouse gas emission reduction and co-products for each scenario

For each of the waste treatment scenarios, the economic analysis in this study accounts for the benefits from both greenhouse gas (GHG) emission reduction and co-products, such as compost and electricity generation. The costs and benefits deriving from such externalities are not usually taken into account; therefore, this study accounts for CO_2 as a GHG emission reduction benefit and for the co-products generated by each waste treatment method, whereas other benefits are neglected. The equation to which the economic analysis is applied is as follows:

$$NPV_{cost} = I + OM + T (1 - (1 + r)^{-t} / r) \tag{12}$$

$$NPV_{revenue} = (R_p + R_{ghg}) \times (1 - (1 + r)^{-t} / r) \tag{13}$$

$$NPV_{benefit} = NPV_{revenue} - NPV_{cost} \tag{14}$$

where

I = the investment cost (USD)

OM = operation and maintenance cost (USD per annum),

T = transportation cost (USD per annum)

R_p = revenue of co-products (USD per annum),

Rghg = revenue from greenhouse gas reduction (USD per annum)

r = discount rate (based on Aye, 2006)

t = project life time.

Greenhouse gas (GHG) emission reductions were calculated in a previous study (Aprilia et al, 2011) in which the GHG emissions of each scenario were compared to the baseline scenario. The carbon price is USD 12 per tonne of CO_2 (UNFCCC, 2009). At the time of that study, the price of grid electricity was on average about IDR 860 per kWh, or USD 0.1 per kWh. A comparisons of the GHG savings externality for each of the waste treatment scenarios is presented in Table 15.

	Scenario 1 (Landfill)	Scenario 2 (Communal composting)	Scenario 3 (Anaerobic digestion)	Scenario 4 (Central composting)	Scenario 5 Landfill gas to energy
CO_2 savings (kg/tonne waste)	0	461,000	498,000	461,300	489,906
Carbon price (USD/tonne CO_2)	0	12			
Project life (year)	20				
Discount rate	6.37%				
NPV cost (USD/tonne)	75	373	4,302	878	876
NPV revenue (USD/tonne)	0	509	1,031	210	768
NPV benefit (USD/tonne)	0	136	-3,271	-668	-108
Revenue:cost ratio	0	1.37	0.24	0.24	0.88

Table 15. Comparison of the economic impact of the scenarios (in USD)

The assumption used for the anaerobic digestion scenario is that the residue is not composted but is placed in a landfill. Regarding the communal composting scenario, voluntary action is assumed. The CH_4 collection efficiency for the landfill gas to energy scenario is 60%.

Based on the economic analysis for each waste treatment scenario, communal composting (Scenario 2) has the highest potential, as it has the highest benefit to cost ratio. However, it should be noted that the communal composting that takes place in Jakarta employs voluntary labour with an average wage below the regular labour wage. The costs for the existing common communal composting sites are also relatively low because simple composting techniques are applied. The costs of construction, equipment, O&M, planning, design, and engineering (which accounts for the total investment cost) are seven up to sixty six times lower than with the other options.

Landfill use for electricity generation (scenario 5) does not generate positive benefit, since its cost per tonne treatment is higher than revenue; however it has better potential rather than anaerobic digestion and central composting scenario. The potential revenue from scenario 5 includes revenue from both GHG emission reductions through the CDM and electricity generation. The price of electricity that can be sold to the grid is currently USD 0.11 per kWh, whereas in 2006 it was USD 0.06/kWh. The implementation of this scenario should be accompanied by financial support by the government, particularly to cover the investment costs of equipment provision and land acquisition.

Centralised composting (scenario 4) is the third-preferred option followed by anaerobic digestion (scenario 3). Both of these scenarios show negative benefit and would need subsidy or financial support to achieve positive benefit. As waste in Jakarta is not sorted, centralised composting becomes labour-intensive, particularly for manually sorting the organic from inorganic waste. The type of machinery used for the centralised composting plant considered in this study is a conventional windrow, which is a manual non-mechanical composting process.

Anaerobic digestion is the least profitable as it requires the highest investment cost for construction and equipment, as well as O&M cost. The revenues obtained from the implementation of this scenario are from the GHG saving with CDM scheme and electricity generation that are sold to the grid, as the case for scenario 5. Scenario 1 has the least cost; however it does not generate any revenues.

All of the options proposed in this study, except for the Scenario 1, require at-source waste sorting by householders. This approach minimises the need for manual and automated sorting within waste treatment facilities and increases the effectiveness of the composting and digestion processes. If plastic and inorganic material is present in urban solid waste during anaerobic digestion or landfill gas to energy generation, the material causes the total amount of gas produced to decrease (Muthuswamy, S. et al., 1990).

6. Conclusions

This study employs socio-economic evaluation to measure household solid waste management scenarios. According to the estimation, communal composting has the highest potential with the highest benefit:cost ratio. Theoretically, composting can be performed at the communal level at temporary storage sites, at composting centres or at the landfill. The costs of processing and transport and the roles, perceptions, and responsibilities of households are arguably different. Despite the potential for communal composting, a high percentage of respondents indicated that there is no neighbourhood composting. Thus, the present composting rates are low compared to the composition of the waste.

There are several possible constraints impacting the further application and expansion of communal composting, such as

1.   Land acquisition

The land being utilised for communal composting usually belongs to a specific entity that dedicated it as a public space, and the land came to be used for communal composting later. For instance, the communal composting that takes place in Rawajati Jakarta uses land that belongs to the Indonesian ground forces and is dedicated to communal composting at no cost. Further application of communal composting throughout other areas would imply the need for open space dedicated to composting. In addition, the limited availability of open space in Jakarta poses particular constraints on the siting of communal composting facilities.

2.   Labour and wage systems

The current communal composting sites in Jakarta employ voluntary labour with a lower waging system. Further application of communal composting would require an appropriate waging system at or above the regional minimum wage. A subsequent issue regards the marketing of compost products and the extent to which compost sales would be able to cover operational costs, such as the provision of income for the labourers. The current practice is that most of the compost produced is used by the community. The tendency of urban residents not to conduct farming practices that require compost and the scarcity of land for farming raise the question of marketing issues such that the marketing of compost might have to be extended to neighbouring areas of Jakarta.

3.   Capacity of composting facilities

The capacity of communal composting facilities is usually much smaller than that of industrialised composting sites, and increasing, their capacity would be a challenge, due to the limited compostable waste feedstock and the limited space for the communal composting facilities.

All of the options proposed in this study, except for the baseline scenario, suggest that at-source waste sorting by householders is necessary. However, the majority of people in Jakarta do not sort their waste, and household waste is a mix of biowaste, inorganic waste, hazardous waste, and bulky waste. Waste sorting tends to take place outside of the home by waste transporters and manual labours at temporary storage sites and waste treatment or disposal facilities.

Despite the current trend of not sorting waste, most of the respondents surveyed for this study agreed to consider waste sorting. The willingness to consider waste sorting by people who have not yet adopted it is more likely if benefits, information, and assistance are provided. Increased transparency from waste service providers and government regarding the modalities of waste treatment and final disposal is expected to increase public awareness and active participation in at-source waste sorting.

At-source waste sorting by householders can be successfully achieved through both voluntary measures and regulatory measures. The current approach to promoting at-source sorting is through voluntary measures, specifically, the introduction of incentives through revenue from sorted recyclables and revenue from home and communal composting.

Although several types of incentives are present, they are not sufficient to encourage the public to sort their waste. Thus, regulatory measures may have to be considered through the formulation of a regulatory framework to mandate sorting at households with the provision of disincentives or penalties for householders that do not properly sort their waste. A regulatory framework for waste sorting would essentially increase composting success rates. For the regulatory measures to prevail, concrete mechanisms would be required, such as the provision of information for proper sorting, trained waste collectors, varied waste collection schedules according to different types of waste, and the provision for purchase of standardised transparent plastic bags to enable checking by responsible officers.

Promoting at-source waste sorting is important; however, appropriate end-of-pipe technologies for the treatment of municipal solid waste are also required. This study identified feasible technologies with cost-efficiency assessments that can be considered for further implementation. Communal composting is found to have the highest potential with the highest benefit:cost ratio and the greatest greenhouse gas savings, but there are challenges, such as land availability, labour and waging systems, and the capacity of composting facilities. The second preferred option is landfill gas to energy scenario, followed by central composting and anaerobic digestion. However it should be noted that the operation of landfill gas to energy, central composting and anaerobic digestion require substantial financial support from the government, particularly to cover investment and O&M costs. The financial support is regarded as the costs for municipal waste treatment that is borne by the government of Jakarta. The imposed subsidy on electricity tariff results in the uncompetitive selling price of electricity from these scenarios. Therefore when it comes to the revenue analysis, scenario 3 and 5 may show better results if the electricity subsidy were lifted. Communal composting would still have high potential as the land acquisition cost very low due to the provisions by the government. If the low-cost land provision were retrieved, communal composting still have good potential since its O&M, construction, equipment and other cost are very low compared to the other scenarios.

Although people displayed a high degree of willingness to sort their household waste, proper monitoring will be required to ensure the success of sorting. Possible criteria that merit further study include social impact analysis and life cycle analysis to determine the environmental impacts of each waste management option. These complementary aspects would complete the analysis within an integrated framework.

Author details

Aretha Aprilia and Tetsuo Tezuka
Department of Socio-Environmental Energy Science, Graduate School of Energy Science, Kyoto University, Yoshida-honmachi, Sakyo-ku, Kyoto, Japan

Gert Spaargaren
Department of Environmental Policy, Wageningen University, Wageningen, the Netherlands

Acknowledgement

The authors would like to express their gratitude for the generous financial support of this research by the Schlumberger Foundation and the Global Centre of Excellence (GCOE) programme of Kyoto University's Graduate School of Energy Science.

7. References

Asian Development Bank (ADB). Urbanization and Sustainability in Asia: Good Practice Approaches in Urban Region Development. Manila: ADB; 2006.

Aprilia, A., Tezuka, T. GHG Emissions Estimation from Household Solid Waste Management in Jakarta, Indonesia. Proceeding of the 4th International Conference on Sustainable Energy and Environment 2011 (In Press).

Aye, L., Widjaya, E.R. Environmental and economic analyses of waste disposal options for traditional markets in Indonesia, Waste Management , 2006; 26:1180–1191.

Berglund, C. The assessment of households' recycling costs: The role of personal motives, Ecological Economics 2006; 56:560– 569.

Bohma, R.A., Folzb, D.H., Kinnamanc, T.C., Podolskyd, M.J. The costs of municipal waste and recycling programs, Resources, Conservation and Recycling 2010;54: 864–871.

BPS Statistics Indonesia. Jakarta in Figure 2009. Jakarta: BPS Statistics DKI Jakarta, 2009.

BPS Statistics Indonesia,. Remuneration Boost Middle Class Indonesia (in Bahasa); 2011, Available from http://us.bisnis.vivanews.com/news/read/197512-naik-kelas-karena-remunerasi. (accessed 15.05.11).

Bruvoll, A., Halvorsen, B., Nyborg, K. Households' recycling efforts. Resources, Conservation and Recycling 2002;36:337–354.

Bruvoll, A. Taxing virgin materials: an approach to waste problems. Resources, Conservation and Recycling 1998;22:15-29.

Cleansing Department. Module 1: Master plan of Waste Management in Jakarta (in Bahasa). Jakarta: Cleansing Department, 2010.

Corotis, R.B. Societal issues in adopting life-cycle concepts within the political system, Structure and Infrastructure Engineering, 2009;5:59- 65.

Dennison, G.J., Dodda, V.A., Whelanb, B. A socio-economic based survey of household waste characteristics in the city of Dublin, Ireland. Resources, Conservation and Recycling Volume 17, 1996; pp. 227-244.

IPCC. IPCC Guidelines for National Greenhouse Gas Inventories, Volume 5 Waste. Prepared by the National Greenhouse Gas Inventories Programme, Eggleston H.S., Buendia L., Miwa K., Ngara T. and Tanabe K. (eds). Kanagawa: IGES, 2006.

Japan Bank for International Cooperation (JBIC). Special assistance for project preparation (SAPROF) for solid waste management project Jakarta Indonesia, final report. Jakarta: JBIC, 2008

JETRO. Feasibility Study for Improvement of Municipal Solid Waste Management in the Metropolitan of Jakarta: Waste to Energy Incineration Facilities. Jakarta: JETRO, 2002.

McCall, C.H. Jr. Sampling and Statistics: Handbook for Research, Iowa, 1982.

McDougall FR, White PR, Franke M, Hindle P. Integrated solid waste management: a life cycle inventory. 2nd ed. London: Blackwell Science Ltd, 2009.

Merkx, F., van der Weijden, I., Oostveen, A.N., van den Besselaar, P., Spaapen, J. Evaluation of Research in Context: A Quick Scan of an Emerging Field; 2007. Available from: http://www.allea.org/Content/ALLEA/WG%20Evaluating/NL_Quick_Scan.pdf. (accessed 9.10.11).

Ministry of Environment. Indonesian Domestic Solid Waste Statistics Year 2008, Jakarta: Ministry of Environment; 2008.

Ministry of Environment. Waste Bank Generates IDR 9 million per month (in Bahasa); 2011. Available from:
http://www.menlh.go.id/home/index.php?option=com_content&view=article&id=5147%3Abank-sampah-hasilkan-9-juta-per-bulan&catid=43%3Aberita&Itemid=73&lang=en. (accessed 18.10.11).

Muthuswamy, S., Nemerow, N.L. Effects of plastics in anaerobic digestion of Urban solid wastes, Waste Management & Research 1990;8:375-378

Oberlin A. The role of households in solid waste management in East Africa capital cities. Dissertation. Wageningen: Wageningen Academic Publishers; 2011.

Oosterveer, P., Spaargaren, G. Meeting Social Challenges in Developing Sustainable Environmental Infrastructures in East African Cities. In: Social Perspectives on the Sanitation Challenge. Dordrecht: Springer; 2010.

Pasang, H., Moore, G., Sitorus, G. Neighbourhood-based waste management: A solution for solid waste problems in Jakarta, Indonesia. Waste Management 2007;27: 1924–1938.

Prasad, N. Stratified Random Sampling. Available from: http://www.stat.ualberta.ca/~prasad/361/STRATIFIED%20RANDOM%20SAMPLING.pdf ;s.a.(accessed 01.12.10).

PT Godang Tua. Recapitulation of the Investment Plan and Infrastructure Improvement Facility Management and Operations of Integrated Waste Management in Bantar Gebang in year 2011 (in Bahasa). Jakarta: Godang Tua; 2011

Purcell, M., Magette, W.L. Attitudes and behaviour towards waste management in the Dublin, Ireland region. Waste Management 2010;30:1997–2006

Rahmawati, Y., Enri, D., 2010. The Usage Pattern of Ink Printer Cartridge and the Recycle Potential in Bandung City, available online: http://www.ftsl.itb.ac.id/kk/air_waste/wp-content/uploads/2010/11/PE-SW2-Yeni-Rahmawati-15305074.pdf. Accessed July 2011.

Reich, M.C. Economic assessment of municipal waste management systems—case studies using a combination of life cycle assessment (LCA) and life cycle costing (LCC). Cleaner Production 2005;13:253–263

Sembiring, E., Nitivatta, V. Sustainable solid waste management toward an inclusive society: Integration of the informal sector. Resources, Conservation and Recycling, 2010;54:802–809

Sonesson, U., Bjorklund, A., Carlsson, M, Dalemo, M. Environmental and economic analysis of management systems for biodegradable waste, Resources, Conservation and Recycling 2000;28:29–53

Sterner, T., Bartelings, H. Household Waste Management in a Swedish Municipality: Determinants of Waste Disposal, Recycling and Composting. Environmental and Resource Economics 1999;13:473–491.

UNFCCC/United Nations Framework Convention on Climate Change. Project Design Document Form of Clean Development Mechanism for the landfill gas to energy project in Bekasi. Bonn: UNFCCC; 2009

UNFCCC. Clean Development Mechanism (CDM); 2011. Available from: http://unfccc.int/kyoto_protocol/mechanisms/clean_development_mechanism/items/2718.php. (accessed 15.12.11)

UNFCCC. PT Navigat Organic Energy Indonesia – Bantar Gebang Landfill Gas Management & Power Generation CDM PDD; 2009. Available from: http://cdm.unfccc.int/UserManagement/FileStorage/5Y1LMNW7I3S4QKDZA0BGF8CV XROTE6. (accessed 20.06.11)

Waste Management Task Force. Integrated Waste Management in Rawajati Ward, Pancoran, South Jakarta (in Bahasa); 2008. Available online: http://gtps.ampl.or.id/index.php?option=com_content&task=view&id=58&Itemid=59. (accessed 09.06.11)

Chapter 5

Environmental Awareness and Education: A Key Approach to Solid Waste Management (SWM) – A Case Study of a University in Malaysia

Asmawati Desa, Nor Ba'yah Abd Kadir and Fatimah Yusooff

Additional information is available at the end of the chapter

1. Introduction

Solid waste is defined as generation of undesirable substances which is left after they are used once [1]. Solid waste can also be defined as the useless and unwanted products in the solid state derived from the activities of and discarded by society. It can be classified into three groups: 1) any materials if they are recycled or accumulated, stored, or treated before recycling, 2) being used in a manner constituting disposal, burned for energy recovery, reclaimed, and accumulated speculatively, and 3) a discarded material that is abandoned, recycled, and inherently waste-like.

Actually, waste can be considered as nothing but useful material at wrong place. There is no material in this world, which is not useful in one-way or the other. Also there is no material, which is created out of nothing. It is man's ignorance that he considers certain things as waste and other thing as useful. Just as types of wastes are changing, so must the attitude of people towards waste must change. People must realize that the solution lies in using waste as a resource rather than to be destroyed. Only due to hazardous to human health, some of these undesirable substances cannot be directly reused.

In relating to change in habits, behaviour and participation, 'what people think about waste' [2] is a significantly important aspect of solid waste management [3]. Studies revealed that 89% of participants considered recycling as an acceptable method for disposing of their waste and 57% agreed with the idea of waste collection being charged per bin or per bag to encourage recycling. Only 34% recycled some waste weekly and 9% recycled four times a year or less [2]. The study further discovered that for those who had children aged 5-14 years old, most information about solid waste management received at school influenced their household. This indicates that school campaigns and focused on recycling can increase awareness and attitudes toward solid waste management among children and their parents. Of relevance on this issue, most of participants agreed that people had a duty to recycle (80%) whilst 60% suggested to avoid buying any goods with too much packaging [4]. Reports on solid waste management recommended that recycling habit needs to be established in relation to sustainability solid waste [5].

2. Background information about Malaysia

Malaysia is located on the South East Asia and there are two distinct parts to this country being Peninsular Malaysia to the west and East Malaysia to the east. The total land area of Malaysia is 329,847 square km (127,350 sq mi) [6]. The Peninsular Malaysia makes up 132,090 square km (51,000 sq mi), or 39.7%, while East Malaysia covers 198,847 square km (76,780 sq mi), or 60.3% of the total land of the country. From the total land area, 1,200 square km (460 sq mi) or 0.37% is made up of water such as lakes, rivers, or other internal waters. Malaysia has a total coastline of 4,675 km (2,905 mi), the two distinct parts of Malaysia, separated from each other by the South China Sea, share a largely similar landscape in that both West (Peninsula) and East Malaysia feature coastal plains rising to hills and mountains. The coastal plains bordering the straits of Malacca are the most densely populated areas of Malaysia, and contains Malaysia's capital, Kuala Lumpur.

In 1992 the World Bank has identified that solid waste is one of the three major environmental problems faced by most municipalities in Malaysia. The amount of solid waste generated went up from 17,000 tons per day in 2002 to 19,100 tons in 2005, an average of 0.8 kilogram per capita per day. Currently, over 23,000 ton of waste is produced each day in Malaysia. However, this amount is expected to rise to 30,000 ton by the year 2020. In the state of Selangor alone, waste generated in 1997 was over 3000t/day and the amount of waste is expected to rise up to 5700t/day in the year 2017 [7].

The amount of waste generated continues to increase due to growing population and increasing development. Modern lifestyle has led to more acute waste problems, convenience products generally require more packaging, careless habits associated with greater affluence lead to greater quantities of waste, as demonstrated by discarded wrappers from the inevitable fast food outlet, and the modern day waste contains a higher proportion of non-degradable materials such as plastics. The waste consists of 45% food waste, 24% plastic, 7% paper and 6% iron. Approximately 95-97% of waste collected is taken to landfill for disposals. The remaining waste is sent to small incineration plants, diverted to recyclers/re-processors or is dumped illegally. However, an alarming 19% of waste ends up in drains, which then causes flash floods and drainage blockage. Today only 5 % of the waste is being recycled, but the government aims to have 22% of the waste recycled by 2020 [8].

Despite the massive amount and complexity of waste produced, the standards of waste management in Malaysia are still poor. These include outdated and poor documentation of waste generation rates and its composition, inefficient storage and collection systems, disposal of municipal wastes with toxic and hazardous waste, indiscriminate disposal or dumping of wastes and inefficient utilization of disposal site space. Litter at the roadside, drains clogged up with rubbish and rivers filled with filthy garbage definitely indicate that solid waste is a major environmental problem in Malaysia [9].

This situation has been and will be reducing our environmental capacity to sustain life. If the present rate of solid-waste production goes on without effective supervision and disposal methods, there will be a substantial negative impact on the quality of our environment. Furthermore, the lack of awareness and knowledge among Malaysian community about solid waste management (SWM) issues, and being ignorant about the effect that improper SWM has to us has definitely worsened the problem.

However, since 2007 environmental awareness is building up within the Malaysian government as well as in consumers' minds. The government has adopted a National Strategic Plan for Solid Waste Management with emphasis on the upgrading of unsanitary landfills as well as the construction of new sanitary landfills and transfer stations with integrated material recovery facilities. A new Solid Waste Management Bill was adopted by parliament in June 2007. The bill is to drastically change the structure of solid waste management in Malaysia and to open up for the development of a completely new business sector. New concessions on domestic waste management will be introduced, as well as recycling, and handling of specific types of solid waste like plastic, paper etc. is highlighted. Solid waste management is a priority area under the 9th Malaysian Plan, as can be seen by the government setting up a Solid Waste Department which is entrusted to enforce the Solid Waste Management Bill.

3. Awareness and education programme towards SWM

Several universities have successfully implemented a 'greening' university campus [3,10,11]; whereby solid waste management programmes were carefully planned based on key focus and waste characterizations. Paper and paper products represent a huge number component of solid waste due to academic and research activities. It is suggested that paper consumption to be reduced and paper recycling is encouraged [12]. It is also advisable for campus community to use refillable cup to replace a single-use beverage containers [13]. The University of Wisconsin-Madison for instance is the first university that initiated this programme.

Students' awareness about environmental problems and solutions can be increased through education [3]. It is expected that solid waste management activities in university campus involve the students as part of their learning process. The particular skills and knowledge gained from environmental education would help in changing human behaviour towards the environment [14]. Students with some knowledge and skills on environmental education are more motivated to take part in environmental protection activities and plans [15] thus would generate new ideas for the solution of environmental problems. Sharing new informations from their activities with families, other adults, and community probably will have some positive implications on solid waste management practices. Although there are a number of literatures on solid waste management in term of intergenerational influence and socialization processes, however the practical impacts of environmental education somewhat has been given little attention [3]. Thus, this study is going to fill this research gap.

4. Theoretical framework

Theory of reasoned action (TRA) and theory planned behaviour (TPB) were used in this solid waste management programme as a framework in understanding, explaining and predicting behaviour. These theories are also useful as a guide for designing intervention strategies to maintain or change a particular behaviour. The theory is based on the assumptions that individual behavioural intentions are directly associated with their attitudes. The theory of reasoned action views an individual's intention to perform or not to perform as an immediate determinant of the action. This behavioural intention has two determinants: 1) attitude towards the behaviour, and 2) the subjective norms. The beliefs related on attitude towards the behaviour are called behavioural beliefs whilst normatif beliefs are for the subjective norms [16]. The theory planned behaviour views an individual's determination is influenced by attitude, social support and perceived behavioural control. Thus, it is best to examine human behaviour when participation decisions are voluntary and under an individual control. Therefore, this theory is suitable to predict a student's intent to participate in a specific behaviour in relation to solid waste management [17].

5. Studies regarding solid waste

Before the awareness and education programme can be conducted, two different researches were carried out. The first study was to identify the current waste collection and waste data. Results showed it is estimated that the National University of Malaysia produce an average collection of about 8 ton of solid waste per day [18]. In order to identify the type of waste produce by the university, waste characterization study was conducted. The method used can be referred from the study by Kian-Ghee Tiew, Stefan Kruppa, Noor Ezlin Ahmad Basri and Hassan Basri (2010) [18] of waste characterization research team from Faculty of Engineering and Built Environmental. After sorting, the waste was store in bins which were labeled for different items and later were weighed to determine waste composition. The study has been successful in highlighting the composition and characteristics of the solid waste produced at the university campus. The main components of the waste are organics (43%), plastics (36%) and paper (17%), which is more than 96% of the total solid waste. The average amount of a sample is 108 kg. Striking is the high plastic and organic content and the third most amounts is paper [18].

Most of the waste collected comprises combustible and noncombustible wastes. The combustible waste consists of materials such as paper, cardboard, furniture parts, textiles, rubber, leather, wood, plastic and garden trimmings. Non-combustible waste consists of items such as glass, discarded tins, aluminum cans and food waste. Characteristics of solid waste can be divided into two: physical and chemical characteristics. The physical characteristics of solid wastes vary widely based on socio-economic, cultural and climatic conditions. The physical qualities of solid waste like bulk density, its moisture content etc., are very important to be considered for the selection of disposal, recycling and other processing methods. Chemical characteristics information of solid wastes such as pH, chemical constituents like carbon content, nitrogen, potassium and micronutrients are important in evaluating processing and recovery options. In addition, the analysis helps in adopting and utilizing proper equipment and techniques for collection and transportation. Identifying both chemical and physical characteristics of solid wastes are important for the selection of proper waste management technology.

Thus both physical and chemical characteristics of the solid waste are important to determine the selection of the final method of waste disposal. Based on this findings the university provide three different bins for the separation at source activities. Each bin is coloured differently, for example green is for organic or biowaste, orange for recyclables and black for residual waste. Before this only one bin is used for all waste.

In the second study, a self-administered questionnaire was used to assess students' awareness, attitudes and perceptions towards the solid waste management. The approach of this research was to analyze problems, create and conduct interventions and then evaluate the effectiveness of interventions. The main tool used in data collection was a structured three part questionnaire specifically designed for this study. The questionnaire covered demographic factors such as year of study and ethnic of the respondent as well as variables related to the respondent's littering attitudes and practices. Examples of statement regarding this variable: 1. I do not care if someone throw litter; 2. I assume waste is not useful and should be thrown away; 3. I do not care if my friends throw rubbish into drains.

Another part of the questionnaire consist of statements regarding the environmental awareness and knowledge of SWM among respondents. Respondents were asked about their knowledge of SWM and programmes conducted by the university in order to create awareness. They were also asked about the source of their information regarding environmental problems.

For the first requirement, simple interactive statistical analysis for size sample calculation was used (Raosoft – sample size calculator) [19] to determine sample size required based on the population size of 5,000 students from the university main campus. Size of sample required at 95% confidence level, a margin of error at 5% was 537. Distribution numbers is estimate for cooperation for questionnaire return at 50%. A much higher number is required to entail sufficient number for survey study. Therefore 600 questionnaires were distributed by convenience sampling.

A total of 589 undergraduate students from eight different faculties at The National University of Malaysia completed the questionnaire forms. There were 458 (77.8%) Malays, 104 (17.7) Chinese, 18 (3.0%) Indians and 9 (1.5%) others. Most of the respondents were first year students (318 = 54%), 180 were second year students, followed by 93 third and fourth year students. Data was analyzed using the Statistical Package for Social Science (SPSS) computer programme version 10 software. Descriptive statistics such as means and ranges were computed. Test of chi square was performed to determine the relationships between attitude and practices and also between facilities and practices.

Results showed that more than half of the students (64%) had high awareness status concerning SWM. But there was still quite a number of them (36%) which have low awareness status. Only 34.1% of the students showed positive attitude towards SWM whereas another 65.9% showed negative attitude. Regarding perceptions concerning SWM only 40% of the subjects' perception status was positive.

Despite the high status of awareness expressed by 64% of the students concerning SWM, it is not consistent with their attitude and perception. The results of this study showed that more than half of the students (65.9%) have negative attitudes towards SWM. Like wise only 40% showed positive perception concerning SWM. Results from the descriptive analysis were supported by the results of Chi Square which showed that there was no relationship between attitude and practice (x^2=2.452, p>0.05), and also between facilities and practice (x^2=1.618, p>0.05).

Although results indicated that majority of the students showed high status of environmental awareness, however, more than half of the subjects showed negative attitude and perceptions concerning SWM. Behavioural problem: not practicing environmentally responsible behaviour (an inconsistent and highly unbalanced strong "knowing" but weak "doing") because: attitudinal problems, lack of enforcement, lack of monitoring and the students did not understand their roles and responsibilities in environment protection. Hvatum and Kelly (2008) [20] labelled the situation as "you know it, but you don't do it"

Results of this study supports some studies that suggest that there is no relation between education and attitude to the environment [21,22]. Findings of previous studies [23-25] and the findings of Hines, Hugerford and Tomera (1986) [26], also suggest that the level of consistency between environmental attitudes and behaviour is affected by a person's knowledge and awareness, public verbal commitment and his/her sense of responsibility. The transfer from attitudes to behaviour can also be affected by lifestyle; many people, while professing to "correct" attitudes to the environment, are not ready to change their lifestyle in ways that might mean sacrificing certain forms of leisure and comfort for the sake of the environment. Other study has also found a weak and inconsistent relationship between environmental attitudes and behaviour; usually attributable to a reluctance to give up the comforts of modern life [27].

This answers the question as to why in certain circumstances individuals with acquired knowledge act on that knowledge to implement changed waste practices, while in other instances, this acquired knowledge does not lead to change. According to Miller and Morris, (1999:74) [28] "there is a commonly held myth that providing individuals or groups with information will lead them to appropriate personal and organizational actions and performance, but this is far from true." According to Pfeffer and Sutton (2000), while information and knowledge are 'crucial to performance', but knowledge of an issue is often not sufficient to cause action: "*there is only a loose and imperfect relationship between knowing what to do and the ability to act on that knowledge*" [29]. The inability to transfer knowledge of what needs to be done into action or behaviour which is consistent with that knowledge, is referred as the 'knowing-doing gap' or the 'performance paradox' [29]. While it was believed that the 'knowing-doing gap' was due to a lack of personal knowledge or skills, research conducted suggests that while personal knowledge is important in ensuring action, it is not as important as having management systems and practices in place [29].

6. Education and awareness programmes implemented at the university

We will now share with the readers, programmes launched by the university to create the students' awareness of environmental and waste problems and to educate them how to practice environmentally responsible behaviours. Programmes were also implemented to promote attitude and intention change and actively pursuing sustainable environmental practices among the students. The attempt to encourage, through education and awareness on managing solid waste in the campus that have been implemented will show good and encouraging results as long as the persons and the organization have the ability to assimilate and interpret the knowledge and to convert this learning into impact (to implement change in managing waste). It is hope that knowledge will act as a precursor to action.

Actually, the best way to create awareness and to educate the students, steps had to be taken to include environmental education in the school educational system. This must be the leading approach to address the environmental problems and engendering sustainable development. Knowledge and understanding of the environment are important since a degraded environment means a lower quality of life for all. It is, therefore, the collective responsibility of all human beings to secure a healthy environment not only for present, but also for future generations, so building environmental curricula on this principle becomes a necessity.

The authors hold the view that there is a clear inadequacy in the environmental education paradigm in Malaysian educational curricula as they pertain to solid waste management. For instance, elements of environmental problems was integrated into the subjects of health education, integrated science, agricultural science and geography among others. These approaches are insufficient if environmental protection is to be undertaken sustainably as presently advocated through environmental awareness and educational programming globally.

Environmental education should, therefore, be a fundamental and integral part of education for all members of society. Modern societies, both developed and developing, need environmental education in its formal and informal aspects. Knowledge of the environment, its conservation and threats must be integrated with the development of sensitivity to, and respect for, the natural environment and the formation of proper attitudes towards it. Fundamental education is therefore the kind of education aimed at realizing a sustainable living for mankind as a whole.

The Malaysian curricula need adjustments to allow for the inclusion of standard environmental education and training at the primary, secondary, tertiary, and informal levels. In so doing, the nations and their peoples would prospectively thwart the on-going environmental damage which is a threat to human survival and sustenance both now and in the future due to the lack of proper management of solid waste. Another aspect that is important to highlight because of its practical consequences on environmental education is teacher education. How do we expect a teacher to teach environmental education if he or she has not received the minimal tools to do so?

For example, environmental education nowadays is included in some way in most of the basic education curricula, but teachers are not qualified to teach it. The pedagogical approach and the teacher's interest in environmental issues seem to affect children's learning processes. A major bottleneck of education in general, and environmental education in particular, is teacher training and sensitivity about environmental matters [30].

In the case of the National University of Malaysia, since the students and most of the workers had never been exposed to any proper environmental education before, we develop an environmental and waste awareness and education programme that will help them. The university is committed to protect the environment by developing practices that are safe, sustainable and environmentally friendly and has developed a practical, staged approach to manage waste in an increasingly sustainable fashion. The programme is known as Integrated SWM UKM Campus towards "Zero Waste Campus". This programme has been developed with the aims of: -

• reducing waste produce;
• increasing and maintaining participation in recycling and composting schemes within the university
• raising and maintaining awareness of waste issues;
• promoting the Waste Hierarchy – reduce, reuse, recycle;
• providing a diverse range of ways of increasing education and awareness within the university;
• giving a message that is consistent;
• linking in with regional and national campaigns;

The first step is to identify the target groups which involve the office staffs, students, lecturers, hostel operators, canteen operators, cleaners inside the building, also out door cleaners (cleaning and landscape workers) and security guards. The next step involve information dissemination through waste awareness and education programme given to the target groups by the researchers with the help of staff from *Alam Flora Sdn Bhd* (an agency responsible for SWM in state of Selangor). For each target group, different approach was used to develop waste recycling systems among them.

1.   Office staffs

Briefing to office staffs were conducted in stages, whereby the importance of proper SWM, recycling and hands-on guide on how to start recycling programme at the office were given.

2. Lecturers and students

Among the lecturers and students, briefing will also conducted in scheduled basis. To create awareness, distribution of leaflets and posters were carried out. Lecturers were encourage to allocate 5 minutes before the start of lecture to explain on SWM issues. Each faculty need to organize recycling week and competition on highest volume of recyclables collected among faculty. Recycling facilities were provided at the faculty building to guide and facilitate the students.

3. The hostels

Briefing were also conducted to the hostels operator and we provide enough facilities at strategic location for separation at source activities. At the end of each semester major spring-cleaning was organized to recover recyclables such as old notes or books.

4. Canteen/cafeteria operators

These operators were explained on how to carry out separation of biowaste and recyclables. They were asked to reduce plastic packaging by encouraging students to bring their own container to pack their food. This program also introduce composting organic waste to the operators by providing them with composters (Best management practices (BMPS)

for incorporating food residuals into existing yard waste composting operation, 2009) [31] and guide how to start composting by staff of *Alam Flora Sdn Bhd*.

5. Cleaners

Briefing to the cleaners was about separation at source activity. We also discussed the effective modus operandi to do collection and storage of recyclables. The incentives in the form of money was also highlighted.

6. Other related activities

Other related activities conducted were e-waste recycling day, no waste day, flora green message to all staffs and exhibitions and awareness talks regarding the greening of the campus. Souvenirs given such as pens, containers and badges were written with wordings such as "think first before throwing" and the 3R logo. All things associated with this program were green in colour.

However, the more important aspect is the ability of the individuals to assimilate and interpret the information gain from education, building knowledge through a process of learning, which would give them the ability to act. In order to transfer the knowledge into practice or good environmental behaviour the students' perceptions and attitude have to be change. It is hope that the knowledge gain from the education and awareness programmes given should at least improve the way in which waste is managed within the university.

Since more than 60% of the university's population are students so another seminar was initiated to increase 'awareness of waste'. By attending this short two-hour training course students will be able to:

- Understand what is waste.
- Reduce, re-use, and recycle waste.
- Know about disposal routes for remaining waste.
- Deal with waste safely.
- Know the legal responsibilities for waste.
- Identify the sources of help for those difficult waste questions

We also initiate the formation of the Zero Waste Club whereby the members of the club act as change agent which provide model actions and were responsible to persuade low willingness students to participate in collection and recycling. They used their personal influence to have their friends join the campaigns and activities associated with zero waste activities. We hope that we are in a position to provide strong leadership and example in the development of sustainable communities, by conducting programmes for the students who will be our future leaders.

7. Conclusion

Any environmental programme at a university must be rooted in the belief that the process of paying attention to the environment will have the greatest impact if it becomes an integral part of the educational mission of the institution. The initiative offers a means to connect what happens in the classroom with what is happening immediately outside. Recycling alone will not earn a campus a clean bill of environmental health. Waste reduction and reuse are far more effective ways of reducing environmental impact, and the goal should be a net reduction in the campus waste stream, not simply more recycling. Yet, recycling is among the most visible, measurable, and enforceable of the environmentally sound practices that a campus can undertake. It is also important to make public the commitment to sustainable waste management since universities assume a special societal responsibility, in that they educate the future decision-makers of society. They take on a multiplier function and therefore a significant responsibility. Environmental protection should be the responsibility of all students and employees. The university will only fulfill this task when as many university members as possible identify with the aims of environmental protection and sustainable development, and actively contribute to the implementation of such aims.

Author details

Asmawati Desa, Nor Ba'yah Abd Kadir and Fatimah Yusooff
School of Psychology and Human Development, The National University of Malaysia, Bangi, Selangor, Malaysia

Acknowledgement

The authors acknowledges the National University of Malaysia for providing fund for research on this topic (UKM-PTS-098-2009).

8. References

[1] Jatput, R., Prasad, G.,& Chopra, A. K. Scenario of solid waste management in present India context. Caspian J. Env. Sci. 2009;7(1) : 45-53.

[2] Watch, W. What people think about waste, attitudes and awareness research into waste management and recycling. 1999. NOP Research Group, London.

[3] Maddox, P., Doran, C., Williams, I.D.,& Kus, M. The role of intergenerational influence in waste education programmes: The THAW project. Waste Management 2011;31:2590-2600.

[4] DEFRA.Survey of public attitudes and behaviors toward the environment. http://www.defra.gov.uk/environment/statistics/wastats/index.htm/ (accessed 2007)

[5] Timlett, R.,& Williams I. D.The impact of transient populations on recycling behavior in a densely populated urban environment. Resources, Conservation and Recycling. 2009; 52: 22-634.

[6] Saw Swee-Hock. The population of Peninsular Malaysia. Singapore: Institute of Southeast Asian Studies. 2007.

[7] Global Environment Center Malaysia. 2000.

[8] Malaysia Environment-Current issues-Geography. 2010.

[9] Cornerstone Content Management System. Solid Waste in Malaysia. 2002.

[10] Smyth, D. P., Fredeen, A, L., & Booth, A. L. Reducing solid waste in higher education: The first step towards 'greening' a university campus. Resources, Conservation and Recycling 2010;54: 1007-1016.

[11] Espinosa, R.M., Turpin, S., Polanco, G., De laTorre, A., Delfin, I., & Raygozs, I. Integral urban solid waste management program in a Mexican university. Waste Management 2008; 28 : S27-S32.

[12] Bolaane, B. Constraints to promoting people centred approaches in cycling. Habitat Int. 2006; 30:731-740.

[13] Ching, R.,& Gohan, R. Campus recycling: everyone plays a part. New Dir Higher Educ. 1992; 77: 113-125.

[14] Ballantyne, R., Connell, S., & Fien, J. Students as catalysis of environmental change: a framework for researching intergenerational influence through environmental education. Environmental Education Research 2006;12(3-4): 413-427.

[15] Tal, R.T. Community-based environmental education – a case study of teacher-parent collaboration. Environmental Education Research 2004; 10(4): 523-543.

[16] Ajzen, I., & Fishbein, M. Understanding attitudes and predicting social behavior. New Jersey : Prentice-Hall; 1980.

[17] Gamba, R., & Oskamp, S. Factors influencing community residents' participation in commingled curbside recycling programs. Environment and Behavior 1994; 26: 587-612.

[18] Kian-Ghee Tiew, Stefan Kruppa, Noor Ezlin Ahmad Basri & Hassan Basri. Municipal Solid Waste Composition Study at Universiti Kebangsaan Malaysia Campus. Australian Journal of Basic and Applied Sciences 2010; 4(12): 6380-6389.

[19] http://www. Raosoft.com/samplesize.htlm

[20] Hvatum, L & Kelly, A. Closing the knowing doing gap. SPA Conference 2008.

[21] Al-Najede, A. 1990. The effect of environmental science curriculum on development of environmental attitudes of in service teachers. Egyptian Associat. Curr. Teach. Methods 1990; 1:40-45.

[22] Lyons, E & G. Breakwell. Factors predicting environmental concern and indifference in 13- to 16 years-olds. Environ. Behaviour 1994; 26: 223-238

[23] Dunlap, R.E., G.H. Gallup & A.M. Gallup. Of global concern: Results of the health planet survey. Environment 1993; 35: 7-39.

[24] Inglehart, R. Public support for environmental protection: Objective problems and subjective values in 43 societies. Polit. Sci. Polit. 1995; 28:57-72.

[25] Olli, E., G. Grendstad & D. Wollebaek. Correlates of environmental behaviours: Bringing back social context. Environ. Behaviour 2001; 33:181-20.

[26] Hines, J.M., H.R. Hugerford & A.N. Tomera. Analysis and synthesis of research on responsible environmental behaviour. Journal Appl. Soc. Psycho.1986; 22: 657-676.

[27] Diekmann, A & P. Preisendorfer. Environmental behaviour-discrepancies between aspirations and reality. Rationality Soc.1998;10:79-102

[28] Miller, W. L., & L. Morris. *4th generation R&D—managing knowledge, technology, and innovation.* John Wiley, New York, New York, USA.1999.

[29] Pfeffer,J & R. I., Sutton. The knowing-doing gap: How smart companies turn knowledge into action. Cambridge: Harvard Business School Press; 2000.

[30] Barraza L. Environmental education in Mexican schools: the primary level. J Environ Educ.2001;32(3):31-/6.

[31] Best management practices (BMPS) for incorporating food residuals into existing yard waste composting operation, The US Composting Council Ronkonkoma, NY;2009.

Chapter 6

Managing Waste Through Managing People

Manfred Fehr

Additional information is available at the end of the chapter

1. Introduction

The status of the landfill

Municipal waste management policies to this date are constructed around the all important central facility: His Majesty The Landfill. Depending on the locally applied technology, there are all kinds of landfills. Some of them receive bio-waste and operate as anaerobic digesters, some collect the methane produced and burn it in flares, and others compress the methane to run gas or steam turbines. In recent times, in many parts of the world, bio-waste has to be treated in windrows before being tipped. In the developing world, a distinction is still made between open dumpsites, controlled landfills and sanitary landfills. Technical experience and local policies are dynamic. Both thrive on scientific accomplishments and are constantly on the move. In a developing country setting, the municipal administrator may be proud of having transformed his or her dumpsite into a landfill that receives all urban garbage without any methane capture. At the same point in time, the municipal administrator in an industrialized country is no longer allowed to tip bio-waste at his or her landfill and has to find next generation treatment or destinations for it. This author has even seen settings where methane capture facilities and steam turbines are being installed now at a landfill site, only to remove the methane originating from past deposits, because tipping of bio-waste has been legally banned.

Apart from the great variety of landfills, there is a great variety of urban waste compositions. Bio-waste represents approximately 30% of city garbage in the industrialized world, whereas it reaches the order of 70% in the developing world. It is easy to see the wide spread of challenges posed by the necessity of adequately "managing" this part of the garbage [1].

Landfill research has been concerned with correct location, construction and operation of the sites [2]. This is the essence of managing waste, because His Majesty is considered eternal.

Taking on the landfill

From simple concepts of material balances, which are the basic package of his chemical engineering background, this author has never been able to accept the existence of landfills as perpetual receivers of matter to the expense of natural resources. They represent a Parallel Planet, which swallows raw material without any return in sight. The battle against landfills finds its driving forces in this material balance upset, coupled with the rising opportunity cost of the huge Parallel Planet.

Inanimate matter does not listen or protest. "Managing" it is very easy. It goes to wherever the "manager" decides it should. It so happens that garbage does not exist in nature. It does not exist by its own will. It takes people to "produce" garbage. So this author set out on the road to "manage" people and induce them to stop producing garbage. If this were to succeed, landfills would become obsolete in the long run. All the scientific and administrative potential trapped in garbage collection and landfill operation would become available to the noble task of "managing" people in order to envisage a future with closed material cycles and without Parallel Planet. Of necessity, the research described here proceeded in Brazil: an emerging economy setting. Replication is thus essentially limited to similar settings inside and outside Brazil.

The top-down concept and its limitations

Governance strategies such as precise targeting for progressive landfill diversion of domestic waste are presently the single most urgent need of municipal administrators in emerging economies. General targets and directives are available from international organizations, but they need to be adapted and applied with local management talent.

The Zero Waste International Alliance (ZWIA) defines a zero waste goal as "to eliminate all discharges to land, water or air that are a threat to planetary health". It encourages the industrial sector to divert more than 90% of its waste from landfills or incinerators, and residential communities to establish benchmarks and timelines to reach 90% diversion within 15 years of adopting a plan [3].

At the turn of the century, the United Nations Organization introduced quantitative targets into global and local sanitation policies. The Millennium Goals of 2000 included environmental sustainability generally [4], and the documents originating from the Johannesburg Summit of 2002 referred to the advancement of sanitation services specifically [5]. The precise target of the Johannesburg Summit was and still is to reduce by one half the proportions of people excluded from sanitary services within a timeframe of 12 years, from 2003 to 2015.

Two unpleasant latent questions have so far eluded a satisfactory answer: When will complete domestic waste collection service become standard practice and when will all domestic waste be diverted from the landfills?

In this top-down environment, the directives are pushed along the pecking order from world summits to national governments and from there to regional and municipal administrations. The latter are considered the end of the line, and results are expected to appear at this geographical entity. The main stakeholder in the waste movement chain is the producer, and not any of the cited government levels. In the case of domestic waste, it is the resident or the residential unit. In the case of school waste it is a particular school community. Neither of them is directly addressed by top-down policy directives and thus, they have remained relatively immune to those directives.

Present hierarchy of waste related policies

Table 1 summarizes the hierarchy for policy directives and decisions relative to urban waste.

Level 1	World Summits emit general and universal directives for the advancement of sanitation services
Level 2	National governments structure their pretensions to improve sanitary services in general and urban waste management in particular through legislation
Level 3	Local administrations are faced with the need for applying the national legislation within their geographical area of responsibility by way of detailing it and defining specific procedures of operation with precise timeframes
Level 4	The community at large is expected to obey the established local procedure, supported by private reverse logistics facilities

Table 1. Levels of waste management hierarchy (source: this research).

In past research [6], [7], this author has addressed levels 1 to 3 of the hierarchy and proposed methods to arrive at suitable timeframes for the evolution of urban waste management in cities. Those timeframes were derived from the 2002 World Summit directives and specific experiments with source separation in Brazil. To illustrate the type of approach, the result of that work is reproduced in Table 2.

Year	Target	% Tipped
2003	local diagnosis of waste movement	100
2015	the universe of residences not served by collection in 2003 has been halved	100
2027	all domestic waste is being collected (turning point)	100
2039	the local management threshold meaning the best achievable source separation result is reached, and all source separated waste is recycled	~33%
2051	the locally defined *zero waste* situation is reached	~0%

Table 2. General target schedule for reaching the stage of *zero waste* in municipalities of emerging economies (source [6]).

The treatment essentially follows the top-down order of directives, but derives its conclusions from experiments at the local level. The management threshold mentioned in Table 2 is the specific result of coaching people in source separation. In the experimental setting of a private initiative [6], it was shown that approximately 33% of domestic waste eluded source separation efforts and ended up in the landfill. The present chapter is dedicated to describing the experimental methods used by the author to tackle this portion and to report on the results obtained. The challenge consisted of working at level 4 of Table 1 by producing precedents in sample communities for progressive landfill diversion of their waste.

The ZWIA (Zero Waste International Alliance) essentially works at level 3 as it attempts to convince city administrations and legislatures to enact the pertinent legal instruments for moving the city towards the final zero waste targets [3]. This author's team has produced material for this level as well. It took the form of a list of temporal and physical commitments by a city administration to reach the *zero waste* target within 15 years, and will be described shortly. Table 3 is anticipated here for illustration [7].

Year 1	Disseminate the program and the specific methods to be used among construction companies and health care institutions.
Year 2	Stimulate construction companies to anticipate the law and develop their treatment and recycling infrastructure. Initiate the collection of dry separated residue from residences by existing reverse logistics operators. Organize pilot units of decentralized composting and collect bio-waste from restaurants and pioneering apartment buildings.
Year 3	Establish voluntary drop-off points for large-volume trash items. Instruct medical establishments to separate their contaminated refuse.
Year 4	Pass the municipal law of solid waste. Irregular or clandestine deposits receive fines. Source separation becomes compulsory and is checked by municipal agents. Bio-waste is collected from residences in closed containers by operators of decentralized composting units.
Year 5	Existing reverse logistics operators absorb all separated dry material.
Year 6	The precarious rubble landfill is closed. Health care residues are collected and treated by private companies against payment.
Year 7	Centralized composting facilities are started by private operators
Year 8	All industries have their waste management programs.
Year 9	Recycled material begins to appear as raw material on the local market.
Year 10	Recycling industries establish themselves in the city.
Year 11	All producers of bio-waste have their management programs. The municipal landfill ceases to accept bio-waste.
Year 12	Barrels for collection of dry and clean packaging material are posted in the streets.
Year 13	Fines are applied to people who do not separate their waste.
Year 14	The municipal collection crews do no longer take away separated residues. They only take refuse, which does not fit the definitions of bio-waste or inert recyclables, to the correct destination to be defined by the city administration.
Year 15	There is no more waste to be tipped at the landfill.

Table 3. Municipal action plan of 15 years to reach a *zero waste* situation (source [7]).

Objectives for bottom-up warfare against the landfill

The following objectives were defined for the author's work on landfill reduction.

Devise methods that address the bottom of the waste movement hierarchy, namely the population at large.

Produce precedents in apartment buildings and schools to be replicated throughout the city in anticipation of municipal policies and laws.

Involve the complete existing reverse logistics chain from residents to wholesalers.

Discover the market forces in the chain and how they can be used to advantage by shrewd managers.

Rewrite waste management theory into people management theory by showing that source separation drives reverse logistics and not vice versa.

Describe and discuss the continuous learning path the research afforded for working with and managing the behavior of people.

Apply and expand the formerly developed theory of target scheduling at the municipal level. This means setting and obeying the timelines mentioned in Table 2.

For purposes of replication by interested communities, describe and discuss the specific infrastructure for waste collection developed in the apartment buildings and schools under study that has to form the transition point between families and reverse logistics operators.

2. Problem statement

The tale of garbage composition is an essential part of the success story.

Recently the author was asked to define *garbage* to lay people, i.e. lay people with respect to environmental management. They were all professionals of some field. So what is *garbage*? It does not exist in nature or on the market. It is produced by families right in the residences. The emphasis is on *produced*! How so? If you have a banana or a potato peel, this is not *garbage*. It is a food residue that can be composted and returned to the cropland. If you have a used sheet of writing paper or some pieces of cardboard, this is not *garbage* either. It is a cellulose residue that can be collected and returned to the paper machine. No *garbage* so far. Now, as soon as the families decide to mix the two items and to put them into a single bag to be thrown away, they *produce garbage* because then neither residue can be used again. *Garbage* then is not a byproduct of manufactured goods or agricultural produce. It is a product of human behavior.

Once *garbage* is produced, what does it consist of? There is a surprising similarity in comparable countries. Since the author lives in the BRICS community (acronym for the emerging economies Brazil, Russia, India, China and South Africa), the pertinent data originate from a group of countries here referred to as BRICS et al. *Garbage*, i.e. raw waste, has been analyzed in many of those countries. How much bio-waste has been found? China 72%, India 71%, Brazil 72%, Nigeria 72%, Nepal 71%. How many plastics? China 11%, India 9%, Brazil 11%, Nigeria 11%, Nepal 12%. In other words: comparable countries – comparable *garbage* – and consequently comparable needs for management [6]. Our world is not as vast and diversified as it may appear to the unsuspecting beholder! We are a global village after all. The apparent similarity may be misleading, though. Raw waste analysis is only the starting point of waste management. The ensuing competition poses to the cited countries the following challenge: Who will first succeed in transforming *garbage* composition into *residue* composition, i.e. raw waste composition into sorted waste composition? What can be recycled is clean *residue*, not raw waste or *garbage*.

Good management practices drive the composition around in a circle with a velocity that depends on local talent. *Garbage* composition opens and closes that circle as it is progressively transformed into residue composition. The secret of management here is to induce people to stop *producing garbage* and instead provide clean residues for recycle. The success may be represented by sequential points on the circle. Take the bio-waste as example. In the *garbage* there were 72% of it, but there was zero residue because *garbage* by definition is mixed matter. As soon as a community sets out on its learning curve and starts to keep its different residues separate, the percentage of bio-waste begins to grow and the *garbage* begins to diminish. This chapter will show separated biodegradable residue to reach 47% in medium size communities and 58% in small communities through sustained effort over long periods of time. The picture of the circle may be obvious now. As the effort proceeds, the bio-waste will eventually move around the circle and up to its limit of 72% of all waste and will thus be available for recycling. This closes the circle. Behavior management will have transformed 72% of biodegradable *garbage* into 72% of clean biodegradable residue for recycling. All other components will follow comparable evolutions, and the tale of *garbage* composition will become the tale of *residue* composition. The nomenclature is still somewhat precarious, but the basic idea of the composition tale is alive and invites for the big competition: Who will first succeed and become the champion of BRICS et al?

The problem may be stated as follows: Top-down directives essentially stagnate at level 3 of Table 1, mainly for lack of management talent at this level. Results have been unsatisfactory in emerging economies simply because level 4 of Table 1 has not been specifically targeted. The population at large has not been induced to modify its attitudes regarding waste, and this will not occur as long as His Majesty The Landfill remains the central piece of waste management strategies. City administrations of emerging economies will face the successive challenges listed in Table 2 over the next four decades. Up to the turning point, where all waste is being collected and tipped, the problem is mainly of technical nature. Raw waste production rate and composition will suffice to design, construct and operate His Majesty and all collection facilities. The demand on people reduces to the classical plastic bag procedure. They put all their garbage into a plastic bag and leave it at the curbside for collection. Picture 1 is provided to illustrate this procedure.

Picture 1. Classical plastic bag method of waste collection

As Table 2 predicts, this will continue until 2027 when top-down directives are expected to turn from tipping to recycling. From then on, the problem changes from managing waste into managing people. Level 4 of Table 1 will have to be specifically addressed. According to the theory of the management threshold mentioned earlier, sorted waste composition will become the all important parameter for setting recycling targets and developing the corresponding procedures. In order to prepare the approximation of levels 3 and 4 of Table 1 needed for successful planning, the research described here initiated the bottom-up direction of waste handling. The idea behind the method was to prepare people to accept and cope with the expected municipal legislation that eventually will make waste sorting at the source compulsory. The timeframe is sufficient for both sides to approach the meeting point. The municipal administration will develop and enact the bylaws, and the residents will practice source separation in anticipation of those bylaws, such that the transition will be rather swift.

3. Application area

As a bottom-up initiative, the research addressed people and their behavior in order to learn management lessons and acquire confidence in the types of results that can possibly be achieved. Apartment dwellers and school communities formed the testing ground. Additionally, existing reverse logistics facilities were included in the study because they are the link between residue producers and recycling industries. The reverse logistics chain provides the means of returning used material to the economy and consists of producers, retailers (recyclers) and wholesalers. The task comprised essentially two lines of action, namely teach residents and students to separate their waste and establish permanent relations with waste retailers in order to keep the separated material moving on to wholesalers.

4. Research course

The research followed its own learning curve by using experience gained in one step to tackle the next step. At the start, no experience was available and consequently, methods found in the literature were copied. They indicated the analysis of raw waste as essential starting point of waste management schemes. It soon became evident that raw waste composition is not a management tool because it cannot be reproduced upon separation at the source. As long as the waste is not separated at the source, reverse logistics has no raw material to work with and cannot grow. The most visible testimonies of this situation are the hydraulic bulk collection vehicles that circulate in the city, load all the waste bags left on the sidewalks and take them to the landfill. In the early stages of this research, the average domestic waste production in the cities studied (Araguari, Uberlândia and Ituiutaba) was measured to be approximately 630 grams per person per day [7], [8], but this number is dynamic. The data were obtained from the landfill operators and thus did not include waste dumped at illegal sites. In order to evaluate the potential for recycling, the research then turned to source separation. This is essentially a people game. Residents had to be coached over long periods of time in the task of separating their waste. This was done by providing oral and written instructions and following up with special visits. As soon as the first results of source separation appeared, reverse logistics entered the scene. This has been the rule of progress of the people game. Continuous contacts with reverse logistics operators have guaranteed that ever increasing quantities of source separated material are moved to appropriate destinations. The operators included informal recyclers and formal wholesalers. Progress of source separation also had to be accompanied by corresponding infrastructure improvements at the apartment building and school levels. The whole game is circular in nature, and this has been the most important experimental result. One action pushes the other, and the initiative may come from any action point in the circle. Source separation stimulates the infrastructure for collection, and suitable infrastructure attracts reverse logistics operators to move the material out. The system also works in the opposite direction. Improved infrastructure stimulates improved separation, and increased demand from reverse logistics pushes the infrastructure.

5. Method used

The strategy has been to experiment with small communities in order to test and approve the approach to "managing people". Pilot scale results are presented, and their extrapolation to the municipal level is outlined. This is a bottom-up procedure based on private initiatives that does not consume public funds.

The classical method of "managing waste" proceeds in the opposite direction. It relies on public funds to implement top-down models of desired waste movement. It starts at the wrong end of the reverse logistics chain, because it invests in recycler facilities without ever considering the residents who are the producers of the waste the recyclers are supposed to collect and recycle. This model has failed to produce any visible result of landfill diversion.

Consequently, this author has developed the theory according to which source separation drives reverse logistics, and not vice versa. Experience has shown that even a perfectly equipped recycler is unable to pick out any significant amount of useful items from mixed garbage. The people game then consists of coaching residents to provide separated residues for collection, instead of mixed garbage. This is the correct starting point for a management scheme because it focuses on the raw material producers of the reverse logistics chain. Separated residues possess an added value in the recycling market, and experience has taught this author that this added value is sufficient to create business opportunities not only for recyclers, but generally for all actors of the chain. There will be no need for social work to keep the residues moving along. Instead, the public administration can concentrate on regulating the market such that the landfill diversion reaches the desired level. Any amount of waste that enters reverse logistics and is recycled, immediately liberates funds from reduced landfill operating costs. Those funds can be redirected to managing people and to raising the added value of certain residues. The fundamental question a city administrator needs to answer in this game is this: How does the opportunity cost of tipping a certain residue compare with the expense required to divert it from the landfill?

The pilot experiments of managing people described here form the starting point of this type of consideration. They provide examples of producing raw material for reverse logistics at no cost to the public budget. The concept of replication refers to taking the message to more communities and teaching them to imitate the examples. A shrewd administrator will know how to proceed.

The basic strategy of the game has been to gradually reduce the degrees of freedom in the residue collection system. How can a resident of an apartment building be induced to deposit a certain residue in a certain container? The answer to this question will lead the way to an operational source separation system, but is not obvious from the start. This author has spent many years to partially reveal the secret. A definitive answer has not been found, but the fraction of success has moved from zero to approximately 0.67, meaning that fifteen years ago the apartment building under study did not divert anything from the landfill, and to-day it permanently diverts 67% of its waste.

The waste compositions studied and measured refer to total municipal waste in city A (Araguari), to unsorted household waste in cities A, B (Uberlândia) and C (Ituiutaba), to unsorted waste in apartment buildings and schools and to sorted waste in apartment buildings in city B. The total municipal waste composition in city A was obtained from the collection reports of city crews assigned to the task of collection and delivery to the landfill. The composition of unsorted household waste was experimentally determined by collecting waste left at the curbside in various town sections and in the town centers. This waste was spread out, sorted and weighed. Pictures 2 and 3 are provided to illustrate the method.

In the case of schools, the waste produced during the day was sorted and weighed by the students under the supervision of a research team. In the case of apartment buildings, the unsorted waste produced at the initial stage of the research was collected daily, spread out, sorted and weighed. Significant amounts of sorted waste became available only after several

Picture 2. Bio-waste sorting

Picture 3. Dry waste sorting

years of coaching. This waste already came from the apartments in divided form. Before weighing, the routine check by building employees moved any items from the wrong bin into the right bin.

6. Status

The zero waste tale for apartments is the happy end of the success story.

It used to be a dream. The Zero Waste International Alliance still calls it a vision. In the case described here for pioneering apartment dwellers, it has become a fact. No more collection and tipping. No more waste sitting on the sidewalk. It really sounds like to-morrow, but it is here to-day for anybody to see. The extrapolation may be restricted, but any determined community in a place with similar idiosyncrasy should be able to replicate the experience.

The point to be made here is this: A situation like the one described does not come about instantaneously through a single person's endeavor. It is a long-term management effort involving various stakeholders. It took fifteen years of team work to reach this fascinating point of a zero waste reality.

Moving on from apartments, a grade school was also chosen for the experiment. The average waste production in the school was measured to be 45 kg/day with the following composition: dry recyclable material 51%, biodegradable material 36% and refuse 13%. With 20 lecture days per month and 9 lecture months per year, this amounts to 8.1 tons/year. Considering that 2139 people frequent the school every day, the waste production reduced to 21 grams per person per day. This amount is an addition to the 630 grams per person per day of residential waste generation. It illustrates the fact that people produce waste in different locations during the day according to their activities. Inspiring extra-curricular activities were proposed to the students that produced compost from bio-waste and delivered inert recyclable material to existing reverse logistics.

This type of practical environmental education in schools takes the message to students' families and thus multiplies the result of landfill diversion. Picture 4 is provided to illustrate compost production by students as practical environmental education activity.

Picture 4. Composting in schools

7. Results

Developing a zero waste project for city A (Araguari)

This particular study was built on the hypothesis that it is possible the reach a *zero waste* situation in a Brazilian city of 100,000 population within a timeframe of 15 years, given administrative dedication and a detailed plan of execution with reasonable progress deadlines. The municipality of A had committed itself to collaborate with data collection. The author's responsibility was to develop the plan and the deadlines for execution [7]

Data were collected on municipal waste and its handling procedure from the administration and through interviews of key stakeholders of the waste business. In all 27 municipal schools, information was collected on environmental education and on school waste handling procedures. The activities of recycler cooperatives operating with inert material in the city were studied. From the publications of ZWIA, existing international experience was extracted. The main operators of reverse logistics facilities were interviewed to collect data on quantity of material commercialized. The main operators were three recycler cooperatives that collect dry recyclable material in the streets. The local prison was operating a composting project where data on quantities were obtained. Information on quantities of material handled in street cleaning activities was secured from the municipal department of public works. All those data were used to construct the diagnosis of the present waste situation.

The author's team followed the movement of all waste in the city. They observed the collection of inert material by autonomous recyclers and by cooperatives. They observed the collection of mixed domestic waste by compression vehicles of the concessionaire. They followed the trail of health care residues and of construction and demolition debris. It became evident e.g. that the trucks used to collect mixed waste are inadequate for the task. They are useful in developed countries where 70% of domestic waste is low density dry material. In that case, the compression is effective in reducing the volume. In emerging countries, only 30% of domestic waste is inert and dry. The rest is bio-waste, which is very humid and by nature is already compact [1]. The compression vehicle does not reduce the volume, but expels leachate, which drops on the pavement and produces bad odors. The study also collected information from the literature on zero waste projects existing elsewhere [3]. Those sources supplied the information on the extensive timeframes necessary for implementation. Personal measurements and data supplied by the public works department allowed the construction of tables with quantities and composition of domestic waste. The fraction of domestic waste collected by recycler cooperatives and thus diverted from the landfill was determined by observation. They operate in large halls where they separate the collected material and prepare it for sale. Information on income and expenses of the public administration with waste management was collected from City Hall archives. The literature survey contributed ideas from cities that already have their *zero waste* projects in place.

The results of the study consist of the quantitative analysis of city waste and of the proposed 15-year management plan. The analysis of waste movement in the year 2007 is shown in Table 4.

Construction debris	172.6 t/d	68.1%
Domestic and commercial residues	54.0 t/d	21.3%
Street cleaning waste	13.6 t/d	5.4%
Health care residues	0.5 t/d	0.2%
Miscellaneous not collected items	12.7 t/d	5.0%
Total	253.4 t/d	100.0%

Table 4. Waste produced in city A in 2007 (t/d = tons per day, source: this research)

Collection covers 95% of demand. The remaining 5% of residences and establishments not covered compose the last item in Table 4.

Construction debris and health care residues are collected and taken to treatment and final destination by the producers. The collection of debris is carried out by three companies who use trucks and mobile recipients of five cubic meters each, and by approximately 1500 autonomous persons who use small pushcarts. The transportation fees are paid by rubble producers. The final destination is a precarious rubble landfill designated by the public administration where all rubble is deposited without treatment or reuse.

A company specialized in pasteurization and incineration of health care residues collects them from the various establishments or residences and treats them against payment. The biologically inert material is then tipped at the landfill.

All other waste is the responsibility of the public administration. It maintains a team of 48 operators who collect and tip domestic waste, and another team of 85 people who take care of street cleaning and park maintenance. The teams use three trucks with hydraulic compression chambers of 10-ton capacity each for domestic waste and one open truck of 14-ton capacity for street cleaning residues. Domestic waste has the characteristics shown in Table 5.

Bio-waste		50.3%
Recyclable dry material		
paper	13.3%	
metals	2.6%	
glass	4.4%	
plastics	12.4%	32.8%
Refuse		16.9%
Total		100%

Table 5. Composition of domestic and commercial waste (including restaurants) in 2007 (source: this research)

There were two recycler cooperatives with 29 members and 180 autonomous collectors in town, for a total of 209 persons in the business of collecting dry recyclable material. Among them, they collected 10.6 t/d of material by visiting residences and by exploring the voluntary drop-off points in town. The 10.6 t/d represented 10.6 / (54.0*0.328) = 59.8% of dry recyclable material in domestic waste. This is a significant quantity. The sale of those components to the wholesale market can earn an average of BRL786.00/t (USD467.86/t). The value obeys the law of supply and demand and varies with time. The best prices encountered between 2007 and 2010 were those indicated in Table 6.

material	wholesale price BRL/t	
cardboard	450	
white paper	530	
ferrous metal	450	
aluminum	2200	
colorless glass	280	
colored glass	280	
rigid plastics	1000	
plastic film	1200	
PET	1200	
Tetrapak type	270	average 786

Table 6. Table 6 – Highest wholesale prices practiced in Brazil, years 2007 to 2010
BRL/t (local currency per metric ton, source: [9])

The 10.6 t/d earn an average of BRL8331.60/d (USD4959.29/d). When shared by the 209 people active in the sector, each one takes BRL39.86/d or BRL956.74/month (USD569.49/month). The conversion factors used here were: 1 month = 24 business days and 1 USD = 1.68 BRL.

As the legal minimum salary in Brazil in 2010 was BRL510.00/month, each recycler was able to earn 1.88 minimum salaries. This is only the starting point. With the action plan proposed hereafter and the consequent compulsory source separation, the collection will become easier and the earnings will grow.

From those data, the study developed the plan to divert from the landfill all waste produced in the city within a timeframe of 15 years. It established specific targets for each one of the 15 years with the respective responsibilities of the public administration who will put the plan into practice. Of the total amount of 253.4 t/d of waste produced in 2007, 12.7 t/d were not collected, 10.6 t/d were absorbed by the recyclers, and the remaining 230.1 t/d were tipped at the landfill. The challenge of the proposed plan consisted in extending the collection to 100% of demand and maintaining it there in spite of the constant demographic expansion.

The research produced the plan shown in Table 3, divided in stages of progress with the corresponding deadlines that still needs to be transformed into a municipal law to become effective. This is the standard procedure adopted by other cities in various countries, although the details vary. In the present case, the chain of events followed this order: oblige source separation using pioneering examples for illustration, take maximum advantage of existing reverse logistics, attract enterprises to process the raw material available in form of source separated waste and apply fines for disrespect of the law. Fifteen years were foreseen for this operation to reach steady state.

In order to dimension the challenge, the waste production was extrapolated 15 years into the future. The existing annual demographic expansion rate of 1.4% was considered constant and taken to represent the increase in waste production. During this period of 15 years, the plan calls for elimination of the miscellaneous not collected items in Table 4 and the refuse in Table 5. As a consequence of complete collection service and compulsory source separation, those items will be distributed to the other entries in the tables.

Tables 7 and 8 show the prospect of waste production and composition at the end of the 15-year period. This result was derived from the premise of compulsory source separation, which is a consequence of managing people. The two tables refer to the waste that is separated at the source and does not need to be taken to the landfill. The collection crews will continue to collect any items that by inspection cannot be recycled. Although the tables show projected values, they can be used by the city administration to dimension diversion efforts over the years. The corresponding investment can be depreciated over the years against reduced collection and tipping costs.

Construction debris	223.8 t/d	71.7%
Domestic and commercial waste	70.0 t/d	22.4%
Street cleaning residues	17.6 t/d	5.6%
Health care residues	0.8 t/d	0.3%
Total	312.2 t/d	100%

The 5% of miscellaneous not collected items in Table 4 were distributed proportionately. Example: $54.0*(1/0.95)*1.014^{15} = 70.0t/d$.

Table 7. Projection of municipal waste production in 2022 (source: this research)

Biodegradable matter	42.4 t/d	60.5%
Dry recyclable material	27.6 t/d	39.5%
Total	70.0 t/d	100%

The 16.9% refuse in Table 5 were distributed proportionately. Example: $50.3\% / [(100 - 16.9)/100] = 60.5\%$ and $70.0 * 0.605 = 42.4$ t/d

Table 8. Projection of composition of domestic and commercial waste in 2022 (source: this research)

Analyzing the existing reverse logistics chain in cities B (Uberlândia) and C (Ituiutaba)

The reverse logistics chains in those cities were studied by observing the actions and reactions of waste retailers and by maintaining contacts with several of them in order to discover the correct destinations for inert as well as biodegradable material.

Compulsory take-back systems are usually limited to specific items that are homogeneous and easily identifiable, such as batteries, tires, light bulbs or bottles. Residents simply take them back to the establishment where they bought them, and the items vanish from sight. The random logistics structure takes care of the large number of heterogeneous items in the waste stream that have to be delivered to their respective recycling industries through unidentified channels. Personal initiative is at a premium here, and the price paid by the industry moves the market. In Northern countries, the city administration usually collects source-separated material, either from households or from voluntary-drop-off areas. In Southern countries, source separation is not yet seriously practiced, and the logistics is left to private initiatives [1], [8]. The people who sift through all the material that was not separated at the source are traditionally referred to as *waste pickers* or *recyclers*. This research has raised their social status and underlined their importance to the municipality by introducing the designation *retailers* in the reverse logistics chain. They collect items of their own choosing from the material left at the curbside by residents, accumulate them and sell them to the wholesalers in the chain who are the intermediaries that stock material bought from the retailers and sell it to the recycling industries. In the cities of emerging economies there usually are a limited number of wholesalers and an impressive number of retailers. The whole system is moved by the price structure of waste material components. The city administration does not intervene as long as no specific landfill diversion target exists.

The waste material retailers limit their attention to dry inert items that may be easily scavenged at the curbside, taken away and sold. This part of the residues is thus well taken care of by private initiatives, but its success depends on effective management models. The research team observed the retailers' activities and identified basic needs for improving the success of this operation. The first problem is source separation. In the absence of source separation, the retailers are obliged to tear apart bags or other containers left at the curbside in order to pick out recyclable items. As a result, landfill diversion is obtained at the cost of dirty sidewalks and personal friction between residents and waste retailers. Institutional waste producers even have the official collection vehicles come to their premises to take away all waste material. In this situation, cardboard e.g., with a market value of 0.27 USD per kg, is mixed with other items and tipped at the landfill. It is like burying money instead of putting it into circulation. The second problem is collection time. The municipal collection vehicles have established timetables for every section of town, and residents leave their unsorted material at the curbside approximately one hour prior to collection. This is the rush hour for retailers. Whatever they are unable to pick up during this hour is fatally taken to the landfill. As a result, landfill diversion is obtained at the cost of uncontrollable intense scavenging activities at the curbside during certain hours of the day, or in some cases at the cost of having material sit on sidewalks all day. The third problem is traffic congestion. Retailers collect their material with pushcarts, bicycles or horse carriages and move along slowly from residence to residence or from building to building. Inevitably, traffic slows down. As a result, landfill diversion is obtained at the cost of slower traffic and the corresponding state of tension of drivers. The management challenge of municipal administrations in this context may be summarized into source separation and traffic control, in order to avoid the stated problems related to landfill diversion.

Moving away from classical waste composition reporting by substance, the new concept of sorted-waste composition was established and quantified by experiment. The experiment consisted of the following steps. A sample community was chosen. Unsorted waste composition was determined experimentally. Source-separation instructions were elaborated and passed on to the families of the community. Source-separation was monitored for eight months with regular analyses of the sorted material. Finally, sorted waste was analyzed and its composition established. This sorted-waste composition was used to establish landfill diversion targets as functions of time, possible to achieve with a modest educational effort.

The community was also used to observe and analyze the reverse logistics operators in the city. Source-separated material was put on display and the quantity of interested spontaneous takers was noted. This procedure was to answer the question whether reverse logistics drives source separation or vice versa. The question was answered by comparing the landfill diversion achieved with the established municipal practices to that achieved with source-separation in the sample community. Realistic landfill life spans were then calculated for various targeting options of recycling initiatives at the municipal level, and the results were evaluated against present Zero Waste initiatives throughout the World. Thus, the research provided the basis of landfill diversion targets as functions of time for the municipal administration.

The research also focused on the biodegradable portion of the waste stream that is presently simply taken to the landfill. It represents approximately 70% of all domestic waste material in Brazil [8]. In analogy to the compulsory take-back system mentioned earlier for specific inert product items, a pragmatic management model could require similar attitudes of food producers. The losses of fresh produce within the commercialization chain have been determined as 18% of turnover in the city, and household scraps amount to 45% of purchased food [10]. Those are experimental data. Environmental education may eventually reduce the amounts, but this would belong to another study. It would be reasonable to expect the farmers to take back those discards for reuse on the cropland. This is a tentative proposal of this research. It sounds futuristic, and the corresponding reverse logistics has to be developed. Even so, an effort has been made to take care of this large portion of waste in the municipal management model. Observations of bio-waste movement originating from restaurants and markets showed that the reuse procedure is already being practiced on a modest scale by strictly private initiative. It remains for the municipal administration to stimulate those initiatives in order to expand them to larger scales.

Two destinations other than the landfill were identified for bio-waste, namely animal feed and compost. Both applications only thrive on pure material, which in turn can only be obtained by source separation. Windrow composting of unsorted waste as applied in mechanical biological treatment stations reduces and neutralizes the mass to be tipped. The method has been tried in Brazil with the wrong objective to commercialize the compost, and has failed. It failed because a large human and financial effort was expanded to sort the collected mixed waste prior to composting, with no result [6]. The author chose to focus on a double-step source-separation in order to obtain pure biodegradable material for recycling, which would thus be diverted from the landfill. Tests of separating bio-waste were performed in sample communities with surprising results [8], [11]. Approximately 80% of residents responded positively to the request for separation. Once the separation was accomplished, compost could be prepared on available premises within the community, and additionally, farmers appeared spontaneously to take the material away for use as animal feed. This was an unexpected result, but it stressed the importance of source-separation in the reverse logistics chain.

The alternative of decentralized composting was also experimented with in sample communities, which included households, service sector enterprises, schools and apartment buildings. The compost produced in all of those places was of excellent quality and found immediate application as soil conditioner in gardening. Open composting bins made of wood or bricks were used, as illustrated on Picture 4. Analyses of compost produced by this research at different locations are presented in Table 9 to illustrate the quality achieved. It is noteworthy that the compost produced in the condominium complex is much drier than that produced in the other communities. This is so because the composting operation had to be conducted close to the residences and any visual or odorous inconvenience had to be avoided. The quality did not suffer.

analysis	Condominium Complex CELT	Residences on Afonso Pena Street	Antonio Nunes de Carvalho School
pH	7.4	8.0	8.1
density	0.40	0.41	0.69
total humidity	8.3%	26.9%	32.4%
total organic matter	46.5%	40.4%	15.0%
total carbon	25.8%	22.4%	8.3%
organic carbon	22.0%	15.7%	7.1%
total mineral residue	44.4%	32.5%	52.9%
total nitrogen	1.2%	1.4%	0.5%
total phosphorus (P_2O_5)	0.6%	3.6%	0.4%
total potassium (K_2O)	2.3%	2.4%	1.5%
total calcium	1.0%	0.3%	1.0%
total magnesium	0.18%	0.3%	0.13%
total sulphur	0.18%	0.1%	0.13%
C / N ratio	20 / 1	16 / 1	17 / 1

Table 9. Analyses of compost prepared from bio-waste collected in three different communities by this research, 2004 - 2007.

With the successful inclusion of bio-waste into the reverse logistics chain, the landfill diversion of this material became viable.

The research moved on to identifying the existing informal reverse logistics operators, in order to determine the landfill diversion potential provided by them but not presently taken advantage of in the municipal waste management model.

The research provided the understanding of reverse logistics in the city, developed and demonstrated the importance of pragmatic sorted-waste composition reports, illustrated the potential tasks of all stakeholders in the effort to reduce the size of the landfill, and developed ideas for solving the management problems inherent in reverse logistics. Apart from source-separation and traffic control, the study also explored unusual means of achieving diversion targets. They refer to promoting or subsidizing the reverse movement of certain waste items, such as multilayer packaging and glass in order to make them attractive to retailers and wholesalers. This idea has not been experimented with heretofore. No literature reports were found. It represents an original contribution to municipal waste management based on market forces.

The waste production rate in municipality C in 2004 stood at 194.3 tons per day with the following origins: builder's rubble 129.1, hospital trash 0.8, industrial 14.4, street cleaning 2, domestic 45 and tires 3 tons per day. This study was concerned exclusively with domestic waste. There were 28844 residential units in the city, with an average occupancy of 3.12 persons per residence. There were also 2985 commercial and 301 industrial establishments, apart from 192 public service points.

In the reverse logistics chain, 29 establishments were identified that buy and sell inert recyclable material in the region. The number of informal collectors could not be determined exactly. An estimate was 220 families.

There existed a modest selective collection program run by the municipal administration through a waste retailer co-operative. This program relied on door-to-door collection and diverted from the landfill 20 tons/month or 0.67 tons/day of dry material items, composed of 59% paper and cardboard, 30% plastics of all kinds, 9% metals and 2% glass of all kinds. The landfill diversion thus achieved was only 1.5% (0.67/45) of domestic waste produced, and even so, there was fierce competition for this small source-separated portion by individual retailers. The remaining 98.5% of the waste, including all bio-waste, was tipped at the landfill, and the operating cost of waste collection and landfilling stood at USD14000 per month in 2004 when the research was initiated in this town. As the ratio of domestic to total tipped waste was 45/194.3 = 0.2316, the cost component pertaining to domestic waste was USD 3242.40 per month (14000 * 0.2316) or USD2.37 per ton (3242.40 USD/month / (45 t/day * 30.4 days/month)). Any recycling scheme would thus be credited with USD2.37 for every ton diverted from the landfill. This is a proactive way of waste management accounting.

The city administration worked with a collected waste density of 0.170 t/m^3, which when compacted in collection vehicles and on the landfill rose to 0.700, estimated the population growth rate at 1.65% per year and the per capita waste production rate at 0,500 kg per person per day (45000 kg/d) / (90000 persons). From those data, and excluding the modest recycling rate, the life span of the existing landfill was estimated to be 20 years and went from 2004 to 2024, for a total projected capacity of 387000 tons or 553000 m3. The domestic waste collection coverage was complete in the city, and 21 vehicles of different kinds were used in this service.

Landfill diversion as a target is not present in the directives of the last World Summit of Sustainable Development [5], but worldwide intellectual movements such as the Zero Waste

International Alliance [3] and the Global Anti-Incineration Alliance [12] work with total diversion targets within timeframes varying from 15 to 25 years depending on specific municipal contexts. The 20-year bulk tipping model encountered in the city under study was thus considered antiquated, and the research turned to finding ways of moving the city closer to worldwide targeting practices.

The basis for a reliable diversion target is a correct waste composition report. The traditional gravimetric composition list for the city was evaluated as stated in Table 10 for a sample community of apartment dwellers. This report results from the analysis by substance of original waste prior to any sorting. In previous research, the method was applied to other sample communities with results similar to those shown in Table 10 [8].

Very soon it was discovered that this way of reporting composition, albeit scientifically correct, was unsuited for reaching management decisions and setting landfill diversion targets. The reason? The fractions indicated in Table 10 are not necessarily pure material, and it remains unknown how much of each fraction could really be separated at the source and consequently, would be absorbed by the reverse logistics chain. In order to establish more reliable reports for targeting, the research developed the concept of *sorted-waste composition*.

The difference between pre-sorting and post-sorting compositions has been established for the first time in this research, and will be explained now.

A source separation program was implemented in sample communities, was supervised and accompanied for eight months. It was learned that source separation has to occur in two steps to be successful. The first step is the separation at the family level. The second step is a complementary separation at the community level, which corrects the errors or flaws that occurred at the family level. In practice, this implies for the case of apartment buildings, e.g., that the building administration runs the second step with its employees who collect the material from the families and pass it on to the retailers only after complementary screening. This is one more lesson learned from the practice of managing people instead of managing waste. Contrary to inanimate waste, people have emotions and are not perfect. Their behavior deviates from the manager's expectations. From this experience, it was possible to produce a new and more reliable composition report, this time for sorted material. Table 11 shows the result.

material	mass %
biodegradable matter	70
plastics of all kinds	13
paper and cardboard	12
glass	3
metals	2

Table 10. Unsorted household waste composition by substance (source: this research)

Upon comparing Table 11 to Table 10, it is apparent that not all bio-waste present in the raw waste was effectively sorted out in the source-separation program. In the particular case under study here, the theoretical amount of 70% shrank to the practical limit of 47%, even with the two-step model. Where did the remainder go? It is hidden in the trash items of Table 11. The expression *educational trash* refers to the amount of material not sorted at the source. It contains bio-waste as well as inert matter in a state of mixture that makes sorting impossible. Here lies the fundamental argument of the composition paradigm: *Unsorted-material composition cannot be reproduced by normal source-separation due to human error and lack of dedication.* Families or house owners, even if instructed to do so, will not necessarily achieve perfect source separation. A realistic management model has to take this fact into account. Finally, the expression *administrative trash* refers to inert items that effectively are sorted out during the source-separation process, but nonetheless are unattractive to reverse logistics, mainly for reasons of pricing. They provide valuable input to a successful management model.

material	mass %
biodegradable matter	47
dry recyclable items	20
educational trash	29
administrative trash	4

Table 11. Sorted household waste composition by destination (source: this research)

The fallacy of the traditional waste composition report by substance can also be deduced from the information obtained at the retailer cooperative. Plastics in the city report of Table 10 represent 13 / 30 = 43.3% of dry waste. This fraction shrinks to 30% in the cooperative's recycling report, indicating that not all plastic material present in the waste stream was sorted at the source. The opposite is true for metals. The unsorted-waste report shows metals to account for 2 / 30 = 6.7% of dry waste (Table 10). In the cooperative's listing, metals occupy 9% of collected dry material, indicating that metals are easier to separate at the source than plastics and thus occupy a higher percentage level in the sorted-waste stream.

Figure 1 shows the material movement in a waste management model developed by this research based on double source-separation, on sorted-waste composition and on decentralized composting. The numbers in Figure 1 are taken directly from Table 11, except for the division of the bio-waste upon composting into bio-gas and compost, which is the result of the team's composting experiments. As the landfill diversion expectance increased from 1.5% to 67% in the new management model, the life span will be much longer than the 20 years originally predicted. The numbers shown in Figure 1 represent the best estimate of the immediate diversion potential in the city, which was not known prior to the execution of this research. It will form the basis of management actions. The challenge is to attain the potential within a stipulated timeframe, which in this case was set as 20 years.

What has been achieved so far? The bulk tipping model existing in 2004 was rejected because it was out of pace with modern waste management philosophies. Experimental sorted-waste composition was shown to indicate a diversion potential of 67% attainable through source-separation strategies. The timeframe of 20 years was fixed to reach this potential.

The challenge to put the model into practice required the study of the existing reverse logistics chain. The first question to be answered was whether the existing waste retailer community had sufficient capacity to face the required increase in inert items recycle. A condominium complex of 48 apartments was used to test the affirmative hypothesis. This experiment was carried out in combination with the sorted-waste composition study referred to earlier. As the source-separation program advanced until finally reaching the situation depicted in Table 11, the separated inert waste items became available for collection. The 220 families who comprise the retailer community randomly cover the whole city with their manual collection vehicles. Their collection efficiency is low because they depend on whatever they find displayed at the curb sides at the time they pass by. Planning is practically impossible. Several of the retailers who covered the street where the condominium complex is situated, were contacted. All of them expressed interest in passing by regularly and take away all inert items available. Those items included everything normally found in household waste and in this case separated from humid material, such as paper and cardboard, all types of plastics and metals, used clothing and wood. This lot has received the label "dry material" because it was and still is displayed in a completely dry state. The result achieved with this experiment was the proven fact that source-separation drives reverse logistics. The retailers to this day take away all dry components regardless of quantity. The arrangement with the selected retailers is to keep the material inside the building's premises for them to pass by during established hours of the day to take it away. There was no more need to have the material sit on the sidewalk. During the short period of loading, the retailers park their vehicles in front of the entrance to the building. Traffic flow is not affected. There is no expense to the city administration for this collection. The retailers take their lot to the wholesalers where it vanishes from sight. The city administration can concentrate its efforts and funds on the improvement of source separation, which according to Table 11 will drive reverse logistics to take care of 20% of all household waste. This experimentally established procedure is one more visible result of people management. Not only the apartment dwellers, but also the reverse logistics retailers had to be coached in order to behave as expected and close the circle of waste movement.

The reverse logistics for bio-waste is still incipient and needs to be developed. This is the other challenge for the administration in the 20-year period under consideration. Centralized and decentralized composting operations may be envisaged, and incentives need to be created for private operators to enter into this business. A collection infrastructure for bio-waste that uses appropriate vehicles without compacting accessories needs to be developed. The collection will attend those residential or institutional units who do not practice decentralized composting. The material will be taken to central composting stations to be processed by private operators. The city administration will provide the correct monetary incentives for this process by using the funds liberated by the reduced collection and tipping rates.

The test community of the aforementioned condominium complex was also used by this research to collect information on existing operators for bio-waste. Farmers from the vicinity were contacted. They expressed interest in the source-separated material for use in their poultry and pig-raising activities. In fact, they covered various points in town where they collected food residues on a daily basis, mainly restaurants and fruit and vegetable stores. They all used motorized vehicles and barrels to move the material. It was only a question of calling upon one farmer at a time to regularly collect the source-separated bio-waste portion of the residues in the test community, which according to Table 11 amounted to 47% of waste produced. The scheme has been operational for several years now and is available for extrapolation or imitation. The precedent has been created without cost to the municipal administration, and as a consequence of people management.

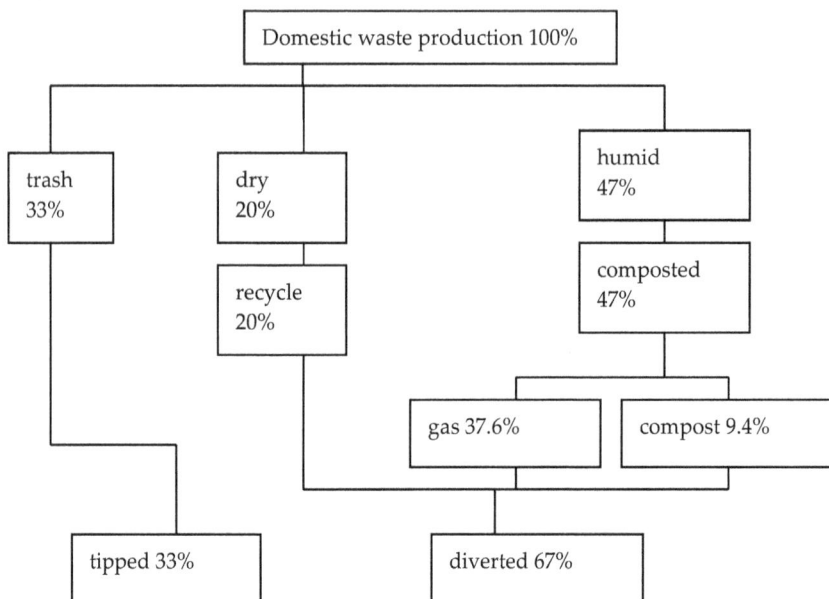

Figure 1. Material movement in the reverse logistics chain.
Management model developed by this research, based on sorted-waste composition.
Landfill diversion rates to be attained within 20 years.

The research also included composting tests. Experiments were carried out on the premises of the condominium complex, on the premises of the city water works and the electrical utility, and in the yards of several grade schools. All of these cases were successful in the sense that compost was prepared from food residues in a decentralized manner with excellent results. The composting operation according to the author's own research yields approximately 80% gas and 20% matured compost with 20% to 30% humidity. The reduction in volume is impressive, and the compost is immediately available for gardening. The experimental data of the decentralized composting operation have been published elsewhere [13].

In this respect, what remains to be done is to create centralized composting facilities in the city and stimulate their operation by private business. This challenge cannot be met by empirical research. It has to be included into the municipal waste management model.

The municipal model will include legal instruments that make source separation compulsory. The logistics for sorted dry material recycling is already in place. The private wholesalers and retailers will adapt their activities to improved source-separation. According to Table 11 and Figure 1 this will take care of 20% of domestic waste material at the end of the period considered. The municipal administration can concentrate its efforts on educational and legal measures to constantly improve source separation until reaching the target diversion of Figure 1. An example of tackling the 47% of bio-waste has been established by interaction with the farming community as illustrated. Additionally, the administration needs to solve the traffic problem created by reverse logistics. To start with, pertinent information has to be provided to the population on the strategic choices to be made. The unsustainable situation of leaving unsorted waste at the curb side to be indefinitely tipped at the landfill has to be confronted with the sustainable situation of presenting sorted material to reverse logistics retailers and exercising patience with temporary traffic slow-downs. A perfect management model is not available at this time. It needs to be constructed, and 20 years is ample time to do this.

What will be left to solve after the 20 years or after the 67% diversion mark? It will be the problem of educational and administrative trash as shown in Table 11. How to do this will depend on the creativity of the municipal administrators. As mentioned before, the most obvious route is to apply savings from reduced landfill operation costs to the creation of new reverse logistics channels for the items in question. What the tables do not show explicitly is the fact that a continuous effort is required to reach the target. If the tenure period of municipal administrations were four years, e.g., the waste management model would have to survive at least five successive administrations. This is the message the research described here conveys.

Developing an environmental management system in school A (EEJIS Public School, city B)

This part of the research confirmed the hypothesis that the concept of environmental management contained in the ISO 14001 norm can be applied in the context of public schools in Brazil. The work with people in this school went beyond waste handling procedures. It carried out an analysis of the school's physical interactions with the municipal infra-structure and indicated specific responsibilities of staff members in a possible environmental management system. The school's population consisted of 1924 students, 175 staff and 40 auxiliary personnel. The analysis of the school's impact on the physical city infrastructure identified water consumption as 3131 m^3/year, energy consumption as 46.1 Mwh/year, solid waste production as 8.1 tons/year, average noise level perceived by passers-by as 65 db, total school area as 6936.5 m^2, water permeable area as 2841 m^2 and impermeable area as 4095.5 m^2. The analysis evolved into the definition of permanent responsibilities of staff members to create a culture of constant improvements necessary for eventual ISO certification. This is an example of managing people to manage not only waste but to manage the environmental impact of a community.

Environmental impact analysis is a common management tool in enterprises of the production and service sectors. Its application in the public school context has not been reported upon, but environmental education practices exist in a variety of forms.

As early as 2004, the International Organization for Standardization (ISO) launched the Kid's ISO 14000 Program with the declared objectives to develop environmental awareness among children, to teach them to implement environmental management in homes and communities and to open them to the value of networking with young people in other schools [14]. The program was developed in Japan and has since spread to many other countries. By 2009 an estimated 210000 children worldwide had participated and achieved a 70000 ton reduction of CO_2 emissions. Reports from 2005 relate efforts to involve school children in Cambodia to protect the architectural, historical and cultural site of Angkor from deterioration by tourism [15].

The ISO 14001 certifiable norm itself details the procedure for any enterprise or community to implement an Environmental Management System (EMS) and apply for certification [16]. Thousands of enterprises worldwide have been certified. The importance of environmental education in the school context is described in [17], but the treatment remains general. No specific procedures for environmental impact analysis or possible certification are given.

The present study concerned itself with the novel idea of inducing a school community to seek certification as a pioneering experience in the Brazilian school universe. It went beyond simple environmental education by establishing and quantifying the impacts a school exerts on its neighborhood and on the municipal environment generally. This type of impact is not usually known to students. Its quantification is a first step toward creating environmental consciousness. From the analysis, proposals were elaborated for the school community as a whole to work towards reducing the impacts with specific physical and temporal targets in mind.

The implementation and continuous operation of an environmental management system is a long term effort that has to be supported by staff members. Students can at best contribute their temporary share during their period of school attendance. Consequently, the proposal defined the responsibilities in the ISO 14000 line of reasoning, namely assignments to specific parts of the community without identifying the persons who temporarily compose those parts. This translated into a management system where e.g. the reduction of energy consumption is the responsibility of the physics teachers and the management of waste is the responsibility of chemistry teachers and successive classes of grade 10 students. The inherent hypothesis stated that once those general responsibilities are routinely attended to with the corresponding result reporting, ISO 14001 certification may be envisaged.

The specific school chosen for the test was a public school in city B, which provides primary and secondary instruction from grades one to twelve. The idea that permeated the study was to create a precedent of an environmental impact analysis to be imitated elsewhere within the school universe.

The literature was searched for similar projects in institutions of higher learning and small businesses, but no precedent to the present study was found.

The analysis of the school's impact on the physical city infrastructure contemplated the topics of solid waste with respect to characteristics, quantity produced and destination, drinking water, storm water, energy consumption, soil available for rain water infiltration, existing vegetation on the premises, food consumption, traffic congestion related to student circulation and noise levels.

Data were collected by direct measurements, from school archives and personal observations. The purchasing files of the school administration yielded data on the consumption of food, light bulbs and plastic drinking cups. The solid waste produced was measured directly and analyzed according to possible destinations.

From the data collected, an environmental management system was proposed with temporal and physical targets as well as assignment of responsibilities. Specific programs were developed for improving the environmental impact derived from on-site vegetation, food quality, water consumption, solid waste production and energy consumption. The proposal also contained administrative infrastructure for continuous monitoring of those programs, a plan for obtaining the involvement of the total school community and for the preparation of monthly evaluations of progress.

Specific situation of solid waste

The average waste production in the school was measured to be 45 kg/day with the following composition: dry recyclable material 51%, biodegradable material 36% and refuse 13%. With 20 lecture days per month and 9 lecture months per year, this amounts to 8.1 tons/year. Considering that 2139 people frequent the school every day, the waste production reduced to 21 grams per person per day. This amount is an addition to the 630 grams per person per day of residential waste generation. It illustrates the fact that people produce waste in different locations during the day according to their activities.

A possible management system

In consequence of the impact analysis, the author's team proposed to the school administrators the participative development of an environmental management system (EMS) within approximately one year by assigning specific tasks to all segments of the school community with physical and temporal targets and demands for result reporting. The proposal indicated the formation of teams consisting of teachers and voluntary groups of students for carrying out each task. A sample task relating to waste management is listed below.

Task A: Solid waste management.

> Scope: establish waste production rates and best practice handling procedures for all waste and compare to production at home.
> Time line: one year.
> Responsible persons: chemistry teachers and 15 students.
> Progress indicator: reduction of percentage of waste taken to landfill

The impact analysis was a pioneering experiment in the Brazilian school context. It represented the starting point for devising the environmental management system. The proposal included the direct involvement of students and thus is expected to provide hands-on environmental education. The task of gardening will transmit notions on esthetics, ecology and utility of vegetation for mitigation of climate change. The task of solid waste management will raise questions on possible landfill diversion, and the idea of comparing generation per person in the school and in students' residences will take the message to the families. The same reasoning applies to water and energy management. Here students will be confronted for the first time with the need to check utility bills. The topic of rain water capture included in the proposal is an excellent engineering challenge to interested students. Topics of innovation and legal scriptures were included in order to provide breeding ground for social involvement of students as extracurricular activity. The topic on traffic congestion will be a challenge for prospective administrators to face, and that of noise levels within the school premises will involve considerations of social behavior. In all, the participation of students in the development of the management system can be an excellent stimulation for general environmental consciousness and personal involvement for many years to come. As students take the message home, they multiply and extrapolate the results of people management obtained in the school itself.

Experimenting with source separation in apartment building A (CELT Condominium, city B)

The evolution of domestic waste management practices in the urban residential condominium complex mentioned earlier is reported on here. A sustained effort over fourteen years has created a benchmark for landfill diversion by private initiative. The project was initiated in 1998 when the prevailing practice was to tip all waste at the landfill. In the presently attained situation, which is available for imitation elsewhere, 67% of all domestic waste produced in the complex is recycled without cost to the municipal administration. Instead of separating the inert recyclables, the effort was turned to separating the bio-waste. The management program derived from waste analyses and the work with people evolved into a two-stage source-separation procedure combined with the participation of handpicked reverse logistics operators. City crews now take to the landfill only 33% of all waste. Although this description is strictly valid only for Brazil, the story in itself might be of wider interest.

The condominium waste management project was created as a cooperation of the condominium administration with a graduate program of the local university. At the time (1998), the literature on urban waste management programs was still in its infancy. The industrialized countries already operated source-separation programs, but in Brazil only a few cities ran selective collection systems. The rule was mixed collection and tipping. Some municipalities had acquired sorting facilities for mixed waste, which at the time were considered modern. However, those facilities did not last because the cost-benefit relation was unfavorable. The cost of acquisition, installation and operation was high, the work environment of the employees was subhuman and the poor quality of the sorted material did not attract reverse logistics operators, i.e., operators of the commercial chain that returns residues to the market. At the end of the nineties, those facilities were gradually closed by environmental government agencies, and a large vacuum was created in the municipal administrations related to financial and managerial aspects of domestic waste [8]. The municipalities that did not have acquired experience with selective collection returned to the old practice of tipping all waste collected.

It was at this moment of pessimism and desperation that the idea of the present project emerged. The author had traveled and inspected management programs in other parts of the world, but had concluded that these programs have very little geographic mobility. The literature shows reports on waste management from various countries. A literature review proceeding from Japan relates the general tendencies for urban waste and the differences in composition and management philosophies from one country to another [18]. A study from Tanzania relates the management philosophies tested in cities there and stressed the importance of popular participation in reverse logistics [19]. The first reports on source-separation efforts proceed from the U.S.A. [20], [21], [22], but did not dwell on the results possible with management of people. The present author visited facilities in Canada and Spain where biogas was captured from landfills and used for electric power generation. This strategy was later challenged by studies on the aggressive impacts of tipped biodegradable matter on soil and ground water [2]. Consequently, industrialized countries started to prohibit tipping of those solids at landfills, and the mechanical-biological pretreatment (MBP) facilities made their appearance and became common in Europe. The author visited them in Austria and Germany, and identified the major impediment to easy technology transfer: significant differences in waste composition. The present author's own research established the "70-30 rule", which states that in industrialized countries domestic waste roughly consists of 70% by weight inert items and 30% by weight bio-waste, and that in Third World countries the opposite is true [1]. The windrows in MBP facilities in Europe handle 30% of domestic waste. In Brazil they would have to handle 70%.

The present project anticipated in Brazil the surge of theories and technologies that, in the first decade of the present century, substantially modified the concept of the landfill, which ceased to be the main component of waste management programs.

The initial step was to produce an analysis of the solid waste situation in the city under study. In accordance with standard practice at the time, the team collected and analyzed several tons of domestic waste. As result, the composition data of Table 10 were obtained.

The following observations are in order. On dividing the quantity collected by official vehicles by the number of inhabitants, the domestic waste production rate was 630 grams per person per day. The relation between bio-waste and inert material in the waste stream according to the results of this study shown in Table 10 was 70 / 30, a piece of information not previously available in the city.

The next step was to refine the results in order to create more confidence in the numbers. Table 10 was constructed from analyses in various city districts. A specific analysis in condominium A resulted in the composition data of Table 12.

material	mass %
biodegradable matter	68
plastics of all kinds	10
paper and cardboard	9
glass	4
metals	2
textiles	3
others	4

Table 12. Unsorted household waste composition in condominium A (source: this research)

The next step was to confirm the new waste composition paradigm. Table 11 represents the best result obtained so far with source separation in residences and indicates the immediate landfill diversion target as 67% of domestic waste. This experimental result is considered a management tool. It provides a precise and realistic target. The timeframe for extrapolation and citywide application will depend on the administrative effort expanded.

In sequence, the destination problem was faced. The team studied the operation of reverse logistics in the city, which is completely informal. This was an additional opportunity to innovate.

The inclusion of reverse logistics in a waste management program had never been considered by the public administration. In the sense understood here, reverse logistics consists of all the operators of the commercial chain that return residues to the market. The main actors are wholesalers of inert material, and individual recyclers who in the new nomenclature are known by the name reverse logistics retailers. There are also cooperatives and operators dedicated to specific residue items like food scraps, or, in the new nomenclature, bio-waste.

At the beginning of the learning process, the research team would leave the source-separated residues at the curbside for anybody to take. Everything always disappeared, but as the passage of the retailers occurred at random, it also happened that the material stayed at the curbside until the municipal collection vehicle came along and took everything to the landfill. With this experience, the team identified the main shortcoming of the municipal waste management program. The activity of the retailers, who are the environmental force in the system, was not part of the program and was even obstructed by the physical impossibility of functioning. In fact, until this day the retailers have at their disposal approximately one hour per day for scavenging recyclable items in the whole city. This is the time period between the moment the residents display their waste at the curbside and the moment the municipal collection vehicle passes by to take everything to the landfill. Quite rapidly, the team succeeded in selecting the most reliable retailers and in establishing a durable partnership with them. This facilitated the job of both the retailers and the municipal collectors and guaranteed the correct destination for all separated items. The recyclable material had to be kept hidden from the municipal collection crew in order to allow for recycling by the retailers. The network of the residue producers (the families), the intermediaries (the condominium administrators) and the receivers of sorted material (the retailers) was closed to the satisfaction of all concerned.

The learning process of this business occupied more than four years. The network of waste mobility was satisfactorily closed. In the last years since then, minor adjustments were still made to the procedure and to the underlying theory, and the system was pushed to the physical limit of recycling capacity.

How was the operational system pushed to the physical limit of recycling capacity?

Again using its own private initiative, the team started to attack the administrative trash. In 2005, the Brazilian National Environment Council passed Directive Number 358/2005, which made producers responsible for the correct destination of health care residues. Although medical establishments began to obey the Directive, residential units continued to place this type of residue into their garbage bins. The city administration simply does not have enough personnel to check on 160000 residences. The condominium administration wanted to set an example, and a contract was celebrated with a local company specialized in sterilizing health care residues. Various families in the condominium have persons under health care, small children or domestic animals that produce this type of residue. An additional container was put at their disposal to deposit their health care residues instead of mixing them with inert material. This type of residue represents up to four percent of total waste, and is collected weekly by the contractor. The condominium pays for collection and sterilization. The cost of this service is negotiated with the company and consists of a flat rate for the first 20 kg per month and an additional rate for every exceeding kg. The containers are supplied and regularly removed by the company. At present, the individual cost to each apartment comes to approximately 2% of the monthly administration charges.

Another contact was made with a company that specialized in processing used light bulbs and broken glass, which amount to two percent of total waste. There is presently no legal obligation in the city to divert this material, and reverse logistics operators do not take it. Once again, the condominium wanted to set an example. The price of the broken glass contributes to the processing cost of the light bulbs. Volunteers in the condominium take care of the transportation. An extra two containers were provided for light bulbs and broken glass, respectively, and the residents were instructed how to use them. The containers are 100-liter plastic barrels, which the condominium acquired as part of the maintenance bill. The collaboration has been satisfactory. The extended period of learning had its virtues. The residents were confronted with new requests for sorting one-by-one, and not all at the same time. This fact explains the satisfactory cooperation. "Satisfactory" means that 80 to 85% of the families adhere to the sorting instructions. As they had never before been exposed to anything similar in terms of behavior, and as the number of families moving in and out of the building is quite high, this degree of collaboration is the maximum to be expected. The building employees complete the sorting task and thus compensate for the families at fault. This task is included in their work contract upon establishing their remuneration. The pails for bio-waste collection from the families and the barrels for curbside display of refuse are part of the condominium's maintenance bill. Accordingly, the project already progressed beyond Table 11. The amount of "administrative trash" has already been reduced by 4% + 2% = 6%.

The proportion of waste presently diverted from the landfill arrived at (47 + 20) % = 67%, is steady and may be confirmed by inspection in the condominium at any time. The remaining 33% of waste is still displayed at the curbside to be taken to the landfill. The waste production in the condominium is on average 100 kg per day. The diversion is 67 kg per day. During the last ten years of operation, the team diverted from the landfill approximately

$$0.067 \text{ t/day} * 365 \text{ days/year} * 10 \text{ years} = 245 \text{ tons of waste.}$$

This is the visible result of people management and is available for replication.

Two more pictures are provided here to illustrate the source separation procedure. Picture 5 shows the basic three recipients used in pioneering apartments for waste disposal, namely one each for bio-waste, for dry recyclable material and for non-recyclable trash. Picture 6 shows the typical contents of the small bag that receives the trash.

Picture 5. Modern source-separation

The people game is a fascinating experience. Just to illustrate the interaction, the collection of batteries is evoked. The law obliges commercial establishments to receive used batteries back for disposal, but people are always looking for the least effort solution. They would throw the used batteries into the refuse bag, which in the apartment receives non-recyclable items. One day, the administration took the initiative of providing a pail at the collection point in the building, especially for batteries. To the surprise of all concerned, that pail kept filling up with batteries at a rate o approximately two kg per month. With a very simple provision, the infrastructure stimulated source separation. This reaction has to be seen in its own context, of course. Had the people not been coached on source separation for a long time, they would not have responded to the provision of the new facility the way they did, which underlines the circular nature of the people game.

Picture 6. Trash to incineration

The interaction between reverse logistics, infrastructure and source separation

The following story pretends to close the report on managing waste through managing people. It optimistically summarizes the success of the people oriented waste administration effort in condominium A. It accounts for the logical cause-effect relations that have been able to lead the way to the surprising limit of zero landfilling.

It used to be a dream. The Zero Waste International Alliance still calls it a vision. In the present case it has become a fact. No more collection and tipping. No more waste sitting on the sidewalk. It really sounds like to-morrow, but it is here to-day for anybody to see. To get the geography straight, this tale is happening in condominium complex CELT in city B with 650000 inhabitants situated in central Brazil. The extrapolation may be restricted, but any determined community in a place with idiosyncrasy similar to this one can replicate the experience.

The point to be made here is this: A situation like the one described does not come about instantaneously through a single person's endeavor. It is a long-term management effort involving various stakeholders, mainly the administrative team of the condominium. It took fourteen years to reach this fascinating point where a zero waste reality can be exhibited.

It started with the classical method of analyzing the waste. There are many forms of analyses for raw as much as for sorted waste. The sorted waste analysis is what matters because it indicates the possible destinations. The next step consisted of studying the existing reverse logistics stakeholders and facilities and of negotiating the waste movement with pertinent operators. This was followed by coaching families to sort their waste in the apartments. As this sorting is not perfect, the condominium employees came into the game to improve it. The source separated waste on average consisted of the items shown in Table 11. The administrative team provided the infrastructure for the waste movement and handed out spreadsheets to show the routing of all items. No miracles. Fourteen years of dedication.

The operational procedure now is such that on the whole, the condominium diverts approximately 67% of its waste from the landfill. This is the result of the overall sorting success of the families, but the present infrastructure allows for much more. So there are a few hardnosed families who have taken up the challenge of really using the infrastructure to its capacity, combined with a perfect sorting procedure at home, to show that "zero waste" can be achieved. Destinations other than the collection vehicle exist right now for all waste items from the condominium.

The point of the story is this: The internal operating procedure works its way down from the community administration to individual families or dwellers. The external scheme constantly negotiates deliveries and destinations with pertinent reverse logistics operators to keep the waste moving out. This is a permanent team effort.

Here is a sample of the pioneering families' present waste destinations. 58% is bio-waste or kitchen scraps that are delivered to a farmer for life stock breading. 30% is dry and clean packaging material like plastics, metals, glass, paper and cardboard. It is taken away by selected reverse logistics operators. 10% is health care or toilet residue taken away by a private company for treatment. This part is the only one with a price tag. The condominium budget covers the treatment. The remaining 2% is what is called refuse or trash. It does not fit into any of the preceding categories and is not attractive to reverse logistics. Examples are used napkins, plastized paper from packaged food, dirt from sweeping the floor, used filter paper from the coffee machine and the like. It is burned in industrial boilers or furnaces around town.

Pictures 7 and 8 are provided to illustrate how sorted waste is being offered to reverse logistics at the moment. The barrel seen in the background contains the 33% trash, which the city crew takes to the landfill. Upon comparing this situation to the starting point shown on Picture 1, the progress can be appreciated. It was slow, but it happened.

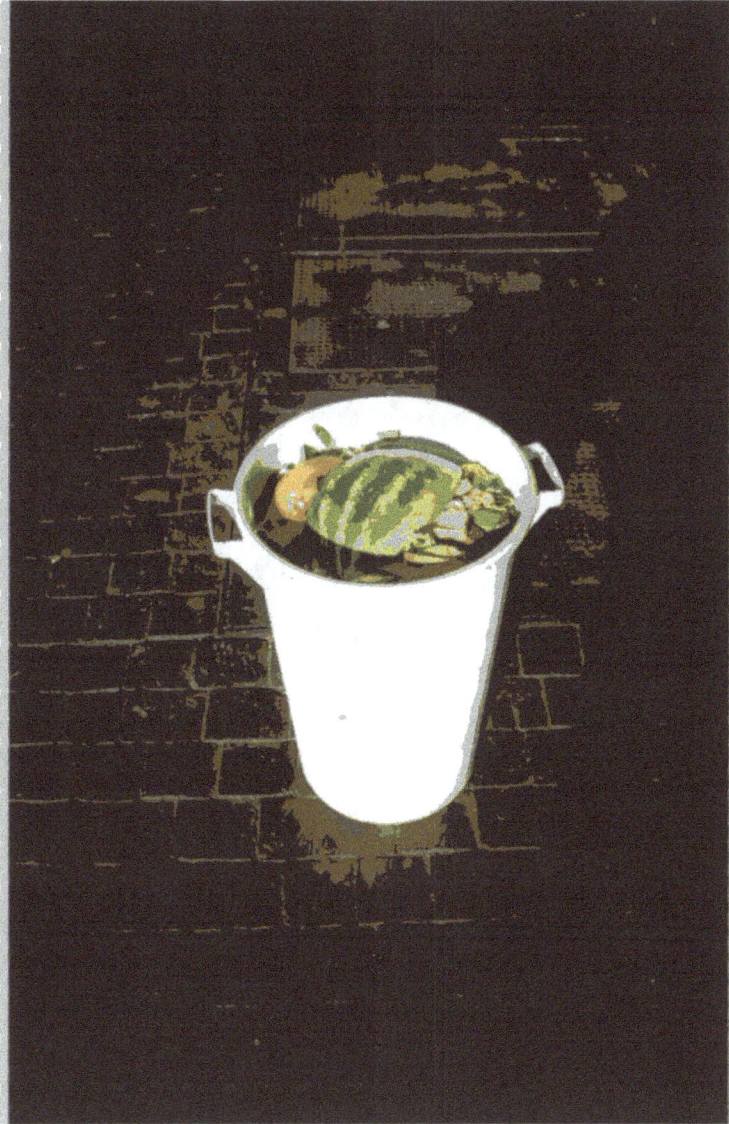

Picture 7. Sorted bio-waste ready for recycle

Picture 8. Sorted dry waste ready for recycle

Only a few families have so far reached this stage of perfect separation that eliminates the need for additional sorting by condominium employees and, above all, the need for collection by city crews. It is "zero waste" in the apartment, thanks to the condominium level infrastructure, and is the logical target for the rest of the families who are still hovering at 33% refuse. No utopia, only hard work.

Admittedly, the bio-waste still represents the biggest challenge, simply because existing reverse logistics operators are not prepared for handling it in the long run. Up to now, selected farmers have taken it away for life stock breading, as a result of continuous negotiations of the condominium administration with the farmers. The extrapolation of the scheme to a great number of condominium buildings will eventually require composting facilities to guarantee a secure and sustainable destination for this waste. This is the only point where the city administration may have to intervene. Incentives need to be created for private initiatives to run the composting facilities. It continues to be a people game, but its scope is beyond the reach of this research. Proposals are welcome.

8. Conclusions

In conclusion, the strategy has so far produced small scale examples on how to fight the landfills. "Zero waste" in apartments and schools has been shown to be achievable through the management of people, simply because waste obeys peoples' orders. Extrapolation of the examples to larger communities will form the challenge of the people game in the near future. This extrapolation will form the base of procedures by city administrations on implementing zero waste programs. His Majesty's days are counted.

Author details

Manfred Fehr
Institute of Geography, Federal University at Uberlândia, Uberlândia, Brazil

Acknowledgement

The Brazilian Research Council CNPq supported the study through grant 301120/2007-2.

9. References

[1] Fehr, M., 2002, The prospect of municipal waste landfill diversion depends on geographical location, The Environmentalist, Springer International, 22 (4): 319-324 ISSN 0251 1088 (print) 1573-2991 (electronic) http://dx.doi.org/10.1023/A:1020710829477

[2] Cossu, R, Lavagnolo, MC, Raga, R, 2000, Role of landfilling in the modern strategies for solid waste management, Wastecon Biennial Conference and Exhibition, Cape Town, September 5-9, proceedings vol. 1 pp. 1-17

[3] ZWIA 2010, Zero Waste International Alliance http://www.zwia.org (2010 12 13)

[4] United Nations 2000, Millennium Goals http://www.un.org/millenniumgoals (2010 11 23)

[5] United Nations 2002, Johannesburg Summit http://www.un.org/events/wssd (2010 11 27)

[6] Fehr, M. 2011, No more need for landfills in emerging economies after 2051, The Open Waste Management Journal, Bentham Open Publishers ISSN 1876 4002, 4: 38-41, http://dx.doi.org/10.2174/1876400201104010038

[7] Fehr, M. and Queiroz, P.C.D. 2012, A pilot project for living without a municipal landfill (in Spanish), Revista AIDIS de Ingeniería y Ciencias Ambientales, AIDIS Publications ISSN 0718 378X, 4 (3) http://www.journals.unam.mx/index.php/aidis

[8] Fehr, M., Calçado, M.R. 2001, Divided collection model for household waste achieves 80% landfill diversion, Journal of Solid Waste Technology and Management, ISSN 1088 1697, 27 (1): 22-26 www2.widener.edu/~sxw0004/abstracts.html

[9] Cempre Informa, São Paulo 2010, 18 (112), 4 http://www.cempre.org.br (2010 12 13)

[10] Fehr, M., Romão, D.C. 2001, Measurement of fruit and vegetable losses in Brazil: a case study, Environment, Development and Sustainability, ISSN 1387 585X, 3: 253-263 http://dx.doi.org/10.1023/A:1012773330384

[11] Fehr, M., 2003, Environmental management by the learning curve, Waste Management, ISSN 0956 053X 23 (5): 397-402 http://dx.doi.org/10.1016/S0956-053X(03)00063-1

[12] Global Anti-Incineration Alliance www.gaia.org (2011 12 09)

[13] Fehr, M. 2007, Confirming decentralized composting as definite option in urban waste management, International Journal of Environmental Technology and Management, special issue on Composting for MSW 7 (3-4): 274-285 ISSN 1741 511X (electronic) and 1466 2132 (print) http://dx.doi.org/10.1504/IJETM.2007.015145

[14] ISO, 2004, Kid's ISO 14000 Program ISBN 92 67 10388 1 http://www.iso.org/iso/pressrelease.htm?refid=Ref895 (2011 06 20) http://www.artech.or.jp/english/kids/envedu/index.html (2011 06 20)

[15] ISO, 2005, Cambodian school children put EMS basics into practice to preserve Angkor site http://www.iso.org/iso/pressrelease.htm?refid=Ref952 (2011 06 20)

[16] ISO 14001 Environmental Management Systems – Specification with guidance for use, 1996, Brazilian Association for Technical Norms ABNT, Rio de Janeiro, 14 pp.

[17] Mamede, F., Leite, A.L.A., 1999, Environmental education for sustainable development (in Portuguese), Ação Ambiental, Federal University at Viçosa Brazil, 2 (8):18-20

[18] Sakai S et al., 1996, *World trends in MSW management*, Waste Management (Elsevier) 16 (5-6): 341-350

[19] Kironde, JML, Ihdego, M, 1997, *The governance of of waste management in urban Tanzania: towards a community based approach*, Resources, Conservation and Recycling 21 (4): 213-226

[20] Dunson, CL, 1997, *Rocking to the right heavy metal*, World Wastes 40 (7): 23-27

[21] Merry, M, Glaub, JC, 1997, *Sticks and stones drive diversion home*, World Wastes 40 (6): 31-35

[22] Mitchell, K, South, G, 1997, *Four-stream collection: education equals results*, Resources Recycling 16 (9): 36-43

Strategies and Applications of Material Recovery and Final Disposal of Solid Waste

Agro-Industrial Waste Management: A Case Study of Soil Fauna Responses to the Use of Biowaste as Meadow Fertiliser in Galiza, Northwestern Spain

Mariana Matos-Moreira, Mario Cunha, M. Elvira López-Mosquera, Teresa Rodríguez and Emilio Carral

Additional information is available at the end of the chapter

1. Introduction

In the recent past, world wide traditional agricultural practice was based on the addition of biowaste, especially manure and slurry, to the soil. This reuse of biowaste allowed the recycling of nutrients and improved the level of organic matter. In the past, the amount of animal waste available was smaller than the amount currently produced, and the environmental impact of such waste application would be consider lower [1]. Over the past 50 years, the intensification of agricultural and livestock breeding activities has produced an increase in the number of livestock and, consequently, in the production and accumulation of large amounts of waste. This increase, associated with the use of mineral fertilisers and pesticides for fodder production, has weakened the complementary relationship between livestock and agricultural production. For this reason, the addition of organic waste to the soil has become a significant problem with potential environmental consequences. This practice can affect watercourses and trophic chains and can contribute to atmospheric pollution.

Agro-food industries are a relevant sector of the economy, and their activity is frequently associated with the production of wastewater. In regions such as Galiza (northwestern Spain), where the primary agricultural activity is the breeding of cattle for milk production, the industries that are dedicated to the processing and packaging of milk constitute a fundamental part of the agri-food sector, generating a significant volume of waste. In recent years, the increase in industrial activity has caused an increase in sewer sludge, and concerns about the economic and environmental impacts of sludge disposal have started to emerge. The recycling of biowaste by incorporating it into agricultural and/or forestry soil is one of the recommended methods for the elimination of this waste because, in addition to being economical, this method benefits the soil as a result of the incorporation of organic matter and nutrients. However, the addition of such waste is not without risk of environmental degradation or negative effects on the health of humans and other animals. Hence, it is fundamental to monitor the use of these materials in the soil. The current legislation requires only the assessment of the nutrient content and the needs of the recipient crop, the heavy metal concentrations in the waste and in the recipient soils, the bacterial content, and the risk of nitrate water pollution [2]. However, given the importance of the biological compartment of the soil in maintaining and sustaining system function, the monitoring of anthropogenic practices, such as the addition of organic waste to soil, should also consider soil biology as a fundamental indicator.

1.1. Organic biowaste

Three primary sources of organic waste exist: i) agricultural and forestry activities, ii) urban activity, and iii) industrial activity [1]. Wastes originating from agricultural and forestry activities include livestock slurry, manure, crop remains, and waste from pruning and from the maintenance of woodlands. Industries generate organic wastes, which include the subproducts of the agri-food industry (e.g., bagasse, coffee dregs, remains from slaughterhouses, subproducts of the fruit and legume industries, and milk serum), wool and skin remains, and cellulose sludge. Such organic waste is increasingly considered not only an environmental problem but also a potential resource whose recovery could lead to important economic benefits. This paradigm shift is powered partly by legislation and partly by market forces.

The addition of such wastes to the soil has several advantages, especially the improvement of the chemical and physical properties of the soil. These wastes (adequately composted) will increase the humus content and, as a consequence, the water retention capacity of the soil. The wastes also improve the soil structure, which is fundamental for root penetration and appropriate drainage and aeration [3]. In addition, organic waste is an important source of nutrients, and its addition to the soils closes the mineral cycle [4,5]. In agricultural areas, where soils are not limited by the organic matter content, as can be the case in Galiza [6], organic waste can help to ameliorate other adverse effects, such as acidity. Several studies indicate that these organic materials, if added to acidic soils, can be effective as acid neutralisers [7,8]; this effect is associated with an increase in fodder crop yield [9,10]. From a biological viewpoint, fertilisation with organic waste also induces an increase in microbial activity, which in turn improves nutrient availability for plants [11]. Similarly, the addition of organic waste reduces the amount of chemical fertiliser needed, thus leading to savings in energy and raw materials, with a concomitant reduction in the greenhouse effect. [12] includes the addition of organic waste among the management practices related to carbon sequestration. [13,14] have observed that organic agriculture systems (no synthetic fertilisers) produce less greenhouse gas than traditional systems.

1.1.1. Associated risks and applicable European legislation

In addition to its role as a source of nutrients, organic waste can also be a source of heavy metals and pathogens (viruses and bacteria) [16, 15]. Organic waste can also add excess nutrients, primarily N and P, that induce the eutrophication of superficial and phreatic watercourses [17]. To minimise such negative effects and regulate the use of organic waste as fertiliser in agriculture, Norm [18] refers to waste and contains the main definitions and principles that govern waste management, emphasising that waste assessment and elimination must be performed without creating risks for water, air, soil, or the flora and fauna.

Considering the great variety of waste types and the specific capacity and sensitivity shown by each soil type to the possible risks represented by the wastes cited above, scientists and researchers criticise the provisions of the current legislation in the field, which are limited to certain chemical analyses [19, 20]. It has been proposed that, in programs for the management of a specific waste type, the effect of that waste on soil should be quantified with parameters that are specific to the recipient soils and whose alteration can lead to the deterioration or the improvement of the soil quality. Moreover, it must be noted that the buffering capacity of soils prevents the detection of the negative consequences of exposure to a contaminant before saturation is reached. For this reason, certain authors propose that chemical analyses should be complemented by the assessment of other types of parameters that permit the collection of information on the bio-available fraction of contaminants and that reflect the effect of other pollutants that have not been identified [21,22]. Using several types of analyses, it will be possible to assess the effect of the waste on soil quality in a concrete and exact manner.

1.2. Soil fauna as a quality indicator

The sensitivity of the soil fauna to environmental disturbances and the roles played by the fauna in physical, chemical, and biological processes are attributes that allow the fauna to be used as an indicator of soil quality [23-29]

The response of the soil fauna to the addition of fertiliser is variable. Generally, the effects of fertiliser application in moderate doses are positive, originating from the modification of microclimatic conditions or from resource availability [4, 30-38].

1.2.1. Quantification of soil fauna

Bio-monitoring allows the identification and quantification of changes over time through the analysis of the following characteristics of the soil fauna: traditional ecological measurements (species abundance and diversity), morphological or behavioural changes, and accumulation in tissues. According to [39], the three levels of interaction between the fauna and soil quality are organisms and populations, communities, and biological processes. The most commonly used parameter in the quantification of the impact of agricultural practices is the assessment of communities because this scale integrates all of the soil factors, including management and pollution effects. At this level, the parameters most commonly used are the abundance of individuals or species, biomass, specific composition, trophic strategies, and the presence or abundance of key species.

Among the organisms that compose the soil fauna, the macrofauna reflects an integration of the processes that occur in the system because the macrofaunal organisms feed on primary decomposers (such as bacteria, fungi, and actinomycetes) and secondary consumers (such as protozoans). Moreover, because macrofaunal organisms are easily collected and because their ecological role is better documented than the roles of the micro- and mesofauna, certain authors view the macrofauna as the most appropriate category for the assessment of the impact of agricultural practices [24].

The data obtained from assessments of the soil macrofauna can be analysed with both univariate and multivariate techniques. Among the univariate techniques, the most general measurements are diversity indices, which synthesise the information on diversity in only one value. These indices are normally distributed and, hence, can be analysed with robust parametric tests such as an analysis of variance. This type of analysis allows rapid comparisons, subject to the statistical test, among the obtained values for different habitats or for the same habitat over time [40].

All multivariate techniques are based on similarity coefficients calculated for each pair of samples. These techniques facilitate the classification or grouping of samples in similar groups, with the distance between a pair of samples reflecting their relative dissimilarity with respect to species composition. Multivariate statistics allow higher resolution (i.e., subtle alterations can be detected) because all of the available information for the community is used. Moreover, by combining community data with soil variables, specific information can be obtained on the factors that are responsible for the alterations [41].

2. Case study: The response of the soil macrofauna to organic waste used as a meadow fertiliser in Galiza

2.1. Agro-industrial organic waste production in Galiza

In Galiza cattle farming produces the most manure, followed by pig and poultry farming (Table 1). This manure contains high levels of organic matter and mineral nutrients.

	Spain (Ton x 10^3)	Galiza (Ton x 10^3)	Galiza (%)
Cattle	42.085,3	6.909,6	16,4
Sheep	12.128,2	98,8	0,8
Goat	1.458,4	18,7	1,3
Pig	25.242,0	907,2	3,6
Horse	2.637,8	242,5	9,2
Broiler	7.695,4	712,7	9,3
Rabbit	407,2	85,7	21,0
Total	91.654,3	8.975,1	9,8

Table 1. Amount of animal manure production for different farm cattle in Galiza and Spain (year 2003)

Nitrogen is present in organic forms and as ammonia, the latter being more abundant in poultry and pig wastes; K is present as highly soluble salts in urine. Potassium is present primarily in organic form (Table 2).

	N (%)	P_2O_5 (%)	K_2O (%)
Cattle	0,35	0,28	0,22
Broiler	1,40	1,00	0,60
Sheep	0,75	0,60	0,30
Pig	0,60	0,45	0,50

Table 2. Content of N, P_2O_5 y K_2O for different animal manure in Galiza (NW Spain)

Manure from broiler chickens

Manure from broiler chickens is the product resulting from the fermentation of poultry manure on a bed that is usually composed of a cellulosic-lignic material, such as straw, sawdust, or rice skin, with a high nutrient content and low humidity. This type of manure contains a high percentage of dry matter and is richer in organic matter and nutrients than other types of manure [42]. Usually, poultry manure is used as a fertiliser for crops of high economic importance, such as corn, soy, hay, and horticultural crops [43].

Cattle manure slurry

Most of the cattle manure slurry produced in Galiza contains a very low percentage of dry matter (less than 6%), which can make the management of the slurry difficult [44]. Table 3 presents estimates of the annual production of nitrogen, phosphorus, potassium, and organic matter. The phosphorus content is considered to be high, and, for this reason, it is not necessary to add this nutrient in its chemical form [45].

	N	P_2O_5	K_2O	CaO	Organic matter
Cattle slurry	65.232	38.251	95.272	44.334	920.919
Pig slurry	8.095	5.980	7.725	5.864	222.685
Total	73.327	44.231	102.998	50.198	1.143.603

Table 3. Fertilizer power from cattle slurry produced in Galiza (NW Spain) (Equivalent Tons/year)

Slurry from dairy-industry purifiers

In Galiza, the sludge generated by the dairy industry is becoming more important given that this autonomous community produces nearly 40% of the country's milk (Table 4).

	Cattle	Sheep	Goat	Total
Galiza	2.300.838	-	-	2.300.838
Spain	6.158.179	414.211	488.746	7.061.136

Table 4. Milk production in Galiza and Spain (2007) (Litres x 1000)

Dairy slurry is generated by the purification of wastewater made up of milk remains and cleaning products such as water, sodium hydroxide, and nitric acid. Generally, effluents from the agri-food industry are easily biodegradable and lack toxins (organic contaminants, heavy metals), making these effluents easy to treat with biological and, especially, microbiological methods [46]. Wastewater can be recycled as part of a closed system in which the dairy slurry produced by purification is used by farmers to fertilise fields in areas near the factory. In general, most research on industrial slurries has focused on products from facilities that purify urban wastewater, although studies were performed on dairy-industry slurry during the 1970s [47]. Likewise, in Australia, certain national programs have attempted to promote a different legislative treatment of slurry produced by the treatment of effluents from dairy factories because heavy metals and chemical contaminants are present at much lower concentrations in these effluents than in slurries from urban purification facilities [48]. In Galiza, research on the dairy industry started a decade ago. [49, 50] determined the optimum application dose for meadow soils and the consequences of this treatment for fodder production. For acidic soils and soils devoted to other uses, [51] concluded that the total concentration of heavy metals was sufficiently low to preclude any environmental risk from this source.

2.2. Materials and methods

Study area

In September 2001, a field trial was performed in which mountain terrain was transformed into a field to provide more land for agriculture. The trial was performed in Goiriz-Lugo-Galiza (northwestern Spain; latitude: 43°19′N; longitude: 7°37′W') on humic umbrisol (FAO, 1998). The mean annual temperature in the area is 11.5 °C, and the mean annual precipitation is 1,084 mm. Most precipitation occurs during the autumn and winter (35% and 29%, respectively), and the amount of rain is the lowest during the spring (22%) and summer (14%). The vegetation on the starting soil was predominantly trees and shrubs, including *Pinus pinaster*, *Castanea sativa*, *Ulex* sp., and *Pteridium aquilinum*. After the soil was fertilised with 3 t ha^{-1} of limestone (CaO 60%), the following mixture was sown: 40 kg ha^{-1} of *Lolium perenne* L. cv. 'Tove', 20 kg ha^{-1} of *Lolium hybridum* Hausskn. cv. 'Texy', and 6 kg ha^{-1} of *Trifolium repens* L. cv. 'Huia'. Different plots were then subjected to different types of fertilisation. The trial consisted of five treatments: Control: low annual doses of mineral fertiliser, equivalent to 1/3 of the dose applied to the Mineral treatment plots, to increase the competitiveness of the sown species over the natural vegetation; Mineral: doses equivalent to 30 kg ha^{-1} of N and 45 kg ha^{-1} of P$_2$O$_5$; Cattle Manure Slurry: 50 m^3 ha^{-1}/year; Dairy Slurry: 120 m^3 ha^{-1}/year; and Broiler Litter: a single application of 4,500 kg ha^{-1} of the dehydrated product. Four randomly distributed replicates were performed for each treatment for a total of 20 experimental units of dimensions 3 x 1.3 m. These units were separated by corridors of 1.65 cm. For more information on the soil characteristics, fertiliser application and field management, see [52].

The primary characteristics of the different organic subproducts are presented in Table 5.

	DM	pH	EC	C^a	N^a %	P^b %	K^b %	Na^b %	Ca^b %	Mg^b	C/N	C/P
Cattle slurry	18,2^c	7,1	4,0	40,0	5,1	2,0	9,6	2,4	0,8	0,7	7,8	20,0
Dairy-industry sludge	20,0^c	7,1	3,4	35,6	6,2	2,1	1,1	3,2	2,2	0,4	5,7	17,0
Broiler litter	89,1^d	7,9	11,1	36,8	4,0	1,6	2,8	1,6	1,9	0,7	9,2	23,0

[a] Determination by CNS2000 auto-analyzer and [b] by Atomic Absorption Spectrometric in HNO₃ 70% extract solution. DM: Dry matter [c] (g L⁻¹) [d] (%). EC: Electrical conductivity (dS m⁻¹). C, N, K, Na, Ca, Mg (%). C/N, C/P, carbon/nitrogen and carbon/phosphorus rate.

Table 5. Physico-chemical characterization of different organic wastes.

Sampling and sample processing

The soil fauna was sampled in 2004 (May and November), 2005 (May and November), and 2006 (May) with pitfall traps [53]. In all, 20 traps were used per sampling season. During each sampling season, traps were collected after four days, and voucher specimens were preserved in 70% alcohol. At the laboratory, specimens were identified to upper taxonomic levels (family/order) using taxonomic keys [54, 55].

Data analysis

Initially, communities were described based on their abundance and taxonomic richness (no. of taxa present), and diversity indices were subsequently calculated based on the method of [39]. The indices calculated for the collected taxa included Simpson's diversity index (1-D), the Shannon-Wiener index (H'), the Berger-Parker index (d), the Simpson evenness index ($E_{1/D}$), and the Smith and Wilson evenness index (Evar). The data obtained were analysed with an analysis of variance (ANOVA). Logarithmic transformations were done when data departures from normal distribution and/or variance homogeneity. After the univariate descriptive analysis, a multivariate analysis was performed to attempt to group the different treatments based on the similarities of the macrofaunal community collected for each treatment. The multivariate analysis was performed with PRIMER 5.0 [55], and three factors were determined:

1. *season*, to separate spring (May 2004, May 2005, and May 2006) and autumn (November 2004 and November 2005).
2. *control*, to differentiate non-fertilised control plots (C) from fertilised plots (M, Mineral fertiliser; CS, Cattle Manure Slurry; DS, Dairy Slurry; and B, Broiler litter).
3. *fertiliser*, to differentiate non-fertilised plots (C) from plots fertilised with mineral fertiliser (M) and plots fertilised with biowaste (CS, DS, B).

In each analysis, taxa with an abundance of fewer than five individuals in all plots were not considered. The square roots of the data values were calculated, and a similarity matrix was calculated based on the Bray-Curtis coefficient [56]. A similarity analysis (ANOSIM) was performed to determine the statistical significance of the in-group discrimination. Afterwards, a SIMPER (*Similarity Percentage Breakdown*) analysis was performed to obtain the percentage contributed by each taxon to the in-group discrimination.

3. Results

A total of 6,496 specimens were captured. These specimens belonged to 42 taxa. The dominant taxonomic groups were Araneae (23.5%), Coleoptera (21.6%), Diptera (19.8%), and Hymenoptera (6.9%). Fewer than five individuals belonging to Diplopoda, Chilopoda, Isopoda, Trichoptera, and Ephemeroptera were observed for each treatment.

The abundance of individuals (N) and the number of taxa captured (S) varied significantly with sampling season. The Control and Mineral plots yielded a greater abundance of individuals. The Cattle Manure Slurry, Dairy Slurry, and Broiler Litter plots exhibited the lowest abundances.. Note that the abundances of individuals of Araneae, Diptera, and Coleoptera may have been influenced by both the treatment and the sampling season, with greater abundances in the spring and lower abundances in the autumn (Table 6).

	Total N		Total S			
	F	p	F	p		
Treatment	13.353	0.000	10.589	0.000		
Date	6,894	0.000	18.082	0.000		
	Aranea (N)		Coleoptera (N)		Diptera (N)	
	F	p	F	p	F	p
Treatment	8.378	0.000	15.336	0.000	6.503	0.000
Date	29.702	0.000	33.735	0.000	23.203	0.000

Table 6. Statically differences in N (abundance) and S (taxon richness) between treatments and sample season.

In the analyses of the distribution of taxon abundance (Figure 1), a better fit to the normal distribution was observed for the plots to which organic waste had been added, and a poorer fit was observed for the communities from the Mineral and Control plots. These results indicate that the addition of organic waste to the soil did not have a severe negative effect on the communities assessed in this study.

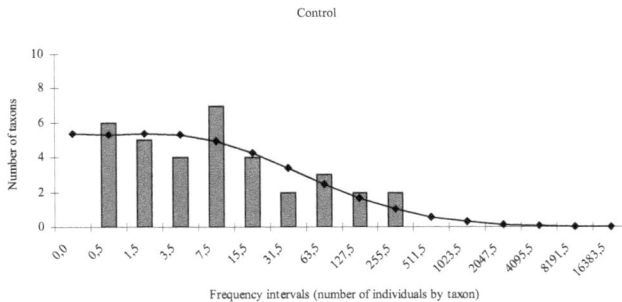

Control

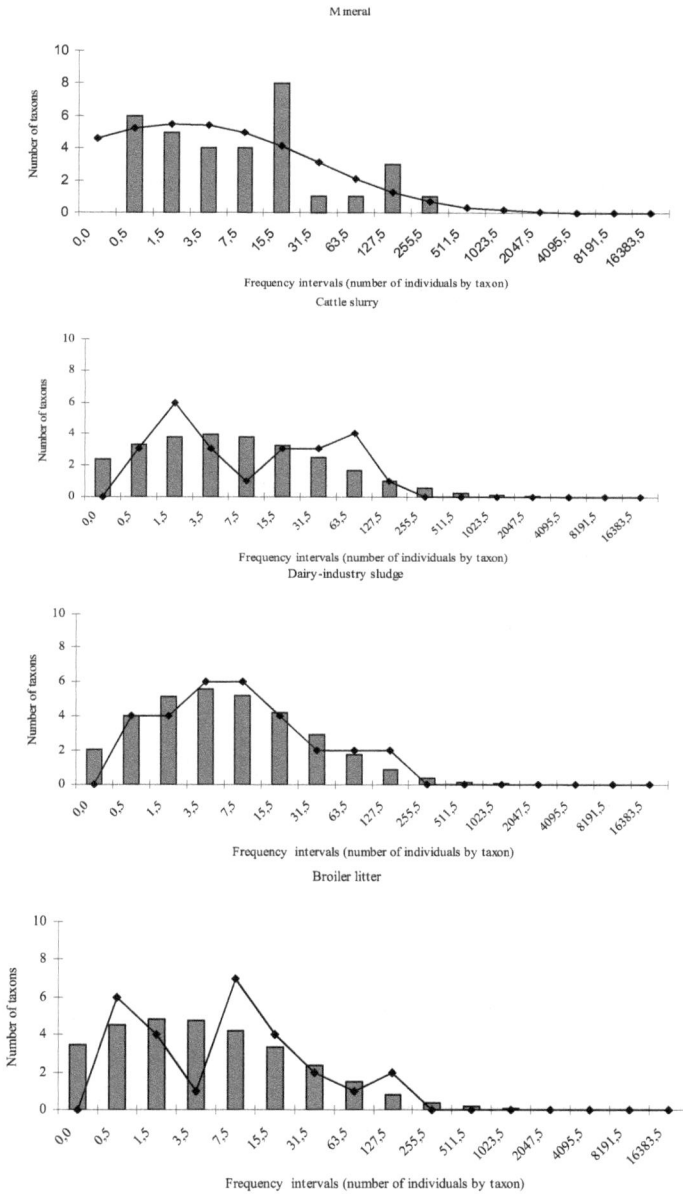

Bars: observed frequency. Dot lines: expected frequency.

Figure 1. Lognormal curves for abundance distribution (individuals/taxon) for each treatment

Indices of ecological diversity

The Simpson (1-D), Berger-Parker (d), and Shannon-Wiener (H') indices were not affected by the fertiliser treatment. According to these indices, the addition of organic waste to the soil did not cause statistically significant changes in the number of taxa or in the abundances of the taxa in the macrofaunal communities. In contrast, the Smith and Wilson and the Simpson evenness indices (Evar and $E_{1/D}$, respectively) were more sensitive to the different fertilisers applied. However, the fluctuations between the sampling seasons were also important (Table 7).

	1-D	H'	Evar	$E_{1/D}$	d
One-way ANOVA					
May-04					
F	1,278	1,141	4,825	5,504	1,822
p	0,316	0,375	0,006	0,003	0,159
Nov-04					
F	0,414	1,241	0,504	0,585	0,233
p	0,796	0,336	0,733	0,678	0,915
May-05					
F	0,688	2,816	3,536	4,749	0,621
p	0,639	0,048	0,021	0,006	0,686
Nov-05					
F	0,821	1,060	5,667	4,647	0,725
p	0,551	0,414	0,003	0,007	0,614
May-06					
F	1,303	0,856	3,644	2,407	1,261
p	0,307	0,529	0,019	0,077	0,323
Two-Way ANOVA					
Date					
F	6,325	14,023	0,435	1,176	6,552
p	0,000	0,000	0,783	0,327	0,000
Treatment					
F	1,807	3,715	5,170	4,294	2,153
p	0,120	0,004	0,000	0,002	0,067
Interaction					
F	0,682	0,505	2,790	2,572	0,600
p	0,826	0,954	0,001	0,002	0,897

1-D: Simpson index, H': Shannon-Wiener index, Evar: Smith-Wilson evenness index, $E_{1/D}$: Simpson evenness index, d: Berger-Parker index

Table 7. ANOVA results for ecological diversity indices.

Certain authors have demonstrated the lack of sensitivity of diversity indices relative to other methods. [57] used seven diversity indices to assess the effect of no-till farming on carabid communities and concluded that these diversity indices and models are not useful for the detection of the possible effects on carabids. [58] concluded that for differences in the values of diversity indices to be observed, the taxonomic level of identification must be deeper. However, identification to lower levels would hinder the use of diversity indices as quality indicators because sampling and identification would be more complex and costly, requiring the aid of specialists knowledgeable about the different taxonomic groups; such high-precision identification contrasts with the indicator characteristics proposed by [59]. The classification of macrofaunal communities to higher taxonomic levels is supported by studies by [60, 61] and has been used in other evaluations of the effect of agricultural practices on soil fauna [62-64]. Note that in [57] study, carabid communities were identified to the species level. However, this level of identification did not aid in the detection of a response of carabids to the disturbance. In this way, is quite difficult establish a real differentiation among treatments using only de ecological diversity indices.

Multivariate analysis

The results from the similarity analysis show that, of the four factors analysed, only the sampling season can differentiate the treatments with statistical significance (r_s = 0.638; p = 0.001) (Table 8).

General analysis			
	Season	Control	Fertilizer
r_s	0,638	0,035	-0,022
p	0,001	0,335	0,551
Pairwise test			
Spring		r_s	p
Fertilizer		0,295	0,015
	C,M	0,111	0,500
	C,O	0,535	0,009
	M,O	0,069	0,300
Fall			
Fertilizer		-0,171	0,780

Season: sampling date, Control: control vs fertilized parcels (both mineral and organic waste application), Fertilizer: control vs mineral fertilization vs organic waste application. Pairwise test: C- control, M- mineral fertilization, O- organic waste application.

Table 8. Similarity analysis results from macro-faunal communities between different sampling date and fertilization treatment

The results from the ANOVA analysis for the factor *season* (Figure 2) show that the differentiation between sampling performed during the autumn and sampling performed during the spring is statistically significant (Stress < 0.1). During the autumn, the number of specimens captured was much lower, a result that is related to the life cycle of soil organisms. During the spring, the populations of most species increase as a consequence of the higher temperature and greater availability of water and food [64, 65].

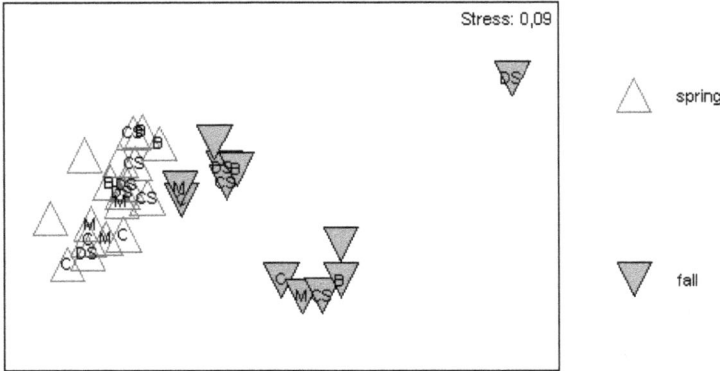

Figure 2. MDS ordination results for all dates, and all experimental parcels. C: Control, M: Mineral, DS: dairy-industry sludge, B: broiler litter, CS: Cattle slurry.

Due to the differentiation according to the sampling seasons described in the previous section, an similarity analysis was performed to separate the data from the spring and the autumn based on the factor *fertiliser*. The results were statistically significant only if the data obtained during the spring were used.

The ordination by MDS tended to separate the Dairy Slurry and Broiler litter plots from the Mineral, Cattle Slurry, and Control plots (Figure 3).

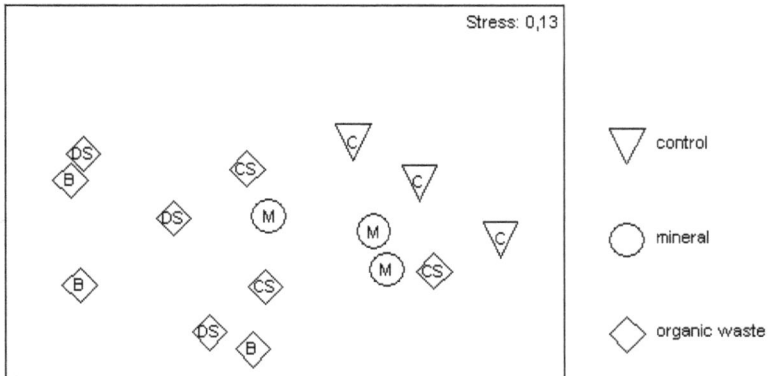

Figure 3. DMS ordination results for spring samples. C: Control, M: Mineral, DS: Dairy-industry sludge, B: Broiler litter, CS: Cattle slurry.

Carabidae and Araneae, with contributions greater than 10%, were the taxa with the greatest ability to separate the communities corresponding to the Control and Mineral plots from those corresponding to the plots treated with organic fertiliser (Table 9). These taxa include polyphagous predators, which have the ability to significantly affect the population dynamics of various phytophagous and saprophagous insects [67, 68]. These results are consistent with those of [69, 70], which demonstrate that the communities of carabids and spiders have a significant bioindicator potential. Similarly, [71] evaluated the effect of altering soil use on populations of coleopterans and spiders. These authors propose that the re-establishment of agricultural processes be monitored using these two groups.

	Average taxon abundance		Percentage of contribution
Control vs. organic waste[1]	Control	Organic waste	
Carabidae	18,33	3,47	11,03
Araneae	33,92	11,00	10,72
Diptera	26,08	9,06	8,60
Formicidae	9,67	3,50	5,43
Mineral vs. organic waste[2]	Mineral	Organic waste	
Araneae	29,25	11,00	13,12
Carabidae	10,67	3,47	10,28
Agrilimacidae	1,75	2,47	6,03
Acrididae	2,08	0,64	5,71
Apionidae	2,50	1,89	5,68
Gryllidae	1,92	0,92	5,54

Average dissimilarity: [1]42, 87%; [2]35, 33%

Table 9. Taxon abundance under different fertilizer application.

Finally, the fertiliser treatments were differentiated based on a two-way crossed similarity analysis based on sampling season and fertiliser treatment (Table 10). This analysis revealed that the presence or absence of fertiliser affected the composition of the macrofauna community. According to this analysis, the effect depends on the type of fertiliser used. For the Mineral and Cattle Slurry treatments, the effects were similar (p = 0.06). The effects of Dairy Slurry and Broiler Litter were equivalent (p = 0.271).

Factor	r_s	p
Season	0,575	0,001
Treatment	0,219	0,001
Pairwise test		
Control vs. Mineral	0,113	0,018
Control vs. Cattle slurry	0,194	0,002
Control vs. Dairy-industry sludge	0,481	0,001
Control vs. Broiler litter	0,315	0,001
Mineral vs. Cattle slurry	0,073	0,060
Mineral vs. Dairy-industry sludge	0,237	0,001
Mineral vs. Broiler litter	0,150	0,019
Cattle slurry vs. Dairy-industry sludge	0,214	0,002
Cattle slurry vs. Broiler litter	0,179	0,006
Dairy-industry sludge vs. Broiler litter	0,030	0,271

Table 10. Two factors cross-way (season x treatment) results for macro-faunal communities similarity analysis between different fertilization treatment

Further research

Based on the results obtained, it is necessary to further evaluate the response of macrofaunal communities to the addition of different types of waste used as fertilisers and/or soil restorers. With this approach, we will be able to analyse the global reaction/regeneration of the edaphic ecosystem beyond concrete and specific responses to the physico-chemical parameters. The extension of this type of research to different types of soil, different crops, and different forms of agricultural management will yield a more thorough view of the biological responses to these different factors, allowing the selection of the most appropriate taxa and indices for the monitoring of the effects of organic wastes. Our results suggest that Araneae and Carabidae should be identified to lower taxonomic levels to obtain better data on species richness and population abundance. This approach will allow a deeper evaluation of waste use.

4. Conclusions

The taxon richness and individual abundance of the soil macrofauna were lower in the plots fertilised with organic waste. However, we cannot conclude that the addition of organic waste has a severe negative effect on the communities studied. Carabidae and Araneae were the most important taxa for the separation of the groups based on the type of fertiliser used, suggesting that the application of organic waste has a positive effect on the total number of predatory arthropods. It is highly probable that this positive effect occurs because these arthropods are polyphagous and, hence, can significantly affect the population dynamics of various phytophagous and saprophagous invertebrates.

Among the organic wastes, dairy slurry and broiler litter had the same effect on the macrofaunal communities. The effect of cattle manure slurry was similar to that of the mineral fertiliser treatment.

The indices of ecological diversity were not effective for detecting differences among the different fertiliser treatments. The multivariate analysis of the macrofaunal communities was more useful, allowing the discrimination of groups and the identification of the taxa responsible for the differences among these groups.

Author details

Mariana Matos-Moreira and Emilio Carral
Department of Cellular Biology and Ecology,
University of Santiago de Compostela, Lugo, Galiza, Spain

Mario Cunha
Department of Mathematics-Research Center for Geo-spatial Sciences,
Porto University, Porto, Portugal

M. Elvira López-Mosquera
Department of Plant Production, University of Santiago de Compostela,
Lugo, Galiza, Spain

Teresa Rodríguez
Department of Zoology and Physic Anthropology,
University of Santiago de Compostela, Lugo, Galiza, Spain

Acknowledgement

Financial supported by Spanish Ministry of Science and Technology (project AGL2000-04-81) and Portuguese Foundation for Science and Technology (Mariana Matos-Moreira pre-doctoral grant: SFRH/BD18486/2004).

5. References

[1] Navarro Pedreño J, Moral Herrero R, Gómez Lucas I, Mataix Beneyto J (1995) Residuos orgánicos y agricultura. Universidad de Alicante. Alicante-Spain.155 p.

[2] European Council - Directive 91/676/EC. L 375: 1-8.

[3] Haynes R J, Naidu R (1998) Influence of lime, fertilizer and manure applications on soil organic matter content and soil physical conditions: a review. Nutrient cycling in agroecosystems. 51(2): 123-137.

[4] Petersen SO, Henriksen K, Mortensen GK, Krogh PH, Brandt KK, Sørensen J, Madsen T, Petersen J, Grøn C (2003) Recycling of sewage sludge and household compost to arable land: fate and effects of organic contaminants, and impact on soil fertility. Soil tillage and research. 72: 139-152.

[5] Antolín MC, Pascual I, García C, Polo A, Sánchez-Díaz M (2005) Growth, yield and solute content of barley in soils treated with sewage sludge under semiarid Mediterranean conditions. Field crops research. 94: 224–237.

[6] López-Arias M, Grau-Corbí JM (2005) Metales pesados, materia orgánica y otros parámetros de la capa superficial de los suelos agrícolas y de pasto de la España peninsular. II: Resultados por provincias. INIA-Ministerio de Educación y Ciencia, Madrid-Spain.383 p.

[7] Hue NV, Licudine DL (1999) Amelioration of subsoil acidity through surface application of organic manures. Journal of environmental quality. 28(2): 623-632.

[8] Materechera SA, Mkhabela TS (2002) The effectiveness of lime, chicken manure and leaf litter ash in ameliorating acidity in a soil previously under black wattle (*Acacia mearnsii*) plantation. Bioresource technology. 85: 9-16.

[9] Hue NV (1992) Correcting Soil Acidity of a Highly Weathered Ultisol with Chicken Manure and Sewage-Sludge. Communications in soil science and plant analysis. 23(3-4): 241-264.

[10] Naramabuye FX, Haynes RJ, Modi AT (2008) Cattle manure and grass residues as liming materials in a semi-subsistence farming system. Agriculture, ecosystems and environment. 124: 136–141.

[11] Marinari S, Masciandaro G, Ceccanti B, Grego S (2000) Influence of organic and mineral fertilisers on soil biological and physical properties. Bioresource technology. 72(1): 9-17.

[12] Lal R (2004) Soil carbon sequestration to mitigate climate change. Geoderma. 123(1-2): 1-22.

[13] Petersen SO, Regina K, Pollinger A, Rigler E, Valli L, Yamulki S, Esala M, Fabbri C, Syvasalo E, Vinther FP (2006) Nitrous oxide emissions from organic and conventional crop rotations in five European countries. Agriculture, ecosystems and environment. 112(2-3): 200-206.

[14] Küstermann B, Kainz M, Hülsbergen KJ (2008) Modeling carbon cycles and estimation of greenhouse gas emissions from organic and conventional farming systems. Renewable agriculture and food systems. 23(1): 38–52.

[15] Venglovsky J, Martinez J, Placha I (2006) Hygienic and ecological risks connected with utilization of animal manures and biosolids in agriculture. Livestock science. 102(3): 197-203.

[16] Zizek S, Hrzenjak R, Kalcher GT, Srimpf K, Semrov N, Zidar P (2011) Does monensin in chicken manure from poultry farms pose a threat to soil invertebrates? Chemosphere. 83(4): 517-523.

[17] Smith KA, Frost JP (2000) Nitrogen excretion by farm livestock with respect to land spreading requirements and controlling nitrogen losses to ground and surface waters. Part 1: cattle and sheep. Bioresource technology. 71(2): 173-181.

[18] European Council - Directive 2006/12/EC. L 114: 9-21.

[19] Moreira R, Sousa JP, Canhoto C (2008) Biological testing of digested sewage sludge and derived composts. Bioresource technology. 99: 8382-8389.

[20] Domene X, Ramirez W, Mattana S, Alcaniz JM, Andres P (2008) Ecological risk assessment of organic waste amendments using the species sensitivity distribution from a soil organisms test battery. Environmental pollution. 155(2): 227-236.

[21] Pandard P, Devillers J, Charissou AM, Poulsen V, Jourdain MJ, Férard JF, Grand C, Bispo A (2006) Selecting a battery of bioassays for ecotoxicological characterization of wastes. Science of the total environment. 363: 114-125.

[22] van Eekeren N, de Boer H, Hanegraaf M, Bokhorst J, Nierop D, Bloem J, Schouten T, de Goede R, Brussaard L (2010) Ecosystem services in grassland associated with biotic and abiotic soil parameters. Soil biology and biochemistry. 42(9): 1491-1504.

[23] Blair JM, Bohlen PJ, Freckman DW (1996) Soil invertebrates as indicators of soil quality, In: Doran JW, Jones AJ, editors. Methods for assessing soil quality. Soil Science Society of America, Madison, Wisconsin, USA. pp. 273-291.

[24] Doube BM, Schmidt O (1997) Can the abundance or activity of soil macrofauna be used to indicate the biological health of soils? In: Pankhurst CE, Doube BM, Gupta VVSR, editors. Biological indicators of soil health. CAB Internacional, Wallingford, Oxon. pp 265-295.

[25] Paoletti MG (1999) Using bioindicators based on biodiversity to assess landscape sustainability. Agriculture, ecosystems and environment.74(1-3): 1-18.

[26] Olfert O, Johnson GD, Brandt SA, Thomas AG (2002) Use of arthropod diversity and abundance to evaluate cropping systems. Agronomy journal. 94: 210-216.

[27] Lavelle P, Decäens T, Aubert M, Barot S, Blouin M, Bureau F, Margerie P, Mora P, Rossi J. P (2006) Soil invertebrates and ecosystem services. European journal of soil biology. 42:S3-S15.

[28] Stork NE, Eggleton P (2009) Invertebrates as determinants as indicators of soil quality. American journal of alternative agriculture. 7 :38-47.

[29] Mazzoncini M, Canali S, Giovannetti M, Castagnoli M, Tittarelli F, Antichi D, Nannelli R, Barberi P (2010) Comparison of organic and conventional stockless arable systems: A multidisciplinary approach to soil quality evaluation. Applied soil ecology.44(2):124-132.

[30] Oliver I, Garden D, Greenslade PJ, Haller B, Rodgers D, Seeman O, Johnston B (2005) Effects of fertiliser and grazing on the arthropod communities of a native grassland in south-eastern

[31] Birkhofer K, Bezemer TM, Bloem J, Bonkowski M, Christensen S, Dubois D, Ekelund F, Fliessbach A, Gunst L, Hedlund K, Mader P, Mikola J, Robin C, Setala H, Tatin-Froux F, Van der Putten WH, Scheu S (2008) Long-term organic farming fosters below and aboveground biota: Implications for soil quality, biological control and productivity. Soil biology & biochemistry. 40 (9): 2297-2308.

[32] Forge TA, Bittman S, Kowalenko CG (2005) Responses of grassland soil nematodes and protozoa to multi-year and single-year applications of dairy manure slurry and fertilizer. Soil biology & biochemistry. 37(10): 1751-1762.

[33] Diez JA, de la Torre AI, Cartagena MC, Carballo M, Vallejo A, Muñoz MJ (2001) Evaluation of the application of pig slurry to an experimental crop using agronomic and ecotoxicological approaches. Journal of environmental quality. 30(6): 2165-2172.

[34] Tomlin AD, Protz R, Martin RR, McCabe DC, Lagace R J (1993) Relationships amongst organic matter content, heavy metal concentrations, earthworm activity, and soil microfabric on a sewage sludge disposal site. Geoderma. 57: 89-103.

[35] Andrés P, Domene X (2005) Ecotoxicological and fertilizing effects dewatered, composted and dry sewage sludge on soil mesofauna: A TME experiment. Ecotoxicology. 14 (5): 545-557.

[36] Andrés P, Mateos, E, Tarrasón D, Cabrera C, Figuerola B (2011) effects of diggested, composted, and thermally dried sewage sludge on soil microbiota and mesofauna. Applied soil ecology. 48: 236-242.

[37] Krogh PH, Pedersen MB (1997) Ecological effects of industrial sludge for microarthropods and decomposition in a spruce plantation. Ecotoxicological environmental safe. 36: 162-168.

[38] van der Wal A, Geerts RHEM, Korevaar H, Schouten AJ, Akkerhuis GAJM, Rutgers M, Mulder C (2009) Dissimilar response of plant and soil biota communities to long-term nutrient addition in grasslands. Biology and fertility of soils. 45(6): 663-667.

[39] Linden DR, Hendrix PF, Coleman DC, van Vleet P (1994). Faunal indicators of soil quality. In: Doran JW, Bezdicek DC, Stewart BA, editors. Defining soil quality for a sustainable environment. Soil Science Society of America, Madison, WI, USA, 91-106.

[40] Magurran AE (2004) Measuring biological diversity. Blackwell Science Ltd. Blackwell Publishing. 260 p.

[41] van Straalen NM (1998) Evaluation of bioindicator systems drived from soil arthropod communities. Applied soil ecology. 9: 429-437.

[42] López-Mosquera ME, Cabaleiro F, Sainz MJ, López-Fabal A, Carral E (2008) Fertilizing value of broiler litter: Effects of drying and pelletizing. Bioresource technology. 99: 5626–5633.

[43] Evers GW (2002) Ryegrass-bermudagrass production and nutrient uptake when combining nitrogen fertilizer with broiler litter. Agronomy journal. 94: 905-910.

[44] Castro J (2002) Estrategia para un manexo sostible da fertilización das terras en Galicia: A reciclaxe do xurro como abono. Cooperación. 60.

[45] Castro J, Mateo E (1999) Ciclos de nutrientes en 12 explotaciones lecheras gallegas: P y K. Actas de la XXXXIX Reunión Científica de la SEEP. 373-378.

[46] Moletta R (2006) Caractérisation des efluentes des industries agroalimentaires. In: Moletta R, editor. Gestion des problèmes environnementaux dans les industries agroalimentaires. Editions Tec & Doc, Collection Sciences & Techniques agroalimentaires, Paris. pp 15-27.

[47] Gras R, Morisot A (1974) Les déchets solides des industries agricoles et alimentaires. Annales agronomiques. 25: 231-242.

[48] Wilkinson KG, Issa, JG, Meehan B, Surapaneni A, Carew M, Palmowski L (2007) Characterisation of selected dairy processing waste streams from Victoria, Australia. Australian Journal of dairy technology. 62 (3): 159-165.

[49] López-Mosquera ME, Alonso XA, Sainz MJ (2001) Short-term effects of soil amendment with dairy sludge on yield, botanical composition, mineral nutrition and arbuscular mycorrhization in a mixed sward. Pastos. 29: 231-243.

[50] Sainz MJ, Matos-Moreira M, Bande MJ, López-Mosquera ME (2006) Forage production in sown meadows under several organic fertilization strategies. Grassland and science in Europe. 11: 700-702.

[51] López-Mosquera ME, Barros R, Sainz MJ, Carral E, Seoane S (2005) Metal concentrations in agricultural and forestry soils in northwest Spain: implications for disposal of organic wastes on acid soils. Soil use and management. 21: 298-305.

[52] Matos-Moreira M, Lopez-Mosquera ME, Cunha M, Oses MJS, Rodriguez T, Carral EV (2011) Effects of Organic Fertilizers on Soil Physicochemistry and on the Yield and Botanical Composition of Forage over 3 Years. Journal of the air & waste management association. 61(7): 778-785

[53] Meyer M (1996) Epigenic macrofauna. In: Schinner F, Öhlinger R, Kandeler E, Margesin R, editors. Methods in Soil Biology. Springer, Berlin, Germany.

[54] Barrientos JA (1988) Bases para un curso práctico de entomología. Asociación Española de Entomología, editor. Salamanca.

[55] Zahradník J (1990) Guía práctica de los coleópteros de España y de Europa. Omega, Barcelona. 570 p.

[56] Clarke KR, Warwick RM (2001) Change in marine communities: an approach to statistical analysis and interpretation. PRIMER-E Ltd., Plymouth, England. 172 p.

[57] Belaoussoff S, Kevan PG, Murphy S, Swanton C (2003) Assessing tillage disturbance on assemblages of ground beetles (Coleoptera : Carabidae) by using a range of ecological indices. Biodiversity and conservation. 12(5): 851-882.

[58] Guerold F (2000) Influence of taxonomic determination level on several community indices. Water research. 34(2): 487-492.

[59] Doran JW, Parkin TB (1994) Defining and assessing soil quality. In: Doran JW, Bezdicek DC, Stewart BA, editors. Defining soil quality for a sustainable environment. Soil Science Society of America, Madison, Wisconsin, USA. pp. 3-21.

[60] Nahmani J, Lavelle P, Rossi JP (2006) Does changing the taxonomical resolution alter the value of soil macroinvertebrates as bioindicators of metal pollution? Soil biology and biochemistry. 38: 385-396.

[61] Biaggini M, Consorti R, Dapporto L, Dellacasa M, Paggeti E, Corti C (2007) The taxonomic level order as a possible tool for rapid assessment of arthropod diversity in agricultural landscapes. Agriculture,ecosystems and environment. 122: 183-191.

[62] Nkem JN, Lobry de Bruyn LA, Hulugalle NR, Grant CD (2002). Changes in invertebrate populations over the growing cycle of an N-fertilised and unfertilized wheat crop in rotation with cotton in a grey Vertosol. Applied soil ecology. 20: 69-74.

[63] Benito NP, Brossard M, Pasini A, Guimarães MF, Bobillier B (2004) Transformations of soil macroinvertebrate populations after native vegetation conversation to pasture cultivation (Brazilian Cerrado). European journal of soil biology. 40: 147-154.

[64] de Aquino A. M., da Silva R. F., Mercante F. M., Correia M. E. F., Guimarães M. F., Lavelle P., 2008. Invertebrate soil macrofauna under different ground cover plants in the no-till system in the Cerrado. European journal of soil biology, 44, 191-197.

[65] Curry J P (1994) Grassland invertebrates. Chapman & Hall, London, UK. 437p.

[66] Rossi JP, Blanchart E (2005) Seasonal and land-use induced variations of soil macrofauna composition in the western Ghats, southern India. Soil biology and biochemistry. 37: 1093-1104.

[67] Ekschmitt K, Wolters V, Weber M (1997) Spiders, carabids, and staphylinids: the ecological potential of predatory macroarthropods. In: Benckiser G, editor. Fauna in soil ecosystems: recycling processes, nutrient fluxes, and agricultural production. Marcel Dekker, Inc., New York, USA. pp 307-362.

[68] Birkhofer K, Fliessbach A, Wise D. H, Scheu S (2008) Generalist predators in organically and conventionally managed grass-clover fields: implications for conservation biological control. Annals of applied biology. 153(2): 271-280.

[69] Rainio J, Niemela J (2003) Ground beetles (Coleoptera : Carabidae) as bioindicators. Biodiversity and conservation. 12(3): 487-506.

[70] Pearce JL, Venier LA (2006) The use of ground beetles (Coleoptera : Carabidae) and spiders (Araneae) as bioindicators of sustainable forest management: A review. Ecological indicators: 6(4): 780-793.

[71] Perner J, Malt S (2003) Assessment of changing agricultural land use: response of vegetation, ground-dwelling spiders and beetles to the conversion of arable land into grassland. Agriculture, ecosystems and environment. 98(1-3): 169-181.

Chapter 8

Perspectives for Sustainable Resource Recovery from Municipal Solid Waste in Developing Countries: Applications and Alternatives

Luis F. Marmolejo, Luis F. Diaz, Patricia Torres and Mariela García

Additional information is available at the end of the chapter

1. Introduction

Municipal solid waste (MSW) is the most complex solid waste stream (Troschinetz & Milhecic, 2009). The search for sustainable development and in particular factors such as fast population growth, land limitations and difficulties associated with finding suitable sites for establishing landfills, as well as the decrease of raw materials, make that practices for the management MSW traditionally used in developing countries such as collection and final disposal, be complemented with recycling as a preferable option for dealing with the solid waste generated. This position has been promoted in international events such as the Johannesburg Summit on Sustainable Development, in which the recycling and reuse of waste were identified as key strategies for the accomplishment of the main objectives and essential requisites for a sustainable development, since it contributes to reduce the negative effects on the environment and increases the efficiency of the use of resources (United Nations, 2002). From this perspective, there are notable efforts made in developing countries such as Tanzania (Mbuligwe et al., 2002), Colombia (Minambiente, 2002) and Botswana (Ketlogetswe & Mothudi, 2005), where policies that give priority to recycling immediately after source reduction have been enforced; however, the application level of these policies is variable and final disposal on the land still remains as the primary option with significant application (Fricke et al., 2001; OPS, 2005; UNEP, 2008).

The data in Table 1 show per capita production and composition of solid waste in different cities in developing countries. In all cases there are significant proportions of putrescible waste in the form of food and yard wastes. The dependency on agriculture for subsistence and economical development of these countries, as well as conservation requirements for soil quality, the productivity and gradual increase of the costs of mineral fertilizers, generate the necessity of using alternative soil amendments. In this sense, the organic matter and nutrients contained in the putrescible fraction of solid waste constitute a viable alternative for this situation (Diaz et al., 2007). Another element that can stimulate the recycling of these wastes is the reduction in the production of greenhouse gases (GHG) compared with traditional techniques of final disposal. Barton et al., (2008) compared the generation of carbon dioxide in final disposal systems at open dump and sanitary landfills (considering three forms of managing emissions) with alternatives such as composting and anaerobic digestion with electricity production, finding that in the open dumpsites and sanitary landfills, emissions varied between 0.09 and 1.2 t CO_2e/t, whereas in composting the emissions were neutral and in the anaerobic digestion plant it was -0.21 tCO_2e/t; however, costs of the last option limit significantly its applicability.

City	PCP (Kg/cap-day)	Putrescible	Paper	Metal	Glass	Plastic, rubber and leather	Textiles	Ceramics, dust and stones
Bangalore, India[1]	0.4	75.2	1.5	0.1	0.2	0.9	3.1	19.0
Manila, Philippines[1]	0.4	45.5	14.5	4.9	2.7	8.6	1.3	27.5
Asunción, Paraguay[1]	0.46	60.8	12.2	2.3	4.6	4.4	2.5	13.2
Mexico City, Mexico [1]	0.68	59.8[a]	11.9	1.1	3.3	3.5	0.4	20.0
Cali, Colombia [2]	0.39	65.54	6.23[b]	1.06	2.56	11.12	1.98	11.51[c]

[a] Includes small amounts of wood, hay, and straw; [b] Includes cardboard; [c] Includes all the others
Sources: [1]UNEP & CalRecovery (2005); [2]DAPM & UNIVALLE (2006)

Table 1. Per capita production (PCP) and composition of solid waste in different cities in developing countries

In the same manner, other materials with reuse potential such as different types of paper, metal, glass and plastics generally represent more than 6% of the MSW generated. Taking into account that most of the developing countries do not have one or more of the raw materials (e.g. iron ore, bauxite or petroleum of importance for its economical development) or other substitute materials, reuse of these materials also is an option (Diaz et al., 2007).

In this article a description and analysis of the application of the options for recovering resources from MSW in different regions of developing countries, identifying common elements that favor or limit the application of these and suggesting alternatives that contribute to the sustainability, is made. The approach is carried out from the Integrated Sustainable Waste Management concept, which proposes to have a vision of the situation that involves the stakeholders, the components of the waste system and sustainability aspects that determine the functioning of the systems to reach technically appropriate,

Perspectives for Sustainable Resource Recovery from Municipal Solid Waste in Developing
Countries: Applications and Alternatives

175

economically viable and socially acceptable solutions that do not degrade the environment (Van de Klundert & Anschutz, 2001). Also, this article focuses in recycling and composting since both of these alternatives constitute the options with the highest degree of application providing an overview of the findings reported by several authors worldwide, with experiences in countries or regions in Africa, Asia and Latin America and the Caribbean (LAC).

2. Discussion

2.1. Recycling

In developing countries it is acknowledged that the recovery of materials such as iron, steel, copper, lead, paper plastic and glass will decrease the investment in importing these materials and save energy (Kocasoy, 2001); however, proper recovery is scarcely applied. Some of the reasons for this situation are: shortage of properly trained professional, absence of appropriate technology, poor public awareness and the relatively high initial capital investments costs required for their implementation. Although a few large-and medium-scale solid waste treatment facilities – imported from industrialized countries, have been built and operated, the intensive mechanical and energetic requirements of these technologies have finally driven most of these facilities to be shut down (Diaz et al., 2002). On the other hand, the prices obtained for some of the recovered materials typically are lower than the segregation / reprocessing costs, which can be even higher than the costs of virgin materials, so that recycling activities usually have to be subsidized, except for materials such as aluminum and paper (Bogner et al., 2007). Table 2 shows the waste recycling rates in some developing countries.

Country/region	Recycling rate of municipal waste (%)				
	Glass	Metal	Paper	Plastic	Total
LAC[1]	0.8	2.1	2.0	3.4	
Brazil[2]	41		30		
China[2]					7- 10
Colombia[3]	10				
Nepal[2]	5				
Thailand[2]	18	39	28	14	15
Turkey[2]	25	30	36	30	
Vietnam[2]	13-20				

[1] Sources: [1]OPS (2005), [2] Troschinetz & Milhecic (2009); [3]Contraloría General de la República (2005)

Table 2. Recycling of waste in developing countries/regions

Waste recovery practices generally are carried out in an informal manner mainly by scavengers on the streets and at final disposal sites, under inadequate working conditions. The formal sector has concentrated on the collection and final disposal; although recycling is viewed as an option, its application is very weak. In the same way, the attitude of the formal waste management sector towards informal recycling often is very negative regarding it as backward, unhygienic and generally incompatible with modern waste management systems (Wilson et al., 2006). Nevertheless, recycling rates reached by the informal sector in several countries are quite high, fluctuating in a range of 20% - 50%, values that are comparable to those achieved by modern waste management systems in industrialized countries (Wilson et al., 2009).

In Africa there are few formal systems for material recovery instituted by public agencies or the private sector. Recovery of materials, including source separation and recycling is carried out mainly by the informal sector. This activity is centered on materials of economic and/or social value; plastic bags, bottles, paper, cardboard and cans are reused before entering the waste chain. A few materials are converted into new products for local use; some examples are the smelting of aluminum cans and scrap metals into household utensils, and paper and plastic residues into products for tourists (Otieno & Taiwo, 2007).

In Kenya, recycling has gained importance due to the increasing costs of raw materials. Initially it was carried out informally by impoverished people, but it is now emerging at an industrial level (Rotich et al., 2006). In Cameroon, governmental policies establish strategies for environmental protection and promotion of conservation of materials through an adequate disposal and recovery of MSW; however, in practice management is focused on, collection and disposal on the land (Manga et al., 2008).

In Nigeria, although SWM is identified as one of the environmental elements to include in the Poverty Reduction Strategy, characterization studies carried out in the central part of the country indicate that the recyclable materials contained in the solid waste do not warrant investment in recycling as a waste management approach (Sha' Ato et al., 2007). In Lagos, recycling and resource recovery exist, but have not received the attention of the government and the waste management authorities. It is estimated that approximately 5- 8% of MSW are recycled through refuse dealers, who separate the materials and sell them to consumers, as well supply them to mills and factories (Kofoworola, 2007).

In Botswana, most of the local companies dedicated to recycling are only in charge of collection, the recovered materials are exported to different countries such as South Africa and Zimbabwe. The amount of recyclable materials recovered in final disposal sites is taken as an indication of the potential for developing a recycling industry on a large scale at the local level (Ketlogetswe & Mothudi, 2005).

In the Southern and Western regions of Asia, industries that deal with repairing items and with used products are important sources of recovery and reuse of waste. In the cities of low or medium income of the East and Pacific regions, informal source separation and recycling have been a common practice for many years; gathering, trading, and reprocessing materials is the work of many people (UNEP, 2008). In the East this activity is generally carried out by medium- scale or household enterprises, and is predicted to grow where it offers economical benefits (Nguyen Ngoc & Schnitzer, 2009).

In Jordan, recycling is carried out by scavengers in final disposal sites and through formal systems managed by the municipalities or NGO's; scavengers usually search for cardboard, tins and plastic bottles. Experiences managed by municipalities include recovery of recyclables from the solid waste stream prior to landfilling and a pilot material recycling facility (Abu Qdais, 2007). In the case of the Gulf Co-operation Council States (Kuwait, Saudi Arabia, Bahrain, Qatar, United Arab Emirates y Oman), Alhumoud (2005) affirms that most of the countries do not report recycling goals or programs and the only comprehensive form of recycling available is the recycling of paper and cardboard; a limited amount of recyclable materials such as cans, metals and cardboard are collected from the waste containers in front of the houses by scavengers. In a research project conducted in seven Palestinian districts there was no evidence of reuse and recycling programs, identifying only the informal recovery of metal scrap from waste collection containers and dump sites (Al-Khatib et al., 2007).

In Pakistan, although in the last four decades the generation of waste has increased significantly, there are not specific recycling programs on a country-wide scale or in the big cities. In Lahore, 1.97 million tons of waste are generated annually, of which 65% to 70% are collected; approximately 0.04 million tons are sold directly to industry; households reuse around 0.054 million and 0.09 million of recyclables are commercialized at junkshops by scavengers and households annually. It is estimated that 21.2% of the total recyclable waste is being used for recycling, generating around US$ 4.5 million a year (Batool et al. 2008).

In Ankara (Turkey), recyclable materials constitute around 18% of the total solid waste generated; Government Statistics Institute estimates that scavengers are collecting up to 50% of these materials. It is considered that the average income from recyclables is between US$ 25.000 - 50.000 per day; however, the income per scavenger is between US$80 and US$100 per month and about US$20,000 is for the owners the recycling system each day (Ali, 2002).

In China, MSW seems to revolve around small scale operations as a result of the application of the principle that the polluter is responsible for treatment and disposal. As a general rule there are not treatment operations only small- scale and inefficient plants (Suocheng et al., 2001); however discarded material is imported from different countries on a large scale for reuse and recycling, generating around 5000 enterprises, about 1.4 million jobs, with recovery rates of 85%, 47%, 25%, 20% and 13% in materials such as iron and steel, rubber, plastics, paper and glass, respectively (Shekdar, 2009).

In the case of Latin America and the Caribbean (LAC), the Pan-American Health Organization PAHO (2005) indicates that the nature of the solid wastes, particularly those generated in residential, commercial and institutional sources, creates a technological problem for the application of recycling since its quality is affected by the mixture of materials. In all the countries of the region informal segregation is common practice and a frequent source of income for the impoverished and unemployed fraction of the population. In Colombia, Mexico, Brazil and Venezuela recycling programs of considerable magnitude have been extended to the communities in order to promote the organization in cooperatives and private associations (Minambiente, 2002). Esteban García et al. (2001) point out that the recycling activity is a well established reality in the region, which reflects in the acknowledgment through names such as "cirujas" in Argentina, "buzos" in Bolivia, "cachureros" in Chile and "pepenadores" in Mexico. They also indicate as an example in

Latin America, that in Mexico the informal sector is composed of people who work in dumpsites and in areas not covered by the collection service, as well as people from the middle class and part -time scavengers (students, retired persons and housewives, who obtain additional income).

McBean et al. (2005) report that through the training and organization of the informal sector, in Tucuman (Argentina) has been possible to recover significant percentages of materials such as paper and newsprint, and plastic (4.3 and 27.2% respectively), making it possible for the people involved in this activity to earn an income 1.75 times the minimum wage in that region. In LAC, more than an impediment, recycling cooperatives represent an opportunity for the private sector and industries to increase the market and level of recycling (Lopes et al., 2007). In an assessment carried out by Do Prado Filho and Sobreira (2007) at 29 recycling-composting plants located in the state of Minas Gerais (Brazil), it identified infrastructure, operation and location conditions that allow to rank 95% of these sites as adequate or acceptable; however, the working conditions of the workers were ranked as regular in 13 and as bad in the rest of them. In the same way, it was also determined that in all of the cases it is feasible to sell the recycled materials.

The topics discussed in the previous paragraphs allow to affirm that recycling of some components of the waste stream constitutes a source of income generation and materials recovery, with market for the products that require the development of programs that integrate the actors involved in recycling and technological advances that take into account the local labor potential to improve the quality of the processes and products.

2.2. Composting

Composting is described as an economically viable method compared with other processes and also effective in contributing to the reduction in the amount of material that should be taken to the landfill (Barreira et al., 2006). However, although successful experiences have been reported in its application (Mbuligwe et al., 2002; Zurbrügg et al., 2005, Bezama et al., 2007) it is acknowledged that there are important limitations (Zurbrügg et al., 2005, Barreira et al., 2006, Körner et al., 2008). Dulac (2001) pointed out that the high organic content of the waste streams of developing countries is ideal for composting, but municipal services operators do not have enough and adequate information and even though they may be familiar with the application of composting in agriculture, it is not considered as a way to solve their urban wastes problems. Bogner et al. (2007) indicate that labor-intensive processes are more appropriate and sustainable for those countries than highly mechanized technological alternatives at large-scale operations.

In Africa, composting has failed in cities such as Dakar (Senegal) y Abidjan (Cote d'Voire), due to the lack of demand for the final product. International NGO's have subsidized small scale composting in countries such as Benin, Cameroon, Kenya, Nigeria and Zambia, without making a significant impact in the reduction of MSW going to the landfill. The problem with composting in African cities is the low quality of the product due to the inadequate segregation of the wastes, which results in a low demand (Otieno & Taiwo,

2007). In Kenya, some groups compost food wastes that are sold to urban farmers and landscapers (Rotich et al., 2006).

In Southeast Asian Nations, composting is not a common practice due to high operation and maintenance costs, the high cost of the final product with respect to commercial fertilizers, and the available market, so that the activity is supported by governments (Nguyen Ngoc & Schnitzer, 2009). Alhumoud (2005) points out that in the Gulf Co-operation Council States in the last 20 years municipalities have concentrated on composting as an alternative for the treatment of MSW in spite of the failure of a large number of plants in the region. This author also affirms that the main problems with these plants have been the poor performance, high operating and maintenance costs, lack of technical support and inefficient management. In Jordan, even with a high fraction of biodegradable organic solid waste generated in the country and the fact that 91% of the country land is arid to semiarid, composting has not been considered as an option for solid waste management (Abu Qdais, 2007).

Hui et al. (2006) indicate that although composting is a widely utilized practice in Western countries, in Chongquing, one of the four largest municipalities in China, it is rarely used due to reasons such as the low application of source separation, low acceptance of compost by farmers, limited usefulness of compost in comparison with chemical fertilizers and strict regulations, monitoring and quality standards of the product. In India, composting is a tradition mainly in rural areas; utilization of large-scale and centralized composting plants during the 1970's had not been economically feasible. Studies have determined that compost is difficult to use because the waste arrives mixed and with high quantities or inorganic materials (Narayana, 2009).

The cooperation with NGO's, the supply of free bins for organic materials and the governmental support for the investment have had a high correlation with the better performance of composting in Thailand (Suttibak & Nitivattananon, 2008).

In LAC, the percentage of recovered waste reaches figures of only 2.2% out of the total and even with the predominance of organic matter; the application of composting is carried out at small scale, reaching only 0.6% of these wastes. The problem does not end with the scarce application of reuse, recovery and recycling technologies, it transcends to the lack of trust and the unsuccessful application of these technologies. From the beginnings of the 70's several initiatives oriented to the establishment of composting plants with diverse imported technologies have failed due to factors related mainly with the inefficient maintenance of equipment, indetermination of markets, inadequate technologies and lack of linking with strategic environmental projects (OPS, 2005), which shows the necessity of conceptual and technological developments contextualized in the reality of the region.

In the case of Cuba, Körner et al., (2008) pointed out than only one facility for composting of MSW is known, and that composting this kind of waste has not been reported officially as a treatment alternative; however its implementation is anticipated by the government.

Fricke et al. (2001) affirm that from 23 composting plants implemented in Brazil only 6 were in operation and that many of those plants were decommissioned after a short period of operation due to reasons such as unsatisfactory operation, high operational costs and low quality of the compost, low materials recovery and nauseous odors. In the monitoring reports of the quality of the compost produced in 20 recycling-composting plants assessed by Do Prado Filho and Sobreira (2007) is brought into focus the presence of heavy metals in variable amounts, which restricts the use compost in the soil. They also indicate that in these reports is rarely included the technical concept conclusive on the figures obtained for each parameter, the analytical methods used and the reference values as to the quality of the material and also that in spite of the presence of trained personnel for the operation of the plants there is considerable difficulty in understanding the results of the analysis of the quality of the products.

In Colombia, the Superintendencia de Servicios Públicos Domiciliarios (SSPD, 2008), determined that there were 28 MSW facilities that dealt with processing the putrescible fraction of MSW, from which 54% carried out composting, 15% vermicomposting and the remaining 31% used both methods. The processing time of the plants varied between 30 and 180 days, obtaining an average production efficiency of 33% (with respect to the material subjected to processing) and although quality control was not carried out, the obtained products were being utilized by farmers of different vegetable species and food without knowing the sanitary risk this could represent.

The scenarios previously presented contrast with the economical, social and environmental potential that the recovery, reuse and recycling of waste has had in regions such as Dhaka - Bangladesh (Zürbrugg et al., 2005), Dar es Salaam –Tanzania (Mbuligwe et al., 2002), Yala - Thailand (Mongkolnchaiarunya, 2005) and Turkey (Metin et al., 2003), where practices such as composting and recycling, besides the sanitary and environmental benefits of the reduction of the amount of materials to be disposed in the land, generate job opportunities and income generally for the sectors with the lowest economical capacity. Reported experiences by Zürbrugg et al. (2005) and Körner et al. (2008), for Bangladesh and Cuba respectively, indicate that the application of composting has had better success at small-scale and in decentralized facilities, but frequent failures are present in the marketing of the product. Drescher & Zürbrugg (2006) suggest that for the large cities, the combination of decentralized small- scale composting systems with medium- scale centralized composting schemes constitute an ideal strategy for the management of organic wastes; at the same time, decentralized composting systems are enough for small municipalities.

The use in agriculture, as soil conditioner or fertilizer, is one of the most usual ways to take advantage of the compost obtained with the processing of MSW; however the quality of the product is subordinated to variables such as the design of the composting facility, type and proportions of feedstock used, composting procedure and maturation period (Hargreaves et al.,2008) the evaluation of alternative techniques for improving and/or facilitating the monitoring of the process and quality of the product (Said-Pullicino et al., 2007, Barrena et al., 2008) that deserve to be assessed in more depth.

2.3. Perspectives for the sustainability of resource recovery from Wastes

Recent studies conclude that amongst the main factors for the sustainability of the reuse and recycling systems are waste collection and segregation, MSW management plan, and a local market for the recycled materials (Troschinetz & Milhecic, 2009). To accomplish the sustainability, it is necessary to develop production and marketing strategies with a marketing vision that acknowledge and integrate the formal and informal sectors, requiring important agreements between stakeholders, the adaptation of educational schemes and technological options, the identification and positioning in the market and the setting of normative references.

At the same time, although the knowledge of the quantity and quality of the materials to be processed is one of the key elements for guiding the industrial vision, it is recognized that the lack of reliable studies on the composition and generation of waste constitute one of the main limitations for the management (Diaz et al., 2002). This situation could be related with the cost and complexity of the methodological procedures used for the execution of such studies (Hristovski et al., 2007). It is necessary to structure alternative methodological schemes that take into account limitations of the trained personnel and the low availability of financial resources, which can be successfully applied. In this aspect, there are some positive experiences in Santiago de Cali in Colombia, where a sampling and characterization program that involved the participation of local stakeholders was structured, adapting a method that utilizes one block as the sampling unit, obtaining results with high confidence levels and low errors, with affordable costs for the local conditions (Klinger et al., 2009).

It is also important to conduct efforts that allow all the stakeholders in the management chain to identify the waste as an element with possibilities of reuse and not as garbage, term associated with problem. At the same time, aspects such as the deterioration of the quality and loss of value of the materials to reuse, which starts at the point of generation and in the collection vehicles, must be avoided. The encouragement of practices such as source separation and separate collection are options with the potential to solve this situation; however, in this way additional costs for the user or for industry can be generated, encouraging the informal recovery which creates conflicts with personnel from the collection system. Analysis and local agreements that take into account these situations and identify a solution must be developed.

In the same way, it is necessary the research and development of technological options that facilitate the transformation in situ, allowing amongst others, the use of locally available resources for the operation and maintenance, as well as the reduction of volumes of materials and the consequent decrease in transportation costs, increasing the added value of the reuse of these products. In this sense, an alternative is the realization of adaptations of the technologies utilized for the recovery of industrial wastes, taking into account the quality and quantity of the MSW. It is also important to put into practice monitoring schemes and quality control of the products since the complexity and costs associated with those traditionally used limit its application.

The previously mentioned situations demonstrate that the sustainability of the reuse of the MSW transcends technical and economical aspects and that it is necessary to think of it as a system. Then, a reuse system is a set-up of parts or interrelated elements that have as a function the efficient reincorporation of recovered materials from solid waste to the economical and productive cycle. Reuse system is maintained in time through technically appropriate and economically and socially feasible strategies, without threatening natural resources for future generations.

The reuse of waste can support dignified work for many people, the conservation of non-renewable natural resources and the reincorporation of products into the productive cycle in developing countries, for such reason its sustainability must be a goal for its human and sustainable development.

3. Conclusions

In developing countries, recovery is an alternative identified and/or applied as a solution to the problems associated with solid waste management. However, it requires that recovery be developed as an option with sustainability potential. For that, it is indispensable the participation of stakeholders and their integration with technologies and markets.

A key element for the sustainability of the reuse systems is the articulation of efforts and the mutual acknowledgement between the formal and informal sectors. The informal sector is one of the main elements of the recovery of paper, plastic, metals and glass, carrying it out generally under precarious working conditions and with inefficient technologies, making necessary the adaptation or development of technologies that favor the employment of local labor, dignifying their working conditions and under economical and sustainable conditions.

The valorization of waste as being an opportunity and not as something to be discarded by the generators is fundamental, so that not only the environmental but also the social and economical benefits associated with the reuse must be recognized, from these benefits also the generator must feel itself as a key part of the solution.

For the implementation of reuse systems from MSW it is important to obtain reliable estimates of the quantity and composition of the materials to take advantage of, as well as the characteristics and markets of the products. In the event of the necessity of a central installation for the reuse operations, siting, design and operation must be adapted to the local conditions.

Author details

Luis F. Marmolejo*, Patricia Torres and Mariela García
Facultad de Ingeniería, Universidad del Valle, Cali, Colombia

Luis F. Diaz
CalRecovery Inc., Stanwell Drive, Concord, CA, USA

---

* Corresponding Author

Perspectives for Sustainable Resource Recovery from Municipal Solid Waste in Developing
Countries: Applications and Alternatives

183

4. References

Abu Qdais, H.A., 2007. Techno-economic assessment of municipal solid waste management in Jordan. Waste Management, Vol. 27, pp. 1666 -1672.

Alhumoud, J.M., 2005. Municipal solid waste recycling in the Gulf Co-operation Council states. Resources Conservation & Recycling, Vol. 45, pp. 142 -158.

Ali, A., 2002. Managing the scavengers as a resource. In: Appropriate Environmental and solid waste management and technologies for Developing Countries, Vol. 1. ISWA.

Al-Khatib I.A., Arafat, H.A., Basheer, T., Shawahneh H., Salaha, A., Eid, J., Ali, W., 2007. Trends and problems of solid waste management in developing countries: A case study in seven Palestinian districts. Waste Management, Vol. 27, pp 1910-1919.

Barreira, L.P., Philippi, A., Rodrigues, M.S., 2006. Usinas de compostagem do estado de São Paulo: qualidade dos compostos e processos de produção. Eng. Sanit. Ambient., Dez , Vol. 11, pp. 385 -393. ISSN 1413 - 4152.

Barrena, R., Vásquez, F., Sánchez, A., 2008. Dehydrogenase activity as a method for monitoring composting process. Bioresource Technology & Recycling, Vol. 99, pp. 905 - 908.

Barton, J.R., Issaias, I., Stentiford, E.I., 2008. Carbon - Making the right choice for waste management in developing countries. Waste Management, Vol. 28, pp. 690 - 698.

Batool, SA; Chaudhry, N., Majeed, K., 2008. Economic potential of recycling business in Lahore, Pakistan. Waste Management, Vol. 28, pp. 294-298.

Bezama, A., Aguayo, P., Konrad, O., Navia, R., Sorber, K.E. 2007. Investigations on mechanical biological treatment of waste in South America: Towards more sustainable MSW management strategies. Waste Management,Vol. 27, pp. 228 - 237.

Bogner, J.M., Abdelrafie, A.C., Diaz, A.F., Hashimoto, G.S., Mareckova, K., Pipatti R., Zhang, T., 2007. Waste Management, In Climate Change 2007: Mitigation. Contribution of Working Group III to the Four Assessment Report of the Intergovernmental Panel of Climate Change. B. Metz, O. R. Davidson, P.R. Bosch, R. Dave, L.A. Meyer (eds). Cambridge University Press, Cambridge, United Kingdom and New York, NY, USA.

Contraloría General de la República de Colombia, 2005. Auditoria Especial al Manejo de los Residuos. República de Colombia, 2005.

DAPM - Departamento Administrativo de Planeación Municipal Santiago de Cali, UNIVALLE -Universidad del Valle, 2006. Caracterización de los residuos sólidos residenciales generados en el municipio de Santiago de Cali - 2006.

Diaz, L.F., Eggerth, G.M., Golueke, C.G., 2002. The role of composting in the management of solid wastes in Economically Developing Countries. In: Appropriate Environmental and solid waste management and technologies for Developing Countries, Vol 2. ISWA.

Diaz, L.F., Savage, G.M., Eggerth L.L., 2007. The management of solid wastes in economically developing countries – major needs. In: Proceedings, Sardinia 2007. Eleventh International Waste Management and Landfill Simposyum. S Margherita di Paula, Cagliary Italy; 1 - 5 October 2007.

Do Prado Filho, F.J., García Sobreira, F., 2007. Desempenho operacional de unidades de reciclagem e disposição final de resíduos sólidos domésticos financiadas pelo ICMS Ecológico de Minas Gerais. Engenharia Sanitária Ambiental. Vol 12, pp 52- 61.

Drescher, S., Zürbrugg Ch., 2006. Decentralised composting: lessons learned and future potentials for meeting the Millennium Development Goals. In: Solid Waste, Health and the Millennium Development Goals. CWG – WASH Workshop 2006, 1 – 5 February in Kolkata, India

Dulac, N., 2001., The Organic Waste flow in Integrated Sustainable Waste Management- The Concept. Waste The Netherlands.

Esteban García, A.I., Muñoz Jofré, J.M., Szantó Nerea, M., Tejero Monzón, I., 2001. The other dimensión in waste management: The informal sector and socio-labour insertion. In: Proceedings, Sardinia 2001. Eighth International Waste Management and Landfill Simposyum. S Margherita di Paula, Cagliary Italy; 1- 5 October 2001.

Fricke, K., Santen, H., Bidlingmaier, W., 2001. Biotechnologicals processes for solving waste management problems in economically less developed countries. In: Proceedings, Sardinia 2001. Eighth International Waste Management and Landfill Simposyum. S Margherita di Paula, Cagliary Italy; 1 - 5 October 2001.

Hargreaves, J.C., Adl M.S., Warman, P.R., 2008. A riview of the use of composted municipal solid waste in agriculture. Agriculture, Ecosystems and Environment, Vol. 123, pp. 1 - 14.

Hristovski, K., Olson, L., Hild, N., Peterson, D., Burge, S., 2007. The municipal solid waste system and solid waste characterization at the municipality of Veles, Macedonia. Waste Management, Vol. 27, pp. 1680 – 1689.

Hui, Y., Li'ao, W., Fenwei, S., Gang, H., 2006. Urban solid waste management in Chongqing: Challenges and opportunities. Waste Management, Vol. 26, pp. 1052 - 1062.

Ketlogetswe, C., Mothudi, T.H., 2005. Botswana's environmental policy on recycling. Resources Conservation & Recycling, Vol. 44, pp. 333 - 342.

Klinger, R.A., Olaya, J., Marmolejo, L., Madera, C., 2009. Plan de muestreo para la cuantificación de residuos sólidos residenciales generados en las zonas urbanas de ciudades de tamaño intermedio. Revista Facultad de Ingeniería Universidad de Antioquia. No. 48 pp. 76-86.

Kocasoy, G., 2001. Solid Waste Management in Developing Countries: Proposed Amendments in the existingsituation. In: Proceedings, Sardinia 2001. Eighth International Waste Management and Landfill Simposyum. S Margherita di Paula, Cagliary Italy; 1 - 5 October 2001.

Kofoworola, O.F., 2007. Recovery and recycling practices in municipal solid waste management in Lagos, Nigeria. Waste Management, Vol. 27, pp. 1139- 1143.

Körner, I., Saborit-Sánchez, I., Aguilera-Corrales, Y., 2008. Proposal for the integration of descentralised composting of the organic fraction of municipal solid waste into the waste management system of Cuba. Waste Management, Vol. 28, pp. 64 - 72.

Lopes, R.F., Shan Muganantha, S., Lippett, R., Keir, A., Mello, R., Hoornweg, D., 2007. Municipal solid waste management in the Latin American and Caribbean regions:Trends and opportunities for improvement. In: Proceedings, Sardinia 2007. Eleventh International Waste Management and Landfill Simposyum. S Margherita di Paula, Cagliary Italy; 1 - 5 October 2007.

Manga, VE; Forton, OT; Read, AD., 2008. Waste management in Cameroon: A new policy perspective?. Resources Conservation & Recycling, Vol 52, pp. 592- 600.

McBean, E.A., del Rosso, E., Rovers, F.A., 2005. Improvements on financing for sustainability in solid waste management. Resources Conservation & Recycling, Vol. 43, pp. 391 - 401.

Mbuligwe, S.E., Kassenga, G.R., Kaseva, M.E., Chaggu, E.J., 2002. Potential and constraints of composting domestic solid waste in developing countries: findings from a pilot study in Dar es Salaam, Tanzania. Resources, Conservation and Recycling, Vol. 36, pp. 45 - 59.

Metin, E., Eröztürk, A., Neyim, C. 2003. Solid waste management practices and review of recovery and recycling operations in Turkey. Waste Management, Vol. 23, pp. 425 – 432.

Minambiente - Ministerio del Medio Ambiente República de Colombia, 2002. Selección de Tecnologías de Manejo Integral de Residuos Sólidos. Guía. ISBN 958-9487-39-4.

Moldes A., Cendón Y., Barral M.T. (2007). Evaluation of municipal solid waste compost as a plant growing media component, by applying mixture design. Bioresource Technology, Vol. 98, pp. 3069-3075

Mongkolnchaiarunya, J., 2005. Promoting a community- based solid-waste management initiative in local government: Yala municipality, Thailand. Habitat International, Vol. 29, pp. 27- 40.

Narayana, T., 2009. Municipal solid waste management in India: From waste disposal to recovery of resources. Waste Management, Vol. 29, pp. 1163-1166.

Nguyen Ngoc, U., Schnitzer, H., 2009. Sustainable solutions for solid waste management in Southeast Asian countries. Waste Management, Vol. 29, pp. 1982-1995.

OPS - Organización Panamericana de la Salud, 2005. Informe de la Evaluación Regional de los Servicios de Manejo de Residuos Sólidos en América Latina y el Caribe. Washington. D.C.

Otieno, F.A.O. & Taiwo, O., 2007. Current state of urban solid waste management in some cities in Africa (2007). In: Proceedings, Sardinia 2007. Eleventh International Waste Management and Landfill Simposyum. S Margherita di Paula, Cagliary Italy; 1 - 5 October 2007.

Rotich K, H., Zhao, Y., Jun, D., 2006. Municipal solid waste management challenges in developing countries -Kenyan case study. Waste Management, Vol. 26, pp. 92 - 100.

Said-Pullicino, D., Erriquens, F.G., Gigliotti, G., 2007. Changes in the chemical characteristics of water- extractable organic matter during composting and their influence on compost stability and maturity. Bioresource Technology, Vol. 98, pp. 1822 - 1831.

Sha'Ato, R., Aboho S.Y., Okentude, F.O., Eneji, I.S., Unazi, G., Agwa, S., 2007. Survey of solid waste generation and composition in rapidly growing urban area in Central Nigeria. Waste Management, Vol. 27, pp. 352 - 358.

Shekdar, A., 2009. Sustainable solid waste management: An integrated approach for Asian countries. Waste Management, Vol. 29 pp. 1438–1448.

Suocheng, D., Tong K. W.,Yuping, W., 2001. Municipal solid waste management in China: using commercial management to solve a growing problem. Utilities Policy, Vol. 10, pp. 7- 11.

SSPD - Superintendencia de Servicios Públicos Domiciliarios. República de Colombia, 2008. Diagnóstico Sectorial Plantas de Aprovechamiento de Residuos Sólidos - Marzo de 2008.

Suttibak,S., Nitivattananon, V., 2008. Assessment of factors influencing the performance of solid waste recycling programs. Resources, Conservation and Recycling, Vol. 53, pp. 45-56

Troschinetz, M., Milhecic, J.R., 2009. Sustainable recycling of municipal solid waste in Developing Countries. Waste Management, Vol. 29, pp. 915-923

United Nations , 2002. Report on the World Summit on Sustainable Development. Johannesburg (South Africa), August 26 to September 4, 2002. In: http://www.unctad.org/sp /docs/aconf199d20&c1_sp.pdf.

UNEP- United Nations Environment Programme, 2008. International Source Book on Environmentally Sound Technologies for Municipal Solid Waste Management. Disponible en Internet: <http://www.unep.or.jp/ietc/ESTdir/Pub/MSW/index.asp.

UNEP - United Nations Environment Programme & CalRecovery Incorporated, 2005. Solid Waste Management, Vol. 1. United Nations Environment Programme (2005). ISBN: 92-807-2676-5.

Van de Klundert, A., Anschütz, J., 2001. Integrated Sustainable Waste Management – The Concept. Waste Advisers on Urban Environment and Development. The Netherlands.

Wang, H., Nie, Y., 2001. Municipal solid waste characteristics and management in China. Journal of Air and Waste Management Association 51, 250–263. Air andWaste Management Association.

Wilson, D., Araba, A.O, Chinwah,K., Cheeseman C.R., 2009. Building recycling rates through the informal sector. Waste Management, Vol. 29, pp. 629-635.

Wilson, D.C., Velis, C., Cheeseman, C., 2006. Role of informal recycling in waste management in developing countries. Habitat Internacional, Vol. 30, pp. 797 – 808.

Zurbrügg, C., Drescher, S., Rytz, I., Maqsood Sinha, A.H.Md., Enayetullah, I., 2005. Descentralised composting in Bangladesh, a win- win situation for all stakeholders. Resources Conservation & Recycling, Vol. 43, pp. 281 - 292.

Chapter 9

Implementation of Recycling Municipal Solid Waste (MSW) at University Campus

Tiew Kian-Ghee, Noor Ezlin Ahmad Basri,
Hassan Basri, Shahrom Md Zain and Sarifah Yaakob

Additional information is available at the end of the chapter

1. Introduction

Conventional waste disposal meets its limits throughout most of the world with increasing waste generation and rising proportions of packaging and toxic compounds in MSW. Landfill of waste leads to pollutant emissions over long periods of time, aggravated with the recent occurring of water pollution caused by open dumping at Sungai Kembong, Bangi (Utusan Malaysia Online, 2010 & David, 2001), thus, requires sophisticated emission control and treatment methods. The consequences are long after-care periods for abandoned landfills. Furthermore, in many countries it is increasingly more difficult to find suitable locations for landfills, which are accepted by the population. In addition, the cost of management and transportation for waste is increasing with population. These circumstances are to be found all over the world and make new strategies for waste management necessary. The promotion of waste minimization and recycling are important components of modern waste management strategies. Nevertheless, even when the minimization and recycling potentials are fully exploited, there is still a residual fraction, which has to be disposed of. The burdens resulting from landfill can be minimized by pre-treating the waste and thus limiting its emission potential. Malaysia, with a population of over 24 million generates 18,000 tones of domestic waste daily. In 2005, the per capita generation of solid waste in Malaysia varies from 0.45 to 1.44kg/day depending on the economic status of an area (Agamuthu et. al 2009). There are now 168 disposal sites but only 7 are sanitary landfills. The rest are open dumps and about 80% of these dumps have filled up to the brim and have to be closed within two years (Mageswari 2005). Fauziah and Agamuthu (2007) reported that the MSW generation rate in Peninsular Malaysia is approximately 1.2 kg/capita.day. The same authors further stated that the current municipal generation rate in Malaysia has reached 1.3 kg/capita.day (Agamuthu and Fauziah, 2008).

The concept of 3R's (Reduce, Reuse, Recycle) is gaining importace as a sustainable and environmental friendly method. Recycling is one way to solving the problems of shortage of sanitary landfill (Chamhuri Siswar et al. 2000). On the other hand, composting is widely being practised at other developing countries (Hoornweg et al. 1999). Domestic waste generated in Malaysia is more than 40% is organic waste and overall in Malaysia, the waste composition is approximately 45% food waste, 24% plastic waste, 7% paper waste, 6% metal waste, 3% glass waste and others (Japan International Cooperation Agency (JICA), 2006). Hence, the 45% food waste can be treated by composters to generate fertilizer and reduce the quantity of food waste directly dumped to the sanitary landfill. Solid waste management (SWM) is a complicated management this is because a lot outside factors will reflected it and normally reflected by culture, economy, food and topography (Shekdar, 2009). Hence, comprehensive study of solid waste management is needed before implementing suitable integrated waste management technology. Therefore, this paper presents the study on waste generation and composition at UKM and the evaluation on the effectiveness of recycling facilities provided being provided to divert waste from landfill.

1.1. Background of case study

Universiti Kebangsaan Malaysia, UKM (the National University of Malaysia) was officially established on 18 May 1970. The main campus of UKM is located at Bandar Baru Bangi, Selangor and consist 12 faculties, 12 centres and 14 institutes. UKM is one of the leading universities in Malaysia with a campus population of approximately 30,000, consists of majority of students, academic and supporting staff. In view of the size of this community, solid waste management represents a major challenge in achieving institutional sustainability. In 2008, a major change took place in the solid waste management (SWM) system in the university. A Memorandum of Agreement (MoA) was signed to upgrade collection of solid waste to a centralized collection system through collaboration between the university and the solid waste company Alam Flora Sdn. Bhd. (AFSB). The collaboration intended to institute integrated solid waste management (ISWM) at the university and enable the university to achieve a Zero-Waste campus objective.

2. Materials and methods

2.1. Waste generation, waste composition and waste characterization

Daily waste generation was carried out based on the weight of solid waste collected by the waste collector company Alam Flora Sdn Bhd (AFSB). The following shows the segregation methods done at the UKM campus:-

2.1.1. Source segregation (24th February 2009 to 2nd March 2009)

The samples were taken from four different categories of generation: (1) faculties (excluding laboratories), (2) dormitories (including cafeteria), (3) offices, and (4) student affairs. These four categories are representative because they cover all the activities carried out at the campus. The characterization of the solid waste was carried out using the modified Methodology for Conducting Composition Study for Discarded Solid Waste (University of Central Florida, 1996). Samples were taken during 7 consecutive days (including Sundays). For the composition analysis, the samples were classified manually into two categories i.e. recyclable items (papers, plastics, metals and glass) and non-recyclable items (food waste, e-waste and others). The results from the solid waste samples were analyzed. The weight percentage for each subcategory was calculated using the following equation:

$$\text{Sub} - \text{category percentage} = \left(\text{Amount of sub} - \text{category, in kg} \, / \, \text{Total weight, in kg}\right) \times 100\%$$

2.1.2. Centralized segregation (27th July 2009 to 23th August 2009)

All baseline characterizations were performed according to the ASTM 5231-92 Standard Test Method for Determination of the Composition of Unprocessed Municipal Solid Waste (ASTM, 2003) and LAGA PN 98 (RFA, 2001). During these four weeks the sampling was done every Monday, Wednesday and Saturday by cone Sampling. At first the samples were taken from a heap and later out of a roro-container. Through the use of a wheelbarrow, public works employees and master course students moved about 450 kg of MSW onto the thin plastic sheet. Ripping and opening of the plastic bags and containers ensued with small knives, and mixing was achieved with shovels and rakes. Then the large sample was quartered so that approximately one fifth of the sample (91–136 kg) was left to form a representative sample. Finally, manual sorting of the waste was done according to ASTM categories i.e. plastic bag, plastic bottle, polystyrene, mixed plastic (consists hard plastic), non-recyclable plastic bag (black plastic bag), paper (newspaper, magazines, cardboard and box), aluminum, rubber and leather, food waste (cooked waste and raw materials), and glass as shown in Fig. 1., and different types of samples were placed into properly labeled buckets.

2.2. Recycling facilities

Recycling facilities at UKM consist of a recycling center, paper recycling boxes in all offices, and 2-bins recycling system at strategic locations throughout the campus.

2.2.1. Paper recycling boxes

Paper recycling boxes are being provided to all offices throughout the campus. The study on the effectiveness of office paper recycling was carried out from April 2010 to October 2010. Monthly data on the collected recycled papers is provided by Alam Flora Sdn. Bhd. Fig. 2. shows paper recycling boxes being located in an office at UKM.

Figure 1. Waste composition categories.

Figure 2. Paper recycling box in an office at UKM

2.2.2. Recycling centre

The recycling centre at UKM is located nearby the main entrance of the university. It had started its operation since April 2010, initially every first Tuesday for the month, but had increased to every Tuesday of the month. Alam Flora Sdn. Bhd. is responsible to buy the recyclable items collected from the Recycling Centre. The Recycling Center is also being used as a centre to educate UKM community on the importance of recycling to reduce the burden on landfills. The study on the effectiveness of the recycling was carried out from April 2010 till October 2010. Monthly data on the recyclable items is provided by Alam Flora Sdn. Bhd. Fig. 3. shows the front view of the Recycling Centre, while Fig. 4. shows the five compartments for different categories recyclable items.

Figure 3. The front view of the Recycling Centre

Figure 4. Five compartment for different categories of recycled items

2.2.3. 2-Bins recycling system

The study on the effectiveness for the 2-bins recycling system at UKM was done at the Faculty of Engineering and Built Environment (FKAB). The orange coloured bin is for commingled recyclable items and the other bin is for non-recyclable waste. The recyclable items in the recycling bins were weighted. Ten waste stations were observed in the four-storey academic building at FKAB. The 2-bins recycling system consists of two types of container, which are a bin for commingled recycling items and a bin for mixed non-recyclable waste as shown in Fig. 5. In addition, the attitudes and acceptance of students were also being evaluated.

Figure 5. 2-bins recycling system located at the FKAB

3. Results and discussions

3.1. Waste generation

The highest monthly waste generation was in July 2009 due to the new semester intake for students. The waste generation in September 2009, which is during a week study period and also the fasting month of Ramadhan for the Muslims is the lowest in comparison to other months. Waste generation in December 2009 and January 2010 was low as well due to the semester break for the students. Results in Table 1 shows the average daily waste generation at UKM campus is approximately 4.76 ton/day.

Month	Total waste generation (ton/month)	Average daily waste generation (ton/day)
July 2009	176.63	5.70
August 2009	156.64	5.22
September 2009	128.09	4.27
October 2009	169.85	5.66
November 2009	145.14	4.84
December 2009	112.22	3.62
January 2010	105.69	3.41
February 2010	145.60	5.20
March 2010	151.28	4.88

Table 1. Waste generation at campus UKM, Bangi

3.2. Waste composition

From the waste composition study at UKM by applying the segregation method at source showed that food waste is the highest portion originating from various locations, i.e. Dormitories (51.6%), Faculties (57.8%), Student Affairs Building (56.4%), and Offices (53.7%) as shown in Table 2. The percentage of plastic items is the second highest for Dormitories 11.8% and Faculties 12.7%, however, at Student Affairs Building is 13.3% and Offices is 10.9%. For both Student Affairs Building and Offices, the second highest amount of recyclable items is papers, i.e. Student Affairs Building, 17.7% and Offices 25.5%.

	Dormitories	Faculties	Student Affairs Bldg	Offices	Average*
Papers	8.3%	12.0%	17.7%	25.5%	15.9%
Plastics	11.8%	12.7%	13.3%	10.9%	12.2%
Metals	1.9%	0.9%	0.4%	1.5%	1.2%
Glass	0.2%	0.6%	1.8%	0.1%	0.7%
Food Waste	51.6%	57.8%	56.4%	53.7%	54.9%
e-Waste	0.1%	0.2%	0.3%	0.3%	0.2%
Others	26.2%	15.8%	10.1%	7.9%	15.0%
Total	100.0%	100.0%	100.0%	100.0%	100.0%

Table 2. Summary waste composition for segregation by source method

In terms of recyclable materials, the study showed that overall approximately 30% of major recyclable materials were disposed of from the campus, this includes mainly papers (15.9%), plastics (12.2%), glasses (0.7%), and metals (include ferrous metals, non-ferrous metals and some aluminium) (1.2%). Offices generated the highest percentage of recyclable items, followed by Student Affairs Building, Faculties and Dormitories.

Waste composition by the centralised method was carried out for a period of one month, and at the end the total number of samples collected was 12. Table 3 showed waste composition results from the centralised method. The food waste constitutes the highest amount in all categories which is 42.95%, i.e. 27.79% for cooked food waste and 15.16% for raw materials of food waste. Currently, approximately 30.55% of major recyclable materials were disposed off at landfills. The recyclable materials consist of mixed papers and plastics, plastic bottles, glasses, and metals. Some other wastes recorded were polystyrene container (5.48%), non-recyclable plastic bags (the black plastic garbage bag) (6.26%) and rubber and leather wastes (1.37%).

Types	Amount (kg)	%
Plastic Bag	14.5	13.39
Plastic Bottle	1.4	1.33
Polystyrene	5.9	5.48
Mixed Plastic	10.3	9.56
Non-recyclable Plastic Bag	6.8	6.26
Newspaper, Magazines, Cardboard and Box (Papers)	18.6	17.18
Aluminum Cans (Metals)	1.9	1.72
Rubber and Leather	1.5	1.37
Cook (Food Waste)	30.1	27.79
Uncook (Food Waste)	16.4	15.16
Glass	0.8	0.76
Total	108.2	100.00

Table 3. Waste composition by centralized method in campus UKM, 2009

The centralised segregation method showed a much higher proportion of plastic items (36.02%) as compared to the source segregation method (12.20%) as shown on Table 4. The moisture content is higher in the centralized segregation method as compared to the sample collected directly from the garbage bins. The higher moisture content is due to the waste samples for the centralised segregation is obtained from compactor trucks, which normally being placed outside buildings for at least one night, thus rainwater may get into the waste during storage. Further, in a compactor truck all types of waste including the wet food waste will be mixed and compacted.

Other waste components showed almost the same percentage, except source segregation method had a higher amount of others waste component, which is 15.2%, while centralized segregation method is nil. Other waste category consists of bulky wastes and some miscellaneous items. In the segregation at source method, everything that was set out as waste was treated as sample, and this item is significant, while the centralized segregation method, the sample was taken from the collection vehicle, which does not collect bulky waste. This is probably the main reason for the difference. The benefit of studying the two methods for waste segregation is to determine the waste composition generated at source and at the end management.

Types	Source Segregation UKM, 2009 %	Centralized Segregation UKM, 2009 %	Average %
Plastics	12.20	36.02	24.11
Papers	15.90	17.18	16.54
Metals	1.20	1.72	1.46
Rubber & Leather	-	1.37	0.68
Food Waste	54.90	42.95	48.92
Glass	0.70	0.76	0.73
Others	15.20	-	7.6
Total	100	100	100

Table 4. Comparison waste composition UKM between segregation by source method and centralized segregation method

3.3. Effectiveness of the recycling facilities

3.3.1. Paper recycling boxes

Paper recycling boxes in offices on UKM campus, Bangi had begun in March 2010. Collection of recyclable papers was conducted in April 2010 to present. A total of 293 paper recycling boxes had been placed in 62 locations, which include institutes, centers, faculties and libraries. The results of recyclable papers collection rate is increasing from month to month. Although in Fig. 6 shows a decline in August (975kg) as compared to July (3,250 kg) with a difference of 2,275 kg because August is semester intake but July is end of short semester. Hence, a lot of materials have been thrown on end of semester compared to semester intake. However, an increasing trend can be seen in comparison to April (809kg) which is first collection of paper recycling program at offices. Detailed recyclable items composition consists of 46% black and white paper, 32% mixed paper and 18% newspapers as shown in Fig. 7.

A total of 11,733 kg paper waste generated by UKM offices had been recycled from April 2010 to October 2010. Coordinators for recycling of office's paper programme were appointed to play important role in the management of paper recycling in offices under the supervision of the Department of Development Management, UKM (JPP).

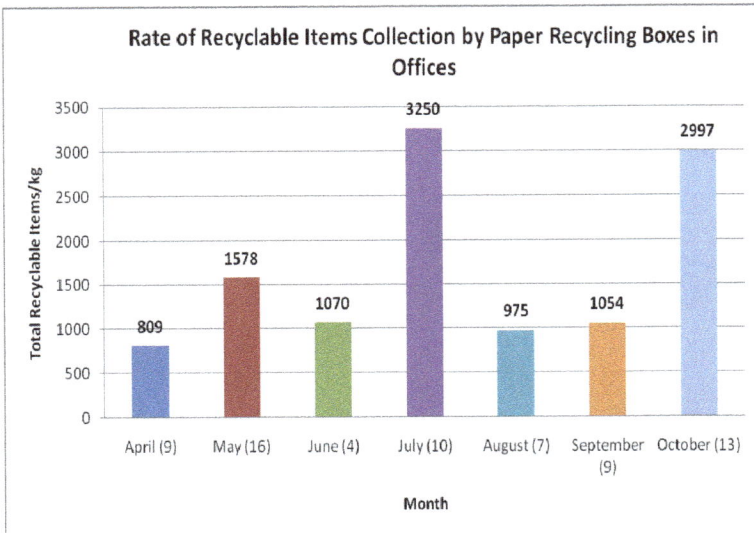

Figure 6. Recycling rate of recyclable papers in paper recycling boxes.

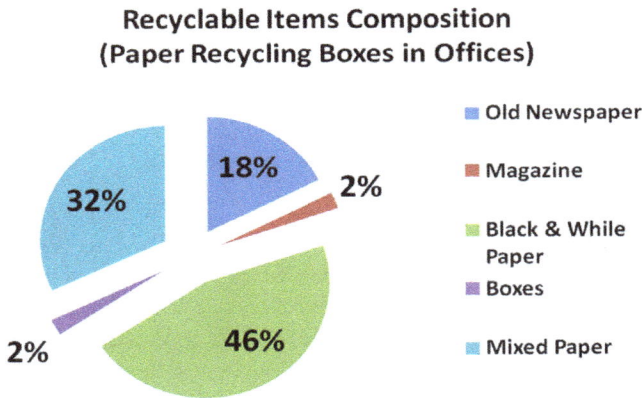

Figure 7. Detailed composition of recyclable papers from paper recycling boxes

3.3.2. Recycling centre

Fig. 8 shows the collection rate at the Recycling Center from April to October 2010. A total of 4,008 kg of recyclable materials from April 2010 to October 2010 was collected. The recyclable items composition is 28.52% newspapers, 27.22% black and white paper, 16.47% mixed paper, 8.43% magazines, 6.29% boxes, 5.41% plastics and 4.17% can, as shown in Fig. 9.

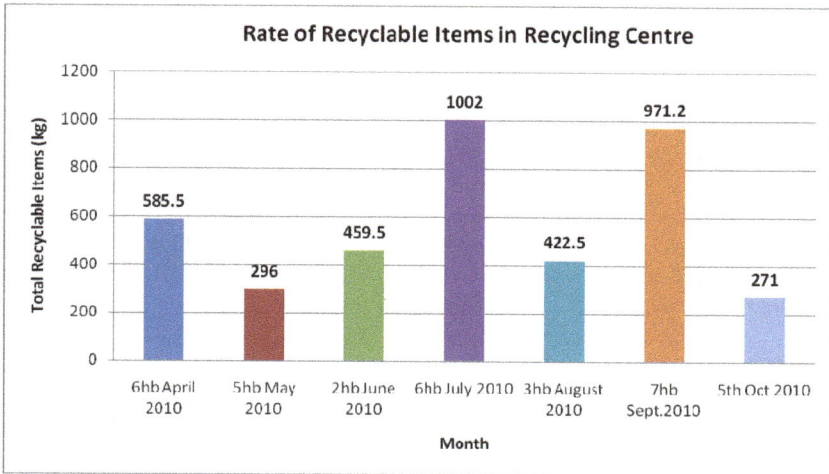

Figure 8. Monthly collection rate of recyclable items at the Recycling Centre

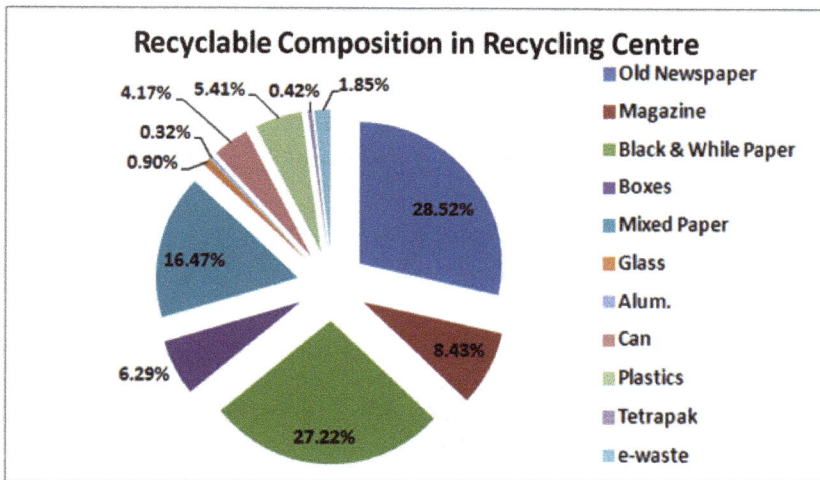

Figure 9. Recyclable items composition at the Recycling Centre

3.3.3. 2-Bins recycling system

From Fig. 10, the 2-bins recycling system at faculties shows that the usage of the recycling bins still needed improvement due to 65.3%, which is slightly more than half of the content in the orange bin (consists of commingled recyclable items) was recyclable items while 34.7% was non-recyclable items. There is still high amount of non-recyclable items were wrongly placed (34.7%) in the orange bin. This shows that, publicity and awareness programs are required to educate the staffs and students on the correct usage of the orange bins. On the other hand, for the mixed waste bins, the content was 58.7% non-recyclable items while the percentage for recyclable items was 41.3%. Thus, this also shows a high percentage of wrong usage for the mixed waste bins (41.3%).

From the study, the 2-bins recycling system is appropriate to ensure students to separate the recyclable items. Many researchers showed that the compliance is increasing when recycling bins are placed closer to users, and the physical features of the bins will also influence the recycling compliance (Sean Duffy & Michelle Verges, 2009). For this study, orange coloured bins with colourful design were specially made for recycling bins at UKM campus and being placed in the areas which easily accessible by the users.

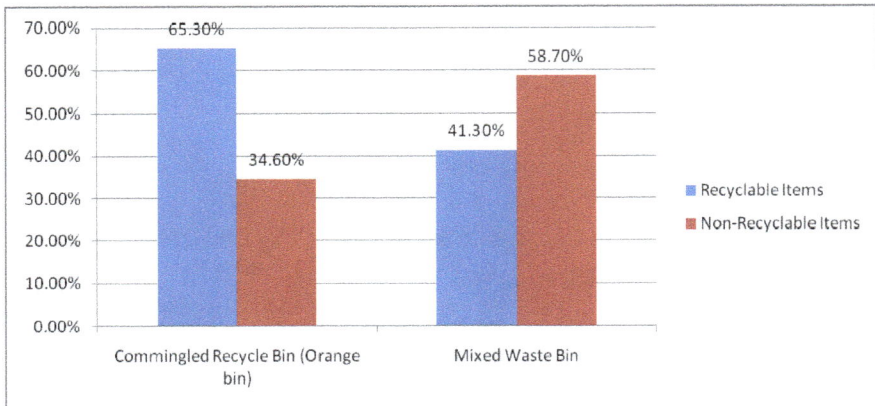

Figure 10. Summary of 2 bins system in FKAB, UKM

4. Conclusions

The results showed that the percentage of food waste generated from UKM campus is the highest, which is 54.9% (source segregation method) and 42.95% (centralized segregation method). Plastics waste is the second highest portion in the waste composition. This is due to the habit of students who prefer to buy food from cafeteria but dine at home, hence, high usage of polystyrene container, drinking plastic bottles and plastic bags used for packaging materials. Papers waste is also in high proportion in the waste composition. This is due to UKM campus being an education institution. From the study, most of the recyclable items could be recovered by providing appropriate recycling facilities along with publicity and the awareness programme. The results showed that waste diversion from landfill is 1.62% by Paper Recycling Boxes at offices and the Recycling Centre. The implementation of recycling MSW at the university campus has the potential to reduce waste disposal to landfills and thus making Universiti Kebangsaan Malaysia a sustainable campus. Finally, sustaining the recycling facilities is vital to ensure the success of continuous recycling programme at UKM campus.

Author details

Tiew Kian-Ghee
Faculty of Engineering and Built Environment, University Kebangsaan, Malaysia

Noor Ezlin Ahmad Basri, Hassan Basri and Shahrom Md. Zain
Faculty of Engineering & Built Environment,
Department of Civil and Structural Engineering, University Kebangsaan, Malaysia

Sarifah Yaakob
Alam Flora Sdn Bhd., Malaysia

Acknowledgement

The authors are grateful to Universiti Kebangsaan Malaysia for funding the project (Grant code: UKM-PTS-007-2009) and Alam Flora Sdn Bhd for the kind cooperation during the project.

5. References

Agamuthu, P. and Fauziah, S.H., 2008. Solid waste: environmental factors and health. Paper presented at *EU-Asia Solid Waste Management Conference*, EA-SWMC, Ipoh, 29-30 October 2008

Agamuthu, P., Fauziah Shahul Hamid & Kahlil Khidzir. 2009. Evolution of solid waste management in Malaysia: Impacts and implications of the solid waste bill, 2007. *J Mater Cycles Waste Management* 11:96–103.

ASTM (American Society of Testing and Materials), 2003. Standard Test Method for Determination of the Composition of Unprocessed Municipal Solid Waste. Designation: D, pp. 5231–5292.

Chamhuri Siswar, Talylor, D. & Hasnah Ali. 2000. Knowledge, attitudes and perception on minimising household municipal solid waste generation: a case of Petaling Jaya Municipality Council. Kertas kerja Bengkel Dasar Memperbaiki Pengurusan Sisa Pepejal Perbandaran. Anjuran Institut Alam Sekitar dan Pembangunan, Universiti Kebangsaan Malaysia. Bangi, Julai 1999 dan November 2000.

David, S.C. 2001. The McGraw-Hill Recycling Handbook. *Characterization of waste streams,* pg. 3.1-3.36. New York: McGraw Hill.

Fauziah, S.H. and Agamuthu, P. 2007. SWPlan Software Application for Malaysian Municipal Solid Waste Management. *Malaysian Journal of Science*, 26 (1). pp. 17-22. ISSN 13943065

Hoornweg, D., Thomas, L. & Otten, L. 1999. Composting and its applicability in developing countries. Urban Waste Management.
http://www.worldbank.org/htm/fpd/urban/solid_wm/uwp8.pdf [5 Ogos 2009].

Japan International Cooperation Agency (JICA). 2006. The Study On National Waste Minimisation In Malaysia. Yachiyo Engineering Co., Ltd. & Ex Corporation.

Länderarbeitsgemeinschaft Abfall (RFA). 2001. *LAGA PN 98 Richtlinie für das Vorgehen beiphysikalischen, chemischen und biologischen Untersuchungen im Zusammenhang mitder Verwertung/Beseitigung von Abfällen.*

Mageswari, S. 2005. *GIAI global meeting. Penang, Malaysia.* 17-21.

Sean Duffy and Michelle Verges. 2009. It Matters a Hole Lot Perceptual Affordances of Waste Containers Influence Recycling Compliance. Environment and Behavior. Vol. 41, 5, 741-749.

Shekdar, A.V. 2009. Sustainable Solid Waste Management: An Integrated Approach for Asian Countries. *Waste Management 29,* Pages 1438-1448.

University of Central Florida. 1996. Methodology for Conducting Composition Study for Discarded Solid Waste.

Utusan Malaysia Online. 2010. "Loji rawat air Sg. Semenyih tercemar". http://www.utusan.com.my/utusan/info.asp?y=2010&dt=0908&pub=Utusan_Malaysia& sec=Muka_Hadapan&pg=mh_05.htm

Chapter 10

Scenarios for Sustainable
Final Waste Treatment in Developing Country

Christia Meidiana

Additional information is available at the end of the chapter

1. Introduction

Worldwide, the waste management sector contributes approximately 3 – 5 % of total anthropogenic emission in 2005. Compared to the total emission, this percentage is relative minor [1]. Yet, the waste sector is in a state that it moves from being a minor source of global emissions to becoming a major saver of emissions [2]. Emission reduction from waste sector can be achieved through waste hierarchy principles including disposal as the least preferred option for managing waste and avoidance and minimization as the most preferred option waste [3]. The implementation of these waste managements can reduce emissions from other sectors of the economy such as energy, forestry, agriculture, mining, transport, and manufacturing sectors. The emission from waste management sector is mainly sourced from landfill through methane which is produced during waste degradation process [1]. Landfills have been practiced for disposing of the waste in developed and developing countries with different level of technical and safety requirements. In developed countries such as EU member states, there is decreasing trend of landfilling for the EU Landfill Directive requiring the reduction of biodegradable waste disposal in landfill [3]. Mean while, landfill is the most common method in waste disposal in developing countries though continuous efforts to promote other waste disposal methods such us recycling, incineration, mechanical and biological treatment. Unfortunately, many developing countries operate an open dump site instead of a controlled landfill [2]. Open dumping method creates environmental damage. It takes up not only more and more valuable land space, but also causes air, water and soil pollution by discharging green house gas i.e. methane (CH_4), carbon dioxide (CO_2), nitrous oxide (N_2O), and nitrogen oxide (NO_x) into atmosphere and chemicals into the earth and groundwater which can threaten human health, plants and animals.

The practice of open dumping method is quite common in Indonesia. Almost 90% of landfills in Indonesia are open dump site. The minor financial viability of the local governments is the reason why they are not be able to operate a proper solid waste disposal site (SWDS) [4]. The waste disposal in open dump site contributes the major greenhouse gas (GHG) from waste sector. At national scale, emission from waste sector is less compared to other sectors. It amounts to 166.8 Mt CO_2e or 8% of the total national GHG emission which was 1,991 Mt CO_2e in 2005 and the government targets to reduce the total GHG emission by 26% by 2020 from 2009 level [5]. This commitment should be supported by adequate legal framework in related sectors including waste sector. In waste sector, there was no law in national level regulating waste management until 2008. The absence of waste law in national level and the lack of laws controlling municipal waste management in regional level is one of some reasons for poor landfill condition [6, 7]. Therefore, The Waste Law No. 18/2008 is not only an opportunity, but also a challenge for the local governments to provide the community with better waste management. The enactment of The Waste Law no. 18/2008 obliges the local governments in Indonesia to implement environmentally sound waste management practices including a safe final disposal site. Article 22 defines this clearly by intending the implementation of environmentally friendly technology for final waste treatment, whereas Article 44 intends the requirement of safe landfill practices [8]. Local government of Yogyakarta as waste authority and landfill operator is also required to meet this law. The municipality will close the old landfill (*Bendo landfill*) in 2012 and construct a new landfill in a new site not so far from the old landfill. Exerting full implementation of the Waste Management Law 18/2008 by constructing a sanitary landfill for environmentally sound landfill is not necessarily suitable for the inferior waste management conditions in Yogyakarta such as subordinate infrastructure, financial stringency, and insufficient technology. A controlled landfill is appropriate for the new landfill for some local conditions [9]. In controlled landfill, scavenging activity is allowed and believed as a contribution to waste reduction. Scavengers involve in Bendo landfill to sort the saleable material such as plastic, paper, metal and glass. Scavenging is becoming a main income for most scavengers and can contribute to waste reduction leading to longer landfill's age and lower landfill gas (LFG) emission. However, there are discussions among local decision makers about the involvement of scavengers in the new landfill. Some believe that reducing the waste by treating it as near as possible to the waste source is more effective than allowing the scavengers to sort the waste at the landfill. Composting is another waste treatment method which has been applied in Yogyakarta since 2005. The organic waste from household is processed in community based composting centers involving about 15,000 households. The current composting rate is 10.33% of total biowaste.

Three different scenarios for the final waste treatment are proposed in this study based on the local situation in Yogyakarta. The selection of the best scenario is determined through the environmental parameters including the global warming potential and the emergy indices. The result of the study can be used as a reference for the local decision maker to determine the suitable final waste treatment in Yogyakarta City.

2. Research problems

The study aims to analyze the scenarios for the new controlled landfill in Yogyakarta, Indonesia. The proposed scenarios are assessed based on global warming potential using

IPCC Tier 2 method suggested by [10] and sustainability as well as efficiency using emergy analysis. By assessing these scenarios, it is possible to determine the best choice for appropriate waste treatment in landfill. Considering the general current local conditions of waste management, the study is conducted with the focus on the following problems;

- The new landfill have to meet the Waste Law No. 18/2008 requiring safe final waste treatment method
- Requirements to shift from open dumping methods to other environmentally sound final waste treatment method
- Inferior condition of waste management especially landfill

In order to solve the above research problems, the study focuses on the following objectives;

- To evaluate current municipal waste management situation in Yogyakarta
- To estimate methane emission from the old landfill
- To predict methane emission from the new landfill
- To determine the appropriate scenarios based on the local conditions
- To investigate the multiple scenarios and to evaluate them in terms of environmental assessment.

3. Method

The area of study is Yogyakarta City as a representative of a big city which has a population about 460,000 inhabitants [11]. The municipality plans to close the old landfill (*Bendo landfill*) and will construct the new landfill not so far from the old landfill in 2012. Surveys for primary and secondary data have been carried out twice which includes the aspects related to the waste management in the city. The first survey was conducted in January until March 2010 and the second was in October 2010. Data on municipal solid waste were collected from waste authorities in Yogyakarta to identify the general municipal solid waste including the waste characteristic, the rate of waste generation, waste collection and waste transportation to the landfill. Data on waste were mainly sourced from statistics on waste management in 2004 - 2008, Regency/City Profile, Waste Status Report 2008 - 2009 and earlier studies about waste management in Yogyakarta. The stakeholders associated with solid waste management are the target for the survey. It comprises the local government, private sectors and the community including scavenger. Nevertheless, after the preliminary study, only two respondents were determined to be the main objects for the primary surveys, namely the local government as the landfill owner/operator and the scavengers. The private sector and the community is not the focus of the surveys since they are not much involved and the major concerns within the scope of study. The number of the scavengers is determined using sampling method and the registered scavenger in the old landfill is the population for the sample. The amount of the waste delivered to the new landfill is a function of projected population, current waste generation per capita and level of service on collection whose rate or value was presented in [9].

The goal of the survey in the old landfill was to estimate the waste reduction rate caused by the existing scavenging and composting activity in Yogyakarta City. The result of the survey was used in scenarios as a reference to estimate the net waste disposal in the new landfill. The selection of the best scenario was based on the GWP and the emergy indices. IPCC Tier 2 method was used to estimate the GWP, while emergy analysis was applied to calculate the emergy indices.

4. Sampling method

Survey for primary data was conducted by means of questionnaires to provide more recent data and through interview in order to follow-up the questionnaire answered by the respondents and to get in-depth information related to landfill operation. Questionnaires were distributed to two kinds of respondents. The first respondents were Municipality of Yogyakarta and Yogyakarta Environmental Board representing the stake holders involved in waste management. The second respondents were scavengers in landfill. Standard open ended interview was selected in which the respondents were asked with same open ended questions to get detailed information which is easy to be analyzed and compared. The questionnaire aimed to examine declared waste treatment in landfill, level of service (LoS) on waste collection, performance of existing landfill and to identify the issues that influence the LoS and landfill's performance.

The number of the scavenger respondent was determined using Slovin formula (Equation 1) proposed by [12].

$$n = \frac{N}{N.e^2 + 1} \tag{1}$$

Where:
n : number of required respondents
N : number of population
e : sample error

5. Methane generation calculation

The methane emission during the new landfill time is estimated by means of time series data on waste disposal from 2013 until 2028. Population from 2013 until 2028 is projected using equation 2 [13].

$$Pn = Po(1 + r)^n \tag{2}$$

Pn : Population in the projected period
Po : Population in starting year
r : The average annual population growth rate
n : The projection period (in years)

Methane emission from landfill is calculated through methane generation estimation using Equation 3 suggested by [10].

$$CH_{4Emiss} = \left[\sum_x CH_{4generated_{x,T}} - R_T \right] * \left(1 - OX_T\right) \tag{3}$$

CH_{4Emiss} : CH_4 emitted in year T [ton/yr]
x : number of waste type
T : inventory year
R_T : recovered CH_4 in year T [ton]
OX_T: oxidation factor in year T (fraction)

  The total methane generation is the sum of the annual methane generation. Due to the fact that there is no soil covering and LFG collection system in the old landfill, the terms of recovered methane and oxidation factor is negligible. As a result, the amount of methane emission equals to the amount of methane generation.

6. Assumptions and limitations

The study focuses on analyzing the alternatives for the final waste treatment. The scenarios were made considering the current situations and the Waste Law no. 18/2008 which requires safe final waste treatment method. The result of the study does not necessarily reflect the actual prediction of future situations because these can be affected by changes including in waste composition (which was kept constant in this study). Some default values proposed by [10] were used to calculate the LFG emission. Due to the lack of input data, the following major assumptions were made:

- Currency rate is Rp 9,500 for US $1 which is the average value of the predicted exchange rate of Rupiah from Central Bank ranging between Rp 9000 - Rp 10,000 in 2010.
- Waste density is assumed 400 kg/m³ based on the average domestic waste density in Indonesia proposed by [14]. The assumption is made to convert some waste data which were in volume units to weight units.
- Waste generation rate per person is derived from the average amount of waste generation and number of population from 2004 - 2008.
- Waste percentages are kept consistent over the time period.
- Population growth is the average value over the period and kept consistent for the prediction.
- All material sorted by the scavengers in landfill will be transported for recycling, whereas the scavenging in community level is neglected because of unquantifiable data at present
- The emergy input from renewable and non-renewable resources per year are kept steady.

Some secondary data are required to be processed due to the following limitations:

- The incomplete data of waste tonnage disposed of in the old landfill in 2008 and 2009. Therefore, the calculation is conducted using the percentage from data in 2010.
- The weigh bridge was failure between May and August 2008. The average waste percentage from nearest month is used to calculate the missing data.
- Different waste classification among the references necessitates modification of existing waste classification to make the physical, proximate and ultimate analysis possible.
- The percentage of metal and glass from typical waste composition in Yogyakarta was consequently used due to minimum data obtained from field survey.

7. Scenarios for future landfill operation in Yogyakarta

The results from the observation of old landfill are also used as a reference in determining the alternatives. The scenarios include the calculation of environmental parameters (GWP and emergy values) from final waste treatment. The assumptions mentioned above are conditioned also to the scenario. It is assumed that the waste collection is constant with the base year 2013 although the rate increases proportionally to the waste generation each year. The calculation in emergy analysis is based on yearly inputs and outputs. Consequently, the value from emergy analysis could be different if the growth rate of waste generation is considered. However, since the same assumptions are applied to all scenarios and the scenarios are compared using the same assumptions, it does not mean that the result of the comparison deviates.

The prediction of waste generation is derived from the population projection. The result is used to calculate the waste which will be disposed of in the new landfill using the actual LoS. The assumptions for the parameters related to the waste management including the waste characteristic, waste percentage and waste composition are kept consistent. The physical and geographical properties of the site are assumed remain the same because of the proximity to the old landfill. Like Bendo landfill, the new landfill accepts the waste not only from Yogyakarta but also from other two counties (Bantul and Sleman). The percentage of the waste from these counties is kept consistent over the inventory years. The methane emission from the landfill is estimated using the IPCC Tier 2 method.

Entirely, there are three scenarios for the final waste treatment method in Yogyakarta presented in this study, i.e.;

1. **Scenario 0**: Zero scenario (Business as usual) is a base line scenario where the new landfill will be operated like the old landfill with the current average waste generation growth per year. Waste is delivered to the landfill without any further treatment and actual composting rate done by community is applied. There is no soil covering and LFG collection system. Furthermore, scavengers from the old landfill will be accommodated to sort the waste disposed of in the new landfill.
2. **Scenario 1**: Meet the target of improving the collection system. The Level of Service (LoS) of collection system will be increased according to the local government claim. The composting rate will be increased according to the local target and scavengers are allowed to work in the new landfill. There is soil covering but no LFG collection system.

3. **Scenario 2**: Meet the Waste Law 18/2008 policy Article 22 for environmentally friendly SWDS. The conditions related to LoS and composting rate in Scenario 1 are applied. Soil covering is applied to the landfill and the collected LFG will be flared with the open flaring system. Scavenging is permitted in restricted landfill area, where LFG collection system is not constructed.

8. The calculation of global warming potential

The calculation of global warming potential from landfill is based on the calculation of the uncontrolled and controlled emission of the methane and carbon dioxide. The methane emission is calculated using Equation 3. Though the existence of the regular soil covering (once a month) in Scenario 1 and Scenario 2, the variable of oxidation factor is assumed zero as the default value from IPCC for the managed but not covered with aerated material. The condition of landfill with few frequency of soil covering is assumed to be the same as that of without soil covering. The uncontrolled CH4 (U_{CH4}) and CO2 (U_{CO2}) emission are emitted from the landfill where a collection/flaring system does not present. The uncontrolled methane emission is calculated using IPCC Tier 2 method. Controlled CH4 (C_{CH4}) and CO2 (C_{CO2}) emission in landfill are from collection and flaring system. The purpose of landfill gas flaring conditioned in Scenario 2 is to release the flammable constituents from the landfill safely and to control odor nuisance, health risks and adverse environmental impacts [15]. In this case, the gas flaring system is assumed to be open flares system. Open flare system is applied since it is quite appropriate for the local situation. It is inexpensive and relatively simple, which are very important factors when there are no emission standards. The controlled emissions of CO2 (C_{CO2}) and CH4 (C_{CH4}) are calculated using Equation 4 and Equation 5 respectively [16]. The methane emission is then converted into emissions of CO2 [CO_{2eq}].

$$C_{CH_4} = \left(1 - \eta_{col}\right) * U_{CH_4} \tag{4}$$

$$C_{CO_2} = U_{CO_2} + \left(\eta_{col} * U_{CH_4}\right) \tag{5}$$

U_{CH4} : uncontrolled CH4 emission [ton]
U_{CO2} : uncontrolled CO2 emission [ton]
C_{CH4} : controlled CH4 emission [ton]
C_{CO2} : controlled CO2 emission [ton]
η_{col} : collection efficiency (fraction)

9. The calculation of emergy values and emergy indices

In this study, the emergy of renewable resources, non-renewable resources, goods and services are calculated as the total amount of emergy flows required to treat the solid waste. The emergy flow of each input is then multiplied by suitable transformity to result in solar emergy.

The emergy analysis is applied to evaluate three different scenarios of final waste treatment, since there is a discussion among the decision maker about the appropriate final waste treatment method for Yogyakarta City. The evaluation includes how much investment is needed for each waste treatment method and how much usage is extracted from the methods. These are the emergy investment and emergy recovery. The emergy values are the emergy investment and the emergy recovery describes the emergy cost and emergy benefits from each scenario. The emergy investment is the measures of the solar emergy required for treating a unit (gram) of solid waste, while emergy recovery is the measure of solar emergy gained from the treatment of a unit (gram) of solid waste. Furthermore, some emergy indices are calculated. The emergy indices are the indicators for the performance of each scenario and Equation 6 – Equation 9 are used to calculate the emergy indices. The result of the calculation is evaluated based on the criteria of each index to judge the sustainability and efficiency of each scenario as described in Table 1. The calculation uses the recalculation values of the 1996 solar empower base (9.44E+24 seJ/yr). Therefore all unit emergy values calculated before 2000 is multiplied by 1.68 as the factor increase from 9.44E+24 seJ/yr to 15.83E+24 seJ/yr as the result of the increase in global emergy base [17, 18].

$$\text{EYR} = \text{Emergy recovery/emergy investment} \qquad (6)$$

$$\text{Net Emergy} = \text{Emergy recovery} - \text{Emergy investment} \qquad (7)$$

$$\text{ELR} = \text{NR+NP+RP/RR} \qquad (8)$$

$$\text{ESI} = \text{EYR/ELR} \qquad (9)$$

10. Results and findings

The waste disposed of in the old landfill sources from the municipal solid waste (MSW) in Yoyagkarta and partly from Bantul and Sleman County. The MSWM in the area of study is characterized by the existence of informal waste management in household level (door to door collection), community level (transfer point collection) and city level (separation in landfill site). Only the involvement of scavenger in landfill was taken into account in this study. The composting centers accept approximately 25 ton/day biowaste and can produce up to 8.3 ton/day which equals to 10.33% biowaste reduction in landfill. The rest of the organic waste and other waste constituents are dumped in the landfill as described in Figure 1.

There are 400 scavengers registered in the old landfill and using Equation 1 there should be 200 samples (sample error of 5%). However, during the preliminary survey, it has been identified that only 45 scavengers can be chosen as respondent. Each scavenger separate approximately 54.3 kg/day and 52.05 kg/day for plastics and paper respectively. The amount of glass and metal sorted from waste in Bendo landfill is very small during the observation which amount to 0.036 kg/day and 0.004 kg/day respectively. The waste reduction of the recyclable materials in landfill through 45 scavengers are 7.54%, 12.87%, 0.15% and 0.03% for plastic, paper, glass and metal respectively. The recyclable wastes are sold to the middle man before it is transported to other parties, such as metal manufactures, recycle centers of plastic and paper. The glass bottles are usually transferred to the home industries.

Index	Abbreviation	Formula	Criteria
Renewable Resources (free)	RR		
Renewable Resources (purchased)	RP		
Non Renewable Resources (free)	NR		
Non Renewable Resources (purchased)	NP		.
Emergy investment	EI	Input emergy/unit MSW treated	The lower the value, the lower the cost.
Emergy Recovery	ER	Output emergy/ unit MSW treated	The greater the the value, the higher the benefit
Emergy Yield Ratio	EYR	EYR=ER/EI	The higher the value, the greater the return obtained per unit of emergy invested.
Net Emergy		Net Emergy = ER-EI	The higher the value, the greater benefit extracted
Environmental Loading Ratio	ELR	ELR=NR+NP+RP/RR	The lower the ratio, the lower the stress to the environment.
Emergy Sustainability Index	ESI	ESI=EYR/ELR	The highest the ratio, the more sustainable.

Table 1. Emergy values and emergy indices analyzed in this study

The value gained from the field survey is used as the reference to estimate the total waste reduction done by the scavengers in the new landfill. The waste reduction done by 45 scavengers is shown in Table 2.

Component	Disposal [kg/day]	Reduction [kg/day]	Percentage [%]
Plastics	32,259.0	2,431.0	7.54
Paper	18,300.0	2,355.0	12.87
Glass	1,101.0	1.6	0.15
Metal	615.4	0.2	0.03

Table 2. Waste reduction at Bendo landfill

If all scavengers (400 people) is assumed work, the amount of sorted waste is 42.56 t/day or about 13.14% of the total waste disposed. The involvement of 45 scavengers has reduced waste disposal at the rate of 1.48%. The income generated from scavenging is about $2.51/p/day or total is $62.75/day. Income comes mainly from selling paper and plastic since these both waste can be found in Bendo landfill every day with abundant amount. Selling the metal and glass contributes very little income because metal can not be found every day and most glass ended in landfill is scattered glass which is worthless. Glass is valuable if it is still in the form of a container such as bottle or jar.

Flows : tons/day

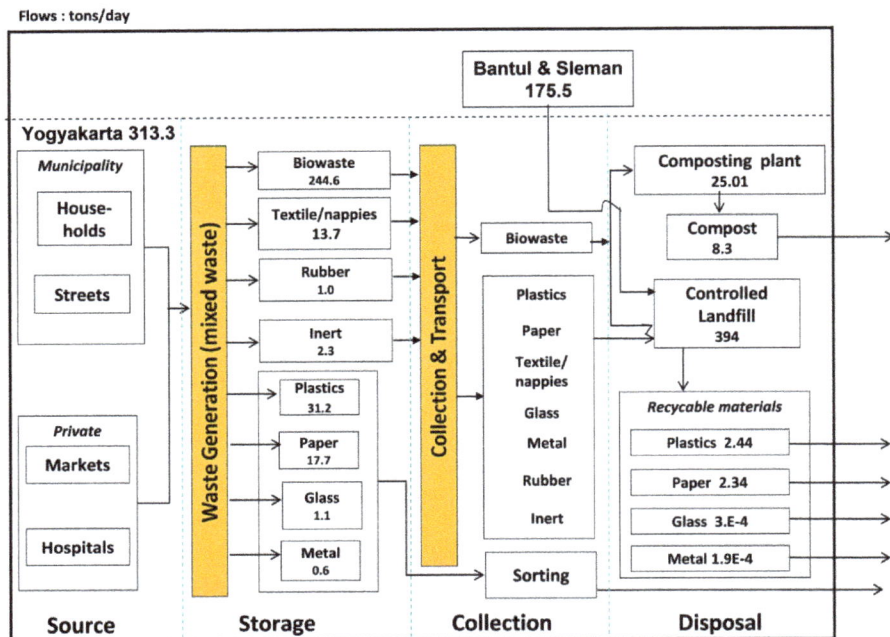

Figure 1. Waste stream in boundary system

Meanwhile, the composting can generate income of $163.5/day if it is assumed that the compost is sold with the current compost price in the market ($0.021/kg) as presented in Table 3.

Materials	Mass	Price		Income	Income
	[kg/p/d]	[Rp/kg]	[US$/kg]	[US$/p/day]	[US$/day]
From Bendo Landfill					
Plastics	54.30	200	0.02	1.14	
Glass	0.036	300	0.03	0	
Metal	0.004	250	0.03	0	
Paper	52.05	250	0.03	1.37	
Total				**2.51**	
total sample(45 scavengers)					62.75
From Composting centers					
Compost [kg/d]	7,766	200	0.021		163.5

Table 3. Income from waste sorting and composting centers in Yogyakarta

The population and waste generation projection is initiated from 2013 using the number of population in 2012. The average population growth rate is 1.51% [19] and the average waste generation rate is 1.61% [20]. The projection is made for 15 years as the landfill will be operated for 15 years (2013 – 2028). Once the population is calculated, the projection of waste generation can be calculated by multiplying it with waste generation per capita. The result is presented in Figure 2.

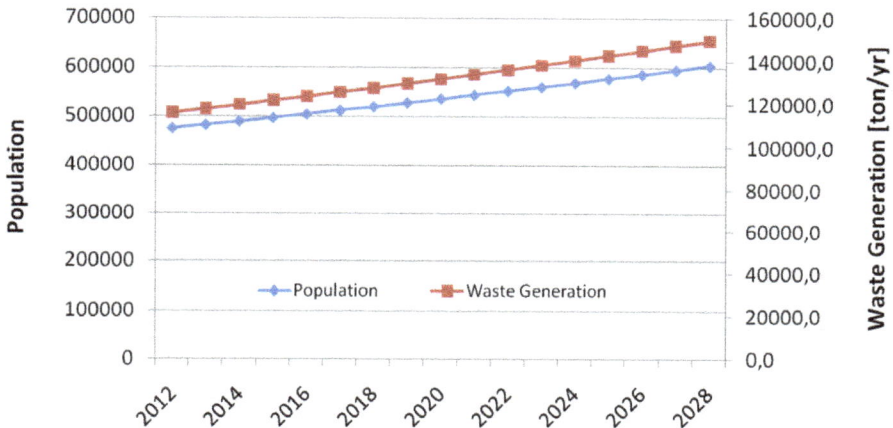

Figure 2. Projection of population and waste generation in Yogyakarta City

The projection of waste disposal in landfill is made referring to the landfill opening year in 2013 and duration for 15 years. Figure 2 shows the projection of waste disposed of in the new landfill with the level of service (LoS) on collection of 70%.

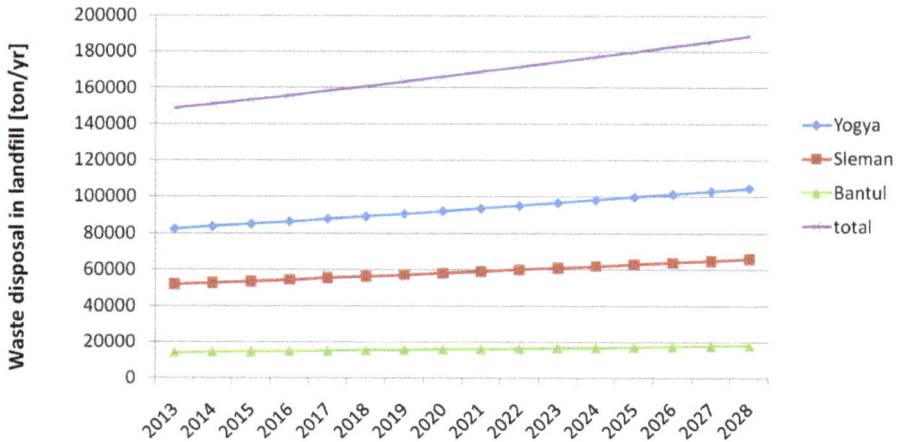

Figure 3. Waste disposal in the new landfill

Mostly waste come from Yogyakarta (64%), while the rest is from Sleman (30%) and Bantul (6%). In the initial year, the waste disposal from these three regions is 148,587 ton. At the last year waste disposal will be 188,811 ton. With the waste disposal growth of 1.61%, the new landfill will totally accept 2.7E+06 ton waste from 2013 until 2028 with the assumption of 70% LoS. If LoS is increased to be 85% (local target), the landfill will accept totally about 3.26E+06 ton waste.

11. Scenarios for final waste treatment method

There are three scenarios in this study to be compared. The scenario reflects the proper alternatives for final waste treatment in Yogyakarta. Each scenario comprises the MSWM stage including collection, landfilling process and composting. All the scenarios are assumed not to affect MSW generation meaning that the amounts and the composition of MSW are considerably the same in all scenarios. The implications of each scenario will be evaluated for its GWP and emergy indices. The GWP is calculated from methane emission from the new landfill. Emission from other facilities of final waste treatment such as composting centre is not taken into account although it is inside the boundary system. In accordance to [21], aerobic decomposition in composting plant results emission of CO_2 and H_2O. Methane can be also generated in anaerobic pockets within a compost pile due to the heterogeneous nature of compost pile [22]. Nevertheless, some studies showed that the majority of methane emission oxidizes to CO_2 in aerobic pockets and near the surface of the compost pile, so that methane emission can be neglected [23, 24]. The methane generation calculation is done with the assumption that methane will be generated for 47 years (2013 – 2060).

The emergy indices are derived from the calculation of emergy input and output within the boundary including the collection, landfill site and composting center.

11.1. Scenario 0: Baseline scenario

Baseline scenario is a reference scenario and assumes that there is no change in the future waste management in Yogyakarta. According to the calculation in the previous sub chapter, 70% of MSW was collected in the landfill and 10.33% of biowaste is treated in the community based composting centers. The composting capacity increases though the constant rate because of the higher average amount of waste collected from 2013 – 2028. There is about 26.8 m^3/day or 36.8 t/day biowastes treated. Waste separation is done by 45 scavengers as the optimal current scavenging activity. It is assumed that they work 8 hours/day from Monday until Friday with the average waste sorting capacity for paper, plastic, glass and metal is 53.34 kg/cap/day, 54.02 kg/cap/day, 0.036 kg/cap/day and 0.004 kg/cap/day respectively. The waste reduction through scavenging is kept constant at 12.87% and 7.54% for paper and plastic respectively.

Calculation of methane emission using Equation 7 – 11 estimate that there will be 1.32E+05 ton CH_4 or 2.78E+06 ton CO_{2eq} emitted from the new landfill during inventory years from

2013 until 2060 if there are no changes in final waste treatment method. If there is no measure of waste reduction through scavenging and composting, the total methane emission is approximately about 3.26E+06 ton CO_{2eq} with LoS of 70%. It means the current practice in waste treatment (scavenging and composting) has reduced the total methane emission about 4.8E+05 ton CO_{2eq} or about 14.71% from total emission in case there is no measure (no scavenging and composting). Figure 4 describes the comparison of the methane emission from the new landfill during its operational time between conditions with waste reduction through scavenging and composting and without it.

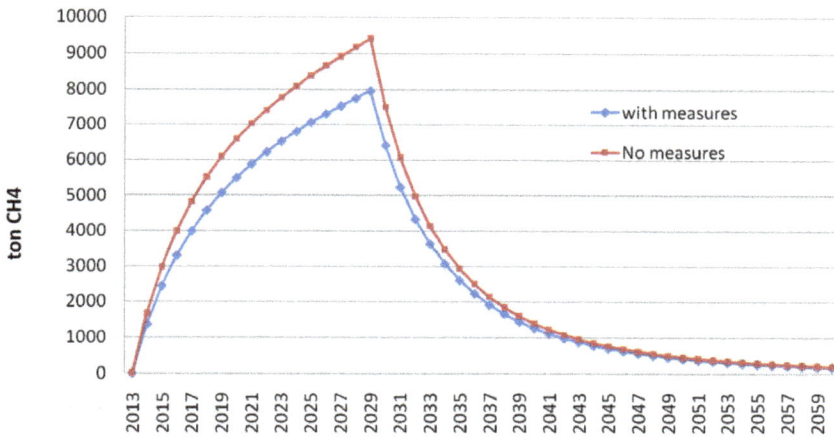

Figure 4. Methane emission from the new landfill based on Scenario 0

11.2. Scenario 1: LoS improvement scenario

As the local government claims that the LoS of collection is 85%, Scenario 1 assumes that LoS will be increased to be 85% meaning that waste volume collected will be more and waste reduction measure is implemented through composting and scavenging. The composting rate will be increased, sum up to 50% to reduce the waste volume delivered to the landfill. 50% is the target of the local government to increase composting rate at the end of year 2011 [25]. Due to this increase, the daily capacity of composting centers will be 230.5 ton/day or almost six fold increase compared to the base case which is 37 ton/day. The target of increasing capacity makes sense as there is abundant organic waste and human resources. However, it requires additional equipment and facilities consequently. The six fold capacity increase requires 31% emergy investment increase as presented later in Table 6 – 8 indicating that it requires relatively restrained investment for the added resources input.

In landfill, 45 scavengers will separate the recyclable materials. Due to the increase LoS, the total amount of the waste collected will increase from 2.69E+06 ton to 3.26E+06 tons. The total amount of methane emission from the new landfill is about 1.02E+05 ton CH_4 or 2.16E+06 ton CO_{2eq}. Figure 5 describes the methane emission from the new landfill during its operational time based on Scenario 2.

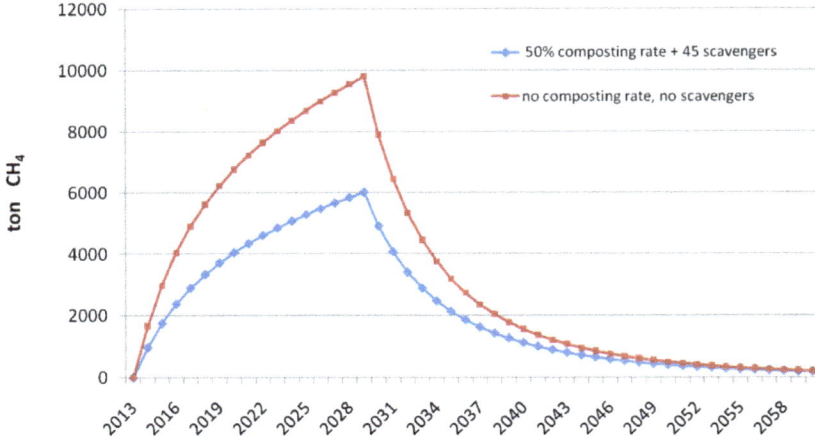

Figure 5. Methane emission from the new landfill based on Scenario 1

11.3. Scenario 2: LFG flaring scenario

In Scenario 2 scenario, scavenging is allowed only in certain area within the landfill site, where LFG collection system is not constructed. It assumed that 200 scavengers will work to separate the recyclable materials. There will be frequent compaction and soil covering (once a month). The composting rate is set to be 50% and the LoS is assumed to be 85%.

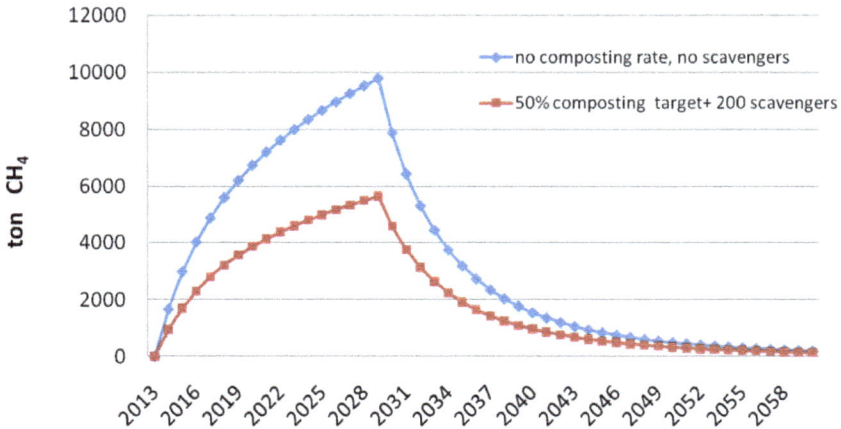

Figure 6. Methane emission from the new landfill based on Scenario 2

The average collection system cost for landfills with flaring system is assumed based on the value proposed by [16] which includes flaring costs. The initial cost for the collection system is US$ 628,000 and the O&M is US$ 89,000/yr. In Scenario 2, the methane emission from landfill will be 2.00E+06 ton CO_2eq as showed in Figure 6. The composting rate is the same as in the Scenario 1. Therefore, composting capacity is 230.5 tons/day and the compost production is 76 tons/day

The global warming potential (GWP in CO_2 equivalent) and specific GHG effect from the scenarios have been compared to the worst condition if there are no waste reduction and the LoS of collection is 85%. The comparison is made to give the overview that the change of the biowaste in landfilled waste changes the global warming potential and specific GHG effect more significantly than that of paper content.

Scenario 0 and Scenario 1 emits more methane and have higher GWP than Scenario 2. Scenario 2 generates the lowest total emission because of the significant reduction of biowaste transported to the landfill and the construction of flaring system in the landfill. Flaring system has converted CH_4 into CO_2 through combustion. The specific GHG emission is calculated for each Scenario and the result shows that in Scenario 0, one ton disposed waste generate the highest specific GHG emission (1,049 kg CO_2eq /t MSW collected). The lowest specific emission is produced in Scenario 2 (613 kg CO_2eq/t MSW collected) as illustrated in Figure 7. The result indicates that Scenario 2 generates the least emission. The graphic implies that the change of composting rate affects the specific GHG emission more considerably than the change of scavenging rate. Scenario 1 and 2 can reduce the impact of GHG on environment about 22% and 28% respectively from the base scenario. The 50% biowaste reduction through composting has decreased the emission considerably.

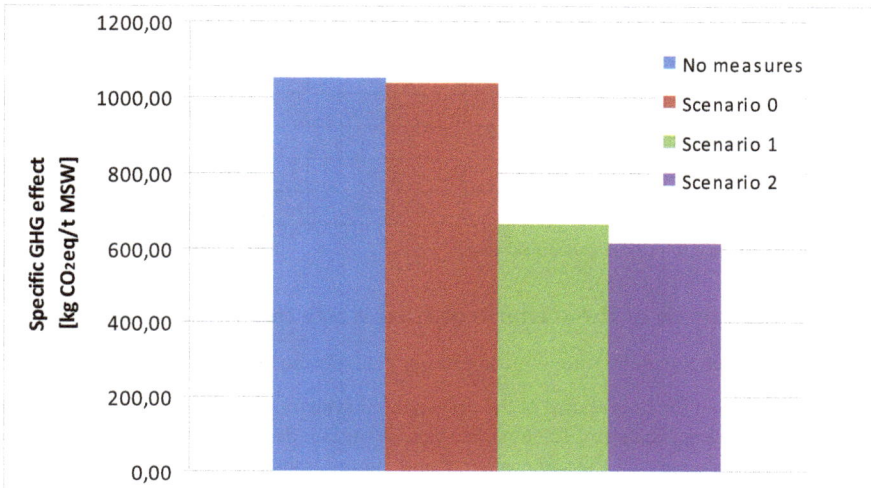

Figure 7. Specific GHG emission comparison of each scenario

The comparison between the three scenarios in terms of GWP demonstrates that the scavenging and composting play role in waste reduction brings the GWP reduction. Therefore, the combination of both measures is the best result as it can minimize the methane emission effectively. Generally, the result of the comparison of all scenarios is summarized in the Table 4 and Table 5.

Parameter	No measures	Scenario 0	Scenario 1	Scenario 2
Input parameters				
LoS collection [%]	85	70	85	85
No. of scavengers	0	45	45	200
Composting rate	0	10.33	50	50
Total waste collected [ton]	3,263,023	2,687,195	3,263,023	3,263,023
Output parameters (calculated)				
CH_4 emission [ton CO_2eq]	3,423,478	2,780,848	2,158,676	2,002,004
CO_2 emission [ton]	2,636,323	2,141,452	1,662,337	1,541,687
Flaring (50% collection) [ton CO_2eq]	-	-	-	1,001,002
Specific GHG effect [kg CO_2eq/ ton MSW collected]	1,049	1,034	661	613

Table 4. Summary of comparison during landfill life

Parameter	Existing (2010)	Scenario 0	Scenario 1	Scenario 2
Input parameters				
LoS collection [%]	70	70	85	85
Composting rate [%]	0	10.33	50	50
Daily collection [ton/d]	313	460	595	595
Biowaste [ton/d]	25.01	36.76	230.5	230.5
Biowaste [g/yr]	9.13E+09	1,43E+10	8,41E+10	8,41E+10
Output parameters				
Compost [ton/d]	8.25	12.9	76.1	76.1
Compost [g/yr]	3.01E+09	4.72E+09	2.78E+10	2.78E+10

Table 5. Summary of comparison for composting

12. Emergy analysis of the scenario of final waste treatment in Yogyakarta

The following steps are undertaken for the emergy analysis during the study:

1. Identification of the boundaries of the investigated system
2. Making an emergy diagram. The emergy system diagram describes the emergy flows into and out of the system in the form material and energy transfers. Hence, it is necessary to identify all variables involved in the process. The main stages, the inputs, the output and the relations between individual elements is presented in emergy system diagram.

3. Calculation of matter and energy flows supporting the scenario All inputs in the system were divided into two groups; renewable resources and non renewable resources. Each group is subdivided into free resource and purchased resources. The calculation of emergy in waste treatment is conducted using Equations 6 – 9. The amount of the available emergy (exergy) is calculated based on the primary and secondary data.

4. Conversion of input matter and energy flows into solar emergy Joules (seJ) by using suitable transformities, recalculated to the new baseline for biosphere (total emergy driving the biosphere: 15.84×10^{24} seJ year [26, 17].

5. Calculation of the emergy cost for safe disposal of one unit of waste (seJ/g).

The values of transformity are presented in the table of emergy evaluation. Some of them are calculated and some are taken from emergy data bases available in the literature.

12.1. Overview of models and flow summary

The final solid waste treatment system in Yogyakarta City is the boundary. The input for the system is waste and renewable, non renewable and services. The input flow of waste assumed to have zero emergy content because mixed waste is not considered as a desired product of human activities, but instead an unavoidable and undesired emission (CO_2, CH_4 and other pollutants) [27]. For the waste material just stored in the landfill, there is no reason for assigning its transformity. The outputs are the products produced during the process including also the good/services that are sold in the market. Compost is the outputs of the process, while emission and recyclable materials are the by products Compost is produced in composting centers, emission is generated from waste degradation process in landfill, and recyclable materials are sorted and sold by scavengers in landfill. The emission from the system is confined to be methane and carbon dioxide emission. The stages involved in final solid waste treatment are collection, waste disposal in landfill and waste treatment in composting centers. Collection includes collection in household level (door to door collection) and collection in community level (transfer point collection). The biowaste collected is distributed to the composting centers spread out in Yogyakarta City. The rest will be transported to the new landfill. The more detailed emergy flow system diagram is presented in each scenario.

The emergy benefits of each scenario are represented by the arrow to the market. Compost and recyclable material from landfill is the emergy benefit for all scenarios. The emergy flow system diagram for scenario 0 and scenario 1 is presented in Figure 8. Meanwhile, Figure 9 describes the emergy flow system diagram for scenario 2. Scenario 0 and Scenario 1 have the same emergy diagram since the process is the same only the amount of the emergy is different caused by the difference inputs. In these figures, the phases including collection, treatment and disposal are shown. Collection is conducted in household level through door to door (DtD) collection and in community level through transfer point collection (TP). In landfilling process, emission is the by-product which is not taken into consideration for the emergy analysis. It has been separately calculated in GWP analysis. Methane emission and carbon dioxide emission is calculated during 47 years as the methane generation is approaching to very small quantity thereafter. Other gases produced from anaerobic process are not considered here as the amount is very little compared to the main LFGs and assumed to be negligible in terms of emergy costs. Surely, the insertion of these little gases would have effect on increasing emergy investment. The emergy benefit from landfilling process is the money flown into the landfill coming from the scavenging activities. The emergy recovery from composting is calculated by transforming the monetary values from the compost sold into emergy units, using the emergy-to-money ratio in Indonesia, 2.06E+13 seJ/$ [28]. As the study is limited to the emission from the landfill, the emission from the WWTP and composting process will not be considered.

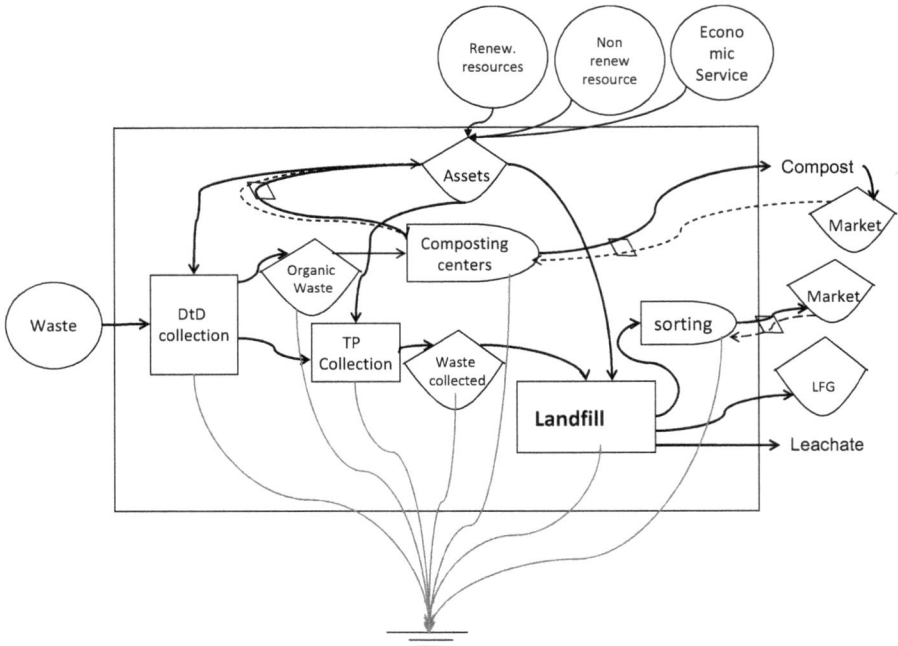

Figure 8. Emergy system diagram of Scenario 0 and Scenario 1

After describing the emergy flow in the diagram, the calculation of the total emergy is conducted and presented in table of emergy. Table 6 – 8 present the results of the emergy values performed in each scenario. The transformities used in this section are based on the value from literatures and from the study self. Each scenario is evaluated for its emergy which is divided into three main parts, namely emergy from the MSW collection, landfilling process and composting. It can be summarized that in terms of emergy investment, the results of the emergy analysis demonstrate a similar trend for all scenarios although the values vary. Landfill requires the highest emergy investment to all scenarios with the percentage ranges between 92% - 97%. Collection ranks in the second place with the percentage of 3% - 9%, while composting invests the smallest percentage of emergy, less than 1% (Figure 10 – 12).

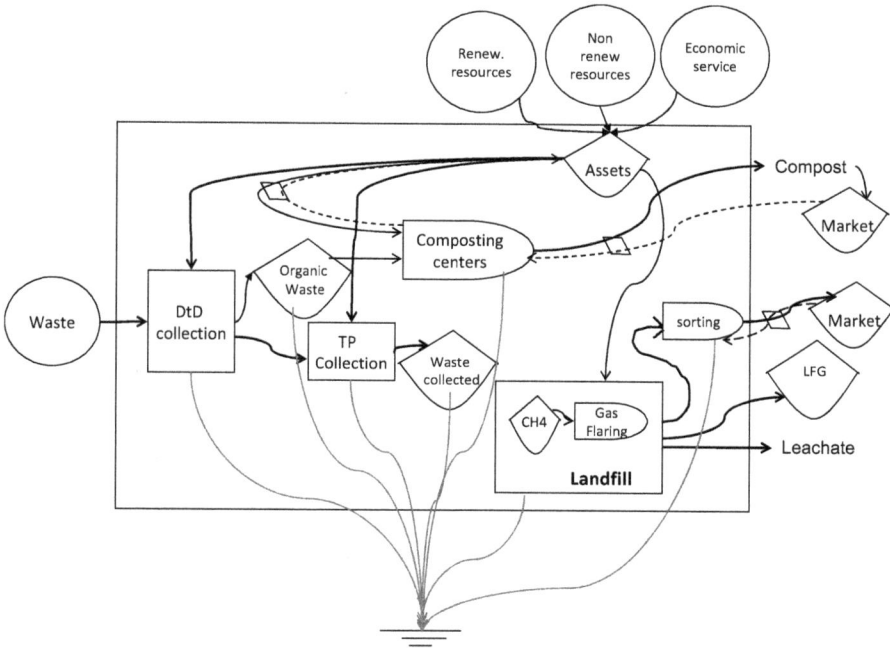

Figure 9. The emergy system diagram of Scenario 2

Table 6 and Figure 10 (Scenario 0) shows that Scenario 0 contributes total solar emergy of 3.30E+23 seJ/yr and needs total emergy investment of 1.84E+12 seJ/gMSW. Most emergy is invested in landfill. The emergy recovery is gained from scavenging in landfill and composting which contributes 4.02E+08 seJ/gMSW. The emergy inputs in Scenario 0 are the lowest. This is because the less amount of disposed waste requires less quantity of equipment, fuel, labor and other capital causes less input of emergy.

Table 7 and Figure 11 (Scenario 1) illustrates that the process contributes total solar emergy of 3.78E+23 seJ/yr and requires emergy investment of 1.74E+12 seJ/gMSW. The emergy investment is mainly from landfill (96.4%). The emergy recovery in Scenario 1 is also from income of scavengers and compost generated in composting centers. The emergy investment in Scenario 1 is the lowest indicating that Scenario 1 has the lowest cost among two others. It means that under Scenario 1, the cost should be provided to manage one unit mass of MSW is lower compared to other scenarios. In this case, the more input in waste treated leading to the more efficiency in waste treatment.

No	Item	Unit	Amount	Transformity [seJ/unit]	References	Solar emergy [seJ/year]	Emergy investment [sej/g MSW treated]
Renewable local resources (RR)							
1	Air (composting)	g	1.96E+08	5.16E+07	[29]	1.01E+16	5.64E+04
2	Scavengers (landfill)	J	1.42E+11	4.63E+06	this study	6.57E+17	3.67E+06
						6.67E+17	**3.72E+06**
Renewable local resources purchased (RP)							
3	Water (landfill)	g	1.10E+10	6.64E+05	[29]	**7.27E+15**	**4.06E+04**
Non renewable resources in collection process purchased (NP)							
4	Handcart	g	1.74E+07	5.91E+09	[30]	1.73E+17	9.67E+05
5	Vehicles	J	951E+11	7.76E+09	[31]	1.24E+22	6.92E+10
6	Fuel	J	2.73E+12	6.60E+04	[32]	3.03E+17	1.69E+06
7	Water	g	3.65E+09	6.64E+05	[29]	2.42E+15	1.35E+04
8	Labor	J	2.90E+12	4.63E+06	this study	1.34E+19	7.50E+07
9	Management cost	$	9.50E+05	2.06E+13	[28]	1.96E+19	1.09E+08
						1.24E+22	**6.94E+10**
Non renewable free (NR)							
10	Material for plant construction	g	9.66E+13	1.68E+09	[32]	2.73E+23	1.52E+12
11	Material for waste final covering	g	6.21E+12	1.68E+09	[32]	1.75E+22	9.79E+10
						3.15E+23	**1.76E+12**
Non renewable input to plant construction, waste management and processing purchased (NP)							
12	Material for plant construction (steel)	g	1.85E+11	4.13E+09	[30]	1.28E+21	7.15E+09
13	Fuel	J	1.35E+12	6.60E+04	[32]	1.50E+17	8.37E+05
14	Electricity	J	3.03E+10	1.60E+05	[17]	4.85E+15	2.71E+04
15	Vehicles	J	8.21E+10	7.76E+09	[31]	1.07E+21	5.98E+09
16	Labor	J	6.31E+09	4.63E+06	this study	2.92E+16	1.63E+05
						2.35E+21	**1.31E+10**
Economic services (NP)							
17	Total cost of landfill plant	$	3.37E+06	2.06E+13	[28]	6.94E+19	3.87E+08
18	Annual O&M cost incl. Labor.	$	1.75E+06	2.06E+13	[28]	3.60E+19	2.01E+08
						1.05E+20	**5.89E+08**
	Average annual disposal of waste	g	1.79E+11				
	Output						
	Total main LFG (CO_2 & CH_4)	**4,92E+12**	g CO_2eq	4.80E+04		2.36E+17	
	Income of scavengers	**4,35E+04**	$	2.06E+13		8.96E+17	5.00E+06
Non renewable input to DtD collection purchased (NP)							
19	Handcart	g	1.99E+03	5.91E+09	[30]	1.98E+13	1.38E+03
20	Labor	J	6.31E+09	4.63E+06	this study	2.92E+16	2.04E+06
						2.92E+16	**2.04E+06**
Non renewable input to composting plant construction, management and processing purchased (NP)							
21	Electricity	J	1.91E+10	1.60E+05	[17]	3.05E+15	2.13E+05
22	Fuel	J	3.25E+10	6.60E+04	[32]	3.60E+15	2.52E+05
23	Labor	J	1.83E+12	4.63E+06	this	8.45E+18	5.90E+08

					study		
						8.46E+18	**5.91E+08**
Economic services (NP)							
24	Investment cost	$	3.45E+03	2.06E+13	[28]	7.11E+16	4.96E+06
25	Management cost	$	5.94E+05	2.06E+13	[28]	1.22E+19	8.55E+08
						1.23E+19	**8.60E+08**
	Annual waste treated	g	1.43E+10	1.39E+11			
	Output						
	Compost	g	**4.72E+09**	4.41E+09			
	Compost Price (Rp 1000/kg)	$/g	0.000105				
	Income	$	**4.97E+05**	2.06E+13		**1,02E+19**	**7,16E+08**
	Total solar emergy (1-25)	3.30E+23 sej/yr					
	Collection	6.94E+10 sej/gMSW	3.76%				
	Treatment in Landfill	1.77E+12 sej/gMSW	96.2%				
	Composting	1.45E+09 sej/gMSW	<1%				
	Total solar emergy investment	1.84E+12 sej/gMSW					

Table 6. Emergy flows of scenario 0

No	Item	Unit	Amount	Transformity [seJ/unit]	References	Solar emergy [seJ/year]	Emergy investment [sej/gMSW treated]
Renewable local resources free (RR)							
1	Air (composting)	g	9.48E+08	5.16E+07	[29]	4.89E+16	2.25E+05
2	Scavengers (landfill)	J	1.42E+11	4.63E+06	this study	6.57E+17	3.02E+06
						7.06E+17	**3.25E+06**
Renewable local resources purchased (RP)							
3	Water	g	1.10E+10	6.64E+05	[29]	**7.27E+15**	**3.34E+04**
Non renewable resources in collection process purchased (NP)							
4	Handcart	g	1.86E+07	5.91E+09	[30]	1.84E+17	8.47E+05
5	Vehicles	J	1.05E+12	7.76E+09	[31]	1.36E+22	6.27E+10
6	Fuel	J	3.14E+12	6.60E+04	[32]	3.48E+17	1.60E+06
7	Water	g	5.48E+09	6.64E+05	[29]	3.64E+15	1.67E+04
8	Labor	J	3.36E+12	4.63E+06	this study	1.56E+19	7.15E+07
9	Management cost	$	1.10E+06	2.06E+13	[28]	2.27E+19	1.04E+08
						1.37E+22	**6.29E+10**
Non renewable resources in landfill free (NR)							
10	Material for plant construction	g	9 66E+13	1.68E+09	[32]	2.73E+23	1.25E+12
11	Material for regular and final covering	g	3 11E+13	1.68E+09	[32]	8.77E+22	4.03E+11
						3.60E+23	**1.66E+12**
Non renewable input to plant construction, waste management and processing purchased (NP)							
	Material for plant construction						
12	(steel)	g	185E+11	4.13E+09	[30]	1.28E+21	5.88E+09
13	Fuel	J	4.06E+12	6.60E+04	[32]	4.50E+17	2.07E+06
14	Electricity	J	3.03E+10	1.60E+05	[17]	4.85E+15	2.23E+04
15	Vehicles	J	2.08E+11	7.76E+09	[31]	2.71E+21	1.25E+10
16	Labor	J	6.31E+09	4.63E+06	this study	2.92E+16	1.34E+05
						3.99E+21	**1.84E+10**

	Economic services (NP)						
17	Total cost of landfill plant	$	3.37E+06	2.06E+13	[28]	6.94E+19	3.19E+08
18	Annual O&M cost incl. Labor.	$	1.94E+06	2.06E+13	[28]	4.00E+19	1.84E+08
						1.09E+20	5.03E+08
	Annual disposal of waste	g	2.18E+11				
	Output						
	Total main LFG (CO$_2$ & CH$_4$)	g CO$_2$ eq	3.82E+12				
	Income of scavengers	$	4.35E+04	8.69E+18		3.78E+23	
	Non renewable input to DtD collection purchased (NP)						
19	Handcart	g	3.98E+03	5.91E+09	[30]	3.96E+13	1.82E+02
20	Labor	J	1.26E+10	2.62E+05	this study	6.14E+17	2.82E+06
						6.14E+17	2.82E+06
	Non renewable input to composting plant construction, management and processing purchased (NP)						
21	Electricity		1.91E+10	1.60E+05	[17]	3.05E+15	1.40E+04
22	Fuel		5.42E+10	6.60E+04	[32]	6.01E+15	2.76E+04
23	Labor		2.34E+12	4.63E+06	this study	1.08E+19	4.99E+07
						1.09E+19	4.99E+07
	Economic services (NP)						
24	Investment cost	$	3.45E+03	2.06E+13	[28]	7.11E+16	3.27E+05
25	Management cost	$	7.65E+05	2.06E+13	[28]	1.57E+19	7.24E+07
						1.58E+19	7.27E+07
	Annual waste treated	g	8.41E+10				
	Output						
	Compost	g	2.78E+10	9.85E+08			
	Compost Price (Rp 1000/kg)	$/g	1.05E-04	**2.06E+13**		6.02E+19	
	Income	$	2,92E+06				
	Total solar emergy (1-25)	**3.78E+23**	seJ/yr				
	Collection	**6.29E+10**	sej/gMSW	3.62%			
	Treatment in Landfill	**1.68E+12**	sej/gMSW	96.4%			
	Composting	**1.26E+08**	sej/gMSW	<1%			
	Total solar emergy investment	**1.74E+12**	sej/gMSW				

Table 7. Emergy flows of the scenario 1

No	Item	Unit	Amount	Transformity [seJ/unit]	References	Solar emergy [seJ/year]	Emergy investment [seJ/g MSW treated]
	Renewable local resources (RR)						
1	Air (composting)	g	9.48E+08	5.16E+07	[29]	4.89E+16	2.25E+05
2	Scavengers (landfill)	J	6.31E+11	4.63E+06	this study	2.92E+18	1.34E+07
						2.97E+18	1,36E+07
	Renewable local resources (RP)						
3	Water	g	1.10E+10	6.64E+05	[29]	**7.27E+15**	**3,34E+04**
	Non renewable resources in collection process (NP)						
4	Handcart	g	1.86E+07	5.91E+09	[30]	1.84E+17	8.47E+05
5	Vehicles	J	2.50E+12	7.76E+09	[31]	3.26E+22	1.50E+11
6	Fuel	J	3.14E+12	6.60E+04	[32]	3.48E+17	1.60E+06
7	Water	g	5.48E+09	6.64E+05	[29]	3.64E+15	1.67E+04
8	Labor	J	3.36E+12	4.63E+06	this study	1.56E+19	7.15E+07
9	Management cost	$	1.10E+06	2.06E+13	[28]	2.27E+19	1.04E+08

						3.27E+22	1.50E+11
Non renewable resources free (NR)							
10	Material for plant construction	g	9.66E+13	1.68E+09	[32]	2.73E+23	1.25E+12
11	Material for regular and final covering	g	3.11E+13	1.68E+09	[32]	8.77E+22	4.03E+11
						3.60E+23	1.66E+12
Non renewable input to plant construction, waste management and processing							
12	Material for plant construction (steel)	g	1.85E+11	4.13E+09	[30]	1.28E+21	5.88E+09
13	Fuel	J	4.06E+12	6.60E+04	[32]	4.50E+17	2.07E+06
14	Electricity	J	3.03E+10	1.60E+05	[17]	4.85E+15	2.23E+04
15	Vehicles	J	2.08E+11	7.76E+09	[31]	2.71E+21	1.25E+10
16	Labor	J	6.31E+09	4.63E+06	this study	2.92E+16	1.34E+05
						3.99E+21	1.84E+10
Economic services							
17	Total cost of landfill plant	$	4.00E+06	2.06E+13	[28]	8.23E+19	3.78E+08
18	Annual O&M cost incl. Labor.	$	1.92E+06	2.06E+13	[28]	3.96E+19	1.82E+08
						1.22E+20	5.61E+08
	Annual disposal of waste	g	2.18E+11				
	Output						
	Total main LFG (CO₂ & CH₄)	g CO₂ eq	3.54E+12	4.80E+04		2.36E+17	
	Income of scavengers	$	8.70E+06	4.56E+16		3.97E+23	
Non renewable input to DtD collection							
19	Handcart	g	3.98E+03	5.91E+09	[30]	3.96E+13	1.82E+02
20	Labor	J	1.26E+10	2.62E+05	this study	6.14E+17	2.82E+06
						6.14E+17	2.82E+06
Non renewable input to composting plant construction, management and processing							
21	Electricity		1.91E+10	1.60E+05	[17]	3.05E+15	1.40E+04
22	Fuel		5.42E+10	6.60E+04	[32]	6.01E+15	2.76E+04
23	Labor		2.34E+12	4.63E+06	this study	1.08E+19	4.99E+07
						1.09E+19	4.99E+07
Economic services							
24	Investment cost	$	3.45E+03	2.06E+13	[28]	7.11E+16	3.27E+05
25	Management cost	$	7.65E+05	2.06E+13	[28]	1.57E+19	7.24E+07
						1.58E+19	7.27E+07
	Annual waste treated	g	8.41E+10				
						Total	3.97E+23	1.83E+12
	Output						
	Compost	2.78E+10	g	9,85E+08			
	Compost Price (Rp 1000/kg)	1.05E-04	$/g				
	Income	2.92E+06	$	2,06E+13		6,02E+19	
	Total solar emergy (1-25)	3.97E+23	sej/yr				
	Collection	1.50E+11	sej/gMSW	8.23%			
	Treatment in Landfill	1.68E+12	sej/gMSW	91.8%			
	Composting	1.26E+08	sej/gMSW	<1%			
	Total solar emergy investment	1.83E+12	sej/gMSW				

Table 8. Emergy flows of the scenario 2

Table 8 and Figure 12 (Scenario 2) demonstrates that the total solar emergy is 3.97E+23 seJ/yr which is the highest value compared to other scenarios. The emergy investment in Scenario 2 is 1.83E+12 seJ/gMSW. The result indicates that the emergy investment depends not only on the emergy input but also the effectiveness of waste collection. In this case, Scenario 1 and 2 with the higher LoS of Collection (85%) and higher emergy inputs than Scenario 0 can reduce the emergy investment because along with the higher emergy inputs, the effectiveness of waste collection is increasing. The more adequate equipment and labor raise the capability of the waste authority to collect the waste leading to lower emergy investment.

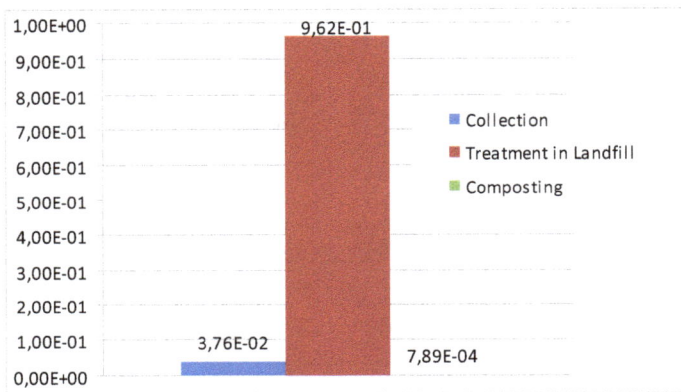

Figure 10. Share of emergy investment in Scenario 0

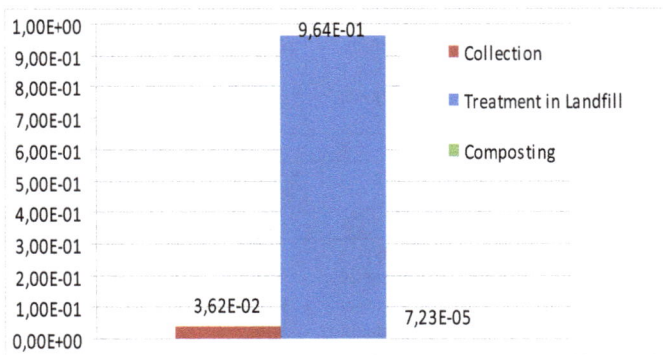

Figure 11. Share of emergy investment in Scenario 1

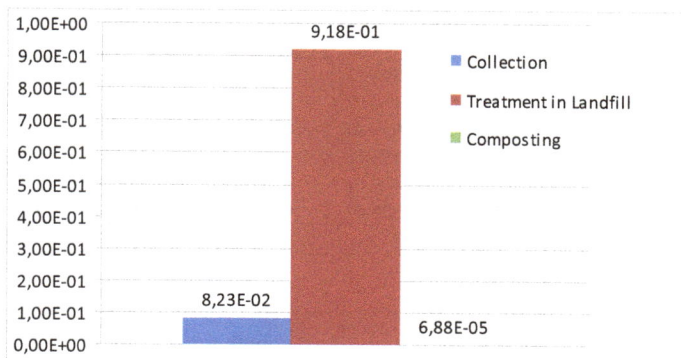

Figure 12. Share of emergy investment in Scenario 2

As mentioned above, scavenging and composting are the source of emergy recovery. Table 9 – 11 describes the emergy recovery from each scenario. The matter/ money recovery is calculated by dividing the product for the amount of waste treated [33]. Emergy recovery is calculated by multiplying energy or matter recovery for the correspondent transformity. The emergy recovery from landfilling is the conversion of the income of the scavengers to the solar emergy by multiplying it to national emergy per unit dollar (2.06E+13 seJ/$). Compost is assumed to have the same content as natural fertilizer with 2.1% nitrogen(N), 1.6% phosphorus (P), 1.1% potassium (K) [34]. The rest is the remaining part assumed as soil [29]. The calculation of emergy in composting uses the transformity of the fertilizer component (N, P, K) and the land cycle from [32].

	Product	Unit	Matter or money recovery	Unit	Transformity [seJ/unit]	Emergy recovery [seJ/g]
Composting	4.72E+09	g	3.30E-01	g/gMSW		4.46E+08
N(2.1%)	2.08E+08	g	1.45E-02	g/gMSW	4.62E+09	3.20E+07
P(1.6%)	5.20E+07	g	3.63E-03	g/gMSW	1.78E+10	9.40E+07
K(1.1%)	8.50E+07	g	5.94E-03	g/gMSW	1.74E+09	6.32E+07
Soil	4.38E+09	g	3.06E-01	g/gMSW	1.00E+09	3.14E+08
Scavenging	4.35E+04	$	2.43E-07	$/gMSW	2.06E+13	5.00E+06
Total						4.51E+08

Table 9. Emergy recovery of Scenario 0

The calculation of emergy recovery presented in Table 9 – 11 clearly shows that composting and scavenging can extract the economic value from waste by generating the flows of money. The highest emergy recovery is produced under Scenario 2 with the value of 1.27E+09 seJ/gMSW. Scenario 0 and Scenario 1 can generate the relative same amount of emergy saving (4.51E+08 seJ/gMSW) although the emergy input in Scenario 0 is higher than Scenario 1. The same scavenging rate and the higher composting rate of Scenario 1 with the higher LoS compared to Scenario 0 cause this, since matter recovery depends not only on the product but also the waste treated.

Product	Unit	Energy/matter recovery	Unit	Transformity [seJ/unit]	Emergy recovery [seJ/g]
Composting	2.78E+10	g	3.30E-01	g/gMSW		4.46E+08
N(2.1%)	1.22E+09	g	1.45E-02	g/gMSW	4.62E+09	3.20E+07
P(1.6%)	3.05E+08	g	3.63E-03	g/gMSW	1.78E+10	9.40E+07
K(1.1%)	5.00E+08	g	5.94E-03	g/gMSW	1.74E+09	6.32E+07
Soil	2.57E+10	g	3.06E-01	g/gMSW	1.00E+09	3.14E+08
Scavenging	4.35E+04	$	2.00E-07	$/gMSW	2.06E+13	4.12E+06
Total						4.51E+08

Table 10. Emergy recovery of Scenario 1

Product	Unit	Matter or money recovery	Unit	Transformity [seJ/unit]	Emergy recovery [seJ/g]
Composting	2.78E+10	g	3.30E-01	g/gMSW		4.46E+08
N(2.1%)	1.22E+09	g	1.45E-02	g/gMSW	4.62E+09	3.20E+07
P(1.1%)	3.05E+08	g	3.63E-03	g/gMSW	1.78E+10	9.40E+07
K(1.8%)	5.00E+08	g	5.94E-03	g/gMSW	1.74E+09	6.32E+06
Soil	2.57E+10	g	3.06E-01	g/gMSW	1.00E+09	3.14E+08
Scavenging	8.70E+06	$	4.00E-05	$/gMSW	2.06E+13	8.24E+08
Total						1.27E+09

Table 11. Emergy recovery of Scenario 2

Scenario 1 and 2 has the same amount of emergy recovery from composting because both scenarios have the same composting rate of 50%. Thus, the value is higher compared to that of Scenario 0 which covers only 10.33% composting rate. The emergy recovery from landfilling of Scenario 2 is the highest compared to other scenarios. The higher scavenging rate involving 200 scavengers is the reason for this.

The analysis of emergy indices is conducted to measure whether one scenario which satisfies the criteria of the above values is really better than any other scenarios. Using these indicators, the evaluation is more comprehensive since it covers not only an assessment from one view of point but also other view of points such as its efficiency and sustainability.

Based on values in Tables 9 – 11, the emergy indices of each scenario is calculated and presented in Table 12.

	S0	S1	S2
Total solar emergy [seJ/y]	3.03E+23	3.78E+23	3.97E+23
Emergy Investment [seJ/g MSW]	1.84E+12	1.74+12	1.82E+12
Emergy recovery [seJ/g MSW]	4.51E+08	4.51E+08	1.27E+09
EYR	2.45E-04	2.59E-04	6.96E-04
Net Emergy [seJ/g MSW]	-1.84E+12	-1.74E+12	-1.82+12
ELR	4.95E+05	5.36E+05	1.34E+05
ESI	4.95E-10	4.84E-10	5.20E-09

Table 12. Emergy evaluation of Scenarios

The results of the analysis demonstrate that Scenario 0 contributes the lowest solar emergy input caused by the lower compliance of landfilling standards and the less amounts of waste disposal and treatment. Scenario 2 demands the highest emergy input because the construction of LFG collection system needs significant additional cost. Nonetheless, the increasing amount of waste collected affects the lower emergy investment compared to Scenario 0. Meanwhile, Scenario 1 needs the lowest emergy investment. The lower emergy input than Scenario 2 for the absence of LFG collection system and the higher amount of waste disposal and treatment than Scenario 0 are the rationales for this. Scenario 2 generates the highest emergy recovery for the higher scavenging rate than Scenario 1 and the higher composting rate than Scenario 0. It shows that the application of LFG collection system has an effect on the entire waste treatment efficiency. The highest EYR is generated by Scenario 2 indicating the most suitable alternative in recovering emergy from MSW though the highest emergy input. All scenarios have the negative value of Net Emergy. It means that none of the scenarios is capable to save the greatest quantity of emergy per unit weight of MSW treated as the emergy investment is higher than the emergy recovery. However, Scenario 1 supplies relatively higher benefits than two other scenarios because it has the highest Net Emergy. Scenario 2 has the lowest ELR reflecting that the pressure on the environment caused by the activities under Scenario 2 is lower compared to other scenarios. The highest EYR and the lowest ELR is the reason for the highest ESI for Scenario 2. The highlighted value in Table 12 is the value that meets the criteria of each parameter.

13. Conclusion

The local government of Yogyakarta in Indonesia will construct a new SWDS not so far from the old landfill. The new SWDS have to be operated as a safe landfill to obey the Waste Law 18/2008 Article 22 and Article 44. Due to the inferior waste management conditions in Yogyakarta, the new SWDS will be a controlled landfill. The existing of the scavengers is also another factor for the option of a controlled landfill. The evaluation of the old landfill showed that scavengers has role in reducing the waste. The involvement of scavengers in the old landfill contributed 7.5% reduction on plastics and 12.8% reduction on paper. Furthermore, they were responsible also for reduction on metal and glass although the percentage was very little (below 0.01%). Using IPCC Tier 2 Method, the methane emission from the old landfill has been calculated. The result demonstrated that the involvement of 45 scavengers in Bendo landfill contributed 0.7% emission reduction. The value was not significant compared to the amount of the degradable waste (paper) sorted since there was no major reduction on organic waste. A considerable biowaste reduction, for example through composting, can effect the methane emission substantially. The increasing number of scavengers was a minor factor compared to the increasing amount of biowaste prevented from disposal in landfill.

Three scenarios of final waste treatment have been evaluated. The evaluation of the scenarios for final waste treatment in Yogyakarta can be used as a reference to determine the appropriate alternative. The cost for the improper final waste treatment and the benefit for better implementation of final waste treatment have been provided in this study. The involvement of scavengers in the new landfill is considered in all scenarios since the evaluation of the old landfill indicates that scavenging has contributed waste and LFG emission reduction. The evaluation includes two environmental parameters; the global warming potential (GWP) and the emergy indices covering some indicators. The estimation of GWP in form of emission of equivalent carbon dioxide shows that the involvement of scavenger in reducing waste in SWDS has less significant contribution in reducing GWP from SWDS. Biowaste reduction through composting affects GWP potential reduction more intensely. Higher percentage of composting in Scenario 1 and 2 contributed the lower GWP from SWDS compared to Scenario 0. Scenario 2 which covers the landfill with open flare system reduces the most GWP.

The application of indicators in emergy analysis such as emergy indices is significant in evaluating the final waste treatment because it enables the assessment of sustainability and efficiency of each scenario. It allows the analysis of environmental cost and benefits of a certain final waste treatment. Therefore, the emergy indices of three scenarios are compared. In all scenarios, landfilling process needs the highest emergy investment which is mainly contributed by emergy input from fuel and plant construction. The positive emergy recovery is contributed by composting and scavenging which generates income. Therefore, the new landfill should not eliminate scavenging totally. The evaluation of emergy indices shows that Scenario 0 contributes the lowest solar emergy input, while Scenario 1 demands the lowest emergy investment and provides the highest Net Emergy. Furthermore, Scenario 2 generates the highest emergy recovery, the highest EYR, the lowest ELR and the highest ESI. Table 13 presents the environmental parameters analyzed in the study.

	S0	S1	S2	Criteria
Global warming potential	-	-	√	lower
Total solar emergy	√	-	-	lower
Emegy Investment	-	√	-	lower
Emergy recovery	-	-	√	higher
EYR	-	-	√	higher
Net Emergy	-	√	-	higher
ELR	-	-	√	lower
ESI	-	-	√	higher

Table 13. The evaluation of the scenarios

According to the value of the environmental parameters analyzed in the study, Scenario 2 shows the best result since it has more environmental parameters which fulfill the criteria. It is characterized by high EYR and low ELR which is an indication of sustainability and the highest emergy recovery implying the efficiency though the relative high emery investment. Hence, it implies that Scenario 2 is the best alternative for final waste treatment scenario in Yoyakarta City.

Author details

Christia Meidiana
Faculty of Engineering, Brawijaya University, Malang, Indonesia

Acknowledgement

The authors would like to express their intense appreciation to Directorate General of Higher Education Department of National Education of the Republic of Indonesia for their important contributions to the development of this work. The author is thankful to the Yogyakarta municipality officers for their kind assistant in the survey works.

14. References

[1] Bogner, J., et.al (2007), Waste Management. In, Metz, B., et. al (eds),. Contribution of Working Group III to the Fourth Assessment Report of the Intergovernmental Panel on Climate Change. Chapter 10 pp. 586 – 618, Cambridge University Press, Cambridge, United Kingdom and New York, NY, USA.

[2] The United Nations for Environmental Program, (2010), Waste and Climate Change: Global Trends and Strategy Framework, UNEP, Division of Technology Industry and Economics, International Environmental Technology Centre, Osaka/Shiga.

[3] European Environment Agency, (2007), The road from landfilling to recycling: common destination, different routes, Brochure No 4/2007., Denmark. Available at http://www.eea.europa.eu/publications/brochure_2007_4.

[4] Susmono, (2009). Urban Environment Sanitation Infrastructure Improvement in Indonesia. Keynote speech at the Conference on Delta Challenges in Urban Areas, Jakarta

[5] The Ministry of Environment (2010), Indonesia Second National Communication under the UNFCC. Climate Change Protection for Present and Future Generation, MoE, Jakarta

[6] Bengtsson, Magnus and Sang-Arun, Janya, (2008). Urban Organic Waste. In, Climate Change Policies in the Asia-Pacific: Re-Uniting Climate Change and Sustainable Development. IGES White Paper, Chapter 6. . IGES (Hayama). p133-157. Available at. http://enviroscope.iges.or.jp/modules/envirolib/view.php?docid=1565.

[7] Meidiana, C. and Gamse, T., (2010). Development of Waste Management Practices in Indonesia. European Journal of Scientific Research, (40), 199 – 210.

[8] The Ministry of Environment (2008), Waste management Act No.18/2008. Ministry of Environment (MoE), Jakarta. Indonesia

[9] Meidiana, C. and Gamse, T., (2011). The New Waste Law: Challenging Opportunity for Future Landfill Operation in Indonesia. Waste Management, 29, (1), 20 – 29.

[10] The Intergovernmental Panel on Climate Change, (2006). 2006 The Intergovernmental Panel on Climate Change (IPCC) Guidelines for National Greenhouse Gas Inventories. Volume 5. Waste. IGES, Japan. Available at http://www.ipcc-nggip.iges.or.jp/public/2006gl/vol5.html.

[11] The Ministry of Public Works, (1994), SNI No. 03-3241-1994. Procedures of Selecting a Landfill Site, Ministry of Public Works (MoPW), Jakarta, Indonesia.

[12] Sevilla, C.G., Jesus A.O., Twila G.P., Bella P.R., Gabriel G.U., (1993). Research Methods, Rex Printing Co. Inc., Quezon City.

[13] Shryock, H.S., and J. S. Siegel, (2004). The Methods and Materials of Demography, 2nd Edition,

[14] Diaz, L.F. and C.G. Golueke (1993), Solid Waste Management in Developing Countries", In, Solid Waste Management, UNEP (2005) Chapter III pp 31 – 49, CalRecovery, Inc., USA

[15] Environment Agency (2002), Guidance on Landfill Gas Flaring. Environment Agency, UK

[16] Jaramillo, P. and Matthews, H., (2005). Landfill-Gas-to-Energy Projects: Analysis of Net Private and Social Benefits. Environmental Science Technology, 39, (19), 7365 – 7373.

[17] Odum, H.T, (2000). Folio No. 2, Emergy of Global Processes. Handbook of Emergy Evaluation. Center for Environmental Policy, Environmental Engineering Sciences, University of Florida, Gainesville.

[18] Brown, M.T., and Ulgiati, S., (2004[a]). Energy Quality, Emergy, and Transformity: HT. Odum's Contributions to Quantifying and Understanding Systems. Ecological Engineering, 178, (1-2), 201 – 213.

[19] Statistics Yogyakarta, (2011), Yogyakarta in Numbers, Statistics Yogyakarta (SY), Indonesia.

[20] Yogyakarta Environmental Board (2010). Report on Environmental Status of Yogyakarta Year 2009. YEB, Indonesia

[21] Lou, X.F. and J. Nair, (2009). The impact of landfilling and composting on greenhouse gas emissions – A review, Bioresource Technology 100, 3792–3798

[22] Brown, S., Leonard, P., 2004. Biosoilds and Global Warming: Evaluating the Management Impacts. Biocycle 45, 54–61.

[23] Brown, S., Subler, S., 2007. Composting and Greenhouse Gas Emissions: a Producer's Perspective. Biocycle 48, 37–41.

[24] Zeman, C., Depken, D., Rich, M., 2002. Literature Review – Research on How The Composting Process Impacts Greenhouse Gas Emissions and Global Warming. Compost Science and Utilization 10, 72–86.

[25] Zudianto, H., (2011). Policies on Environmental Management of Yogyakarta City. Presented in Waste Management Symposium 2011, Singapore

[26] Campbell, D., Maria M. Thomas D., John P., and M. Patricia B., (2004), Keeping the Books for Environmental Systems: An Emergy Analysis of West Virginia. Environmental Monitoring and Assessment 94: 217–230.

[27] Ulgiati S., Bargigli S., Raugei M., (2007). An Emergy Evaluation of Complexity, Information and Technology, Towards Maximum Power and Zero Emissions. Journal of Cleaner Production, (15), 1359 – 1372.

[28] University of Florida, (2000). National Environmental Accounting Database. Indonesia. Available at
http://sahel.ees.ufl.edu/frame_database_resources_test.php?search_type=basic&country=IDN

[29] Wang, L., Ni, W., Li, Z., (2006). Emergy Evaluation of Combined Heat and Power Plant Eco-Industrial Park (CHP plant EIP). Resources, Conservation and Recycling, (48). 56 – 70.

[30] Buranakarn, V, (1998). Evaluation of Recycling and Reuse of Building Materials Using Emergy Analysis Method. University of Florida, FL. In, Emergy Analysis of West Virginia - Data Sources and Calculations,. Available at
http://www.epa.gov/aed/research/TechReptApp.pdf.

[31] Odum, H.T., (1998). Emergy Evaluation. Unpublished paper presented at the International Workshop on Advances in Energy Studies: Energy flows in ecology and economy, Porto Venere, Italy, May 27.

[32] Odum, H.T., (1996). Environmental Accounting: Emergy and Environmental Decision Making. John Wiley and Sons, NY

[33] Marchettini, N., R. Ridolfi, M. Rustici, (2006), An environmental analysis for comparing waste management options and strategies, Waste Management 27, 562–571

[34] Herity, L. (2003). A study of the quality of waste derived compost in Ireland. Available at http://www.bvsde.paho.org/bvsacd/cd43/jlorra.pdf

Chapter 11

Plastics Recycling –
Technology and Business in Japan

Yoichi Kodera

Additional information is available at the end of the chapter

1. Introduction

The Japanese government promotes 3R policies in the country and to developing countries [1]. However, there are many obstacles and discrepancies between the idea and reality. After the straggles in technologies and consumer movements over thirty years, a legislative approach to waste-plastic recycling started in Japan in the year 2000. The first target was plastic containers and packaging from household wastes. Approximately ten years have passed, and we still face many problems within the country: high recycling costs, low quality of recycled resin with respect to the market value, and so on. Some of our challenges and achievements, or fact data itself would be good material for people that have an interest in waste plastics recycling. Many discussions and data in this field are not published in academic journals, and the reports of the national government, municipalities and companies are mostly written in Japanese. The fact data, commercial technologies, and businesses for recycling waste plastics in Japan are reviewed in this chapter.

2. Generation of waste plastics, and the legislation for their management

2.1. Generation of waste plastics

The Ministry of Environment of Japan announces the current status of generation and treatment of general wastes [1,2] and industrial wastes every year. Municipal wastes of the total collection amount of 46.3 million tons include those produced from households (25.6 million tons) and those from small businesses (13.3 million tons). Waste plastic contents in municipal wastes are regularly monitored by each municipality. Depending on separation categories decided by municipalities, waste plastics are incinerated as mixed burnable wastes, or recycled like plastic containers and packaging. Some plastic materials such as those in toys and in electronic devices go to landfill. A composition survey of mixed wastes that were brought into the Chuo incineration plant in Tokyo gave 16.0 wt% as a mean value of waste-plastics content in burnable wastes from households (fiscal year 2010) [3]. This mean value is an average of four times a year. The detailed composition and property of burnable wastes is shown in Table 1. For industrial wastes, waste plastic is one of 20 collection criteria of industrial wastes, and the recent survey shows the generation of 5.67 million tons in 2009 [4].

Category		Mean value
	Burnable components	98.45
Composition, wt%	paper	45.00
	kitchen waste	22.92
	plastics	15.97
	wood and grass	5.74
	textile	4.61
	rubber and leather	0.51
	others	3.70
	Inflammable contaminants	1.55
	metal	0.46
	glass	0.32
	stone and ceramics	0.13
	others	0.64
Proximate analysis, wt%	Moisture content	38.69
	Combustible matter	54.58
	Ash content	6.75
Elementary analysis, %	Carbon	28.5
	Hydrogen	4.24
	Nitrogen	0.34
	Oxygen	21.33
	Combustible sulfur	0.02
	Volatile chlorine	0.17
	Higher calorific value, kJ/kg	12,263
	Lower calorific value, kJ/kg	10,335
	Bulk density, kg/L	0.130

Table 1. Composition and property of burnable wastes in Chuo incineration plant

The world plastics production is 265 million tons (2010) [5]. Production of typical synthetic resins in Japan (2010) is 12.2 million tons (Table 2) [6], and it accounts for 4.6% of the world production. The generation and the macro flow of the waste plastics in Japan are published annually by the plastic waste management institute. The latest data of the year 2010 are shown in Figures 1 though 3 [7]. The total generation of waste plastics increased until 2004 (10.13 million tons), and gradually decreased to 9.12 million tons (2009) due to the shrinking economy in Japan. In 2010, the total generation increased to 9.45 million tons because of the recovery from the economic crisis in 2008 [7].

Synthetic resin		Production, tons
Thermoplastics		
PE	HDPE	1,704,076
	LDPE	1,015,260
	EVA	244,231
PP		2,709,023
PS	For molding	698,113
	For expanded PS	123,560
PVC		1,749,046
PET		631,101
ABS		454,109
PC		369,270
Other thermoplastics		1,311,767
Subtotal of thermoplastics		11,009,556
Thermosetting plastics		
Phenolic resin		284,152
Epoxy resin		187,565
Urethane foam		180,152
Other thermosetting plastics		352,841
Subtotal of thermosetting plastics		1,004,710
Other plastics		227,861
Total		12,242,127

Table 2. Production of typical synthetic resins in Japan (2010)

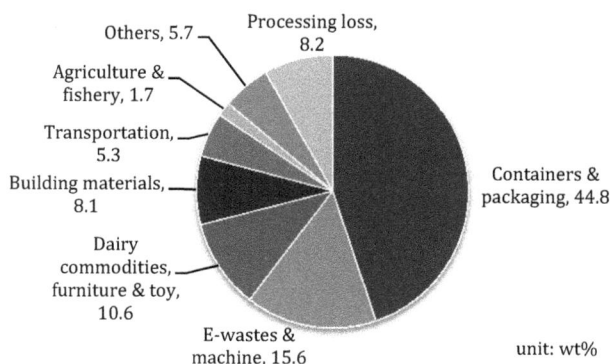

Figure 1. Waste plastics generation by user's application (Total generation 9.45 million tons, 2010)

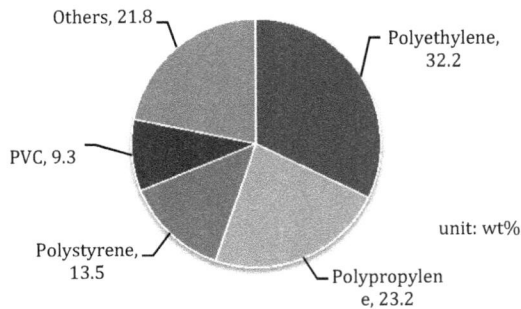

Figure 2. Waste plastics generation by the types of plastics (Total generation 9.45 million tons, 2010)

Entry	Type of waste plastics / waste source	Generation amount ton/year
1	Total PET / Total domestic sales of PET resin as bottles	594,689
	Estimated amount of PET resin from bottles in general wastes	290,000
	Estimated amount of PET resin from bottles in industrial wastes	305,000
2	Mixed plastics / containers and packaging in general wastes	1,040,658
3	Plastic parts in Automobiles	399,000
	Plastics of non-polyurethane in ASR	193,000
	Polyurethane in ASR	93,000
4	Electric and electronic equipment	
	Home appliances under the recycling law	181,884
	Other E-wastes	20,000
5	Agriculture	
	Polyethylene film	62,778
	PVC film	42,852
	Other	9,588
6	Business	
	Chemical industry sector	91,504
	Manufacture of plastic products	77,229
	Manufacture of rubber products	45,425
	Manufacture of electric equipment and machinery sector	14,037
	Manufacture of equipment for transportation	18,753
	Manufacture of pulp, paper or paper products	35,929
	Food processing factory	5,285
	Publishing and printing workshop	3,986
	Steel manufacture	32,939
	Non-ferrous metal manufacture	12,259
	Manufacturer of metal products	4,229
	General machinery	3,768
	General construction	27,430
	Electrical and mechanical service	462

Table 3. Waste plastics generation from various sources in Japan

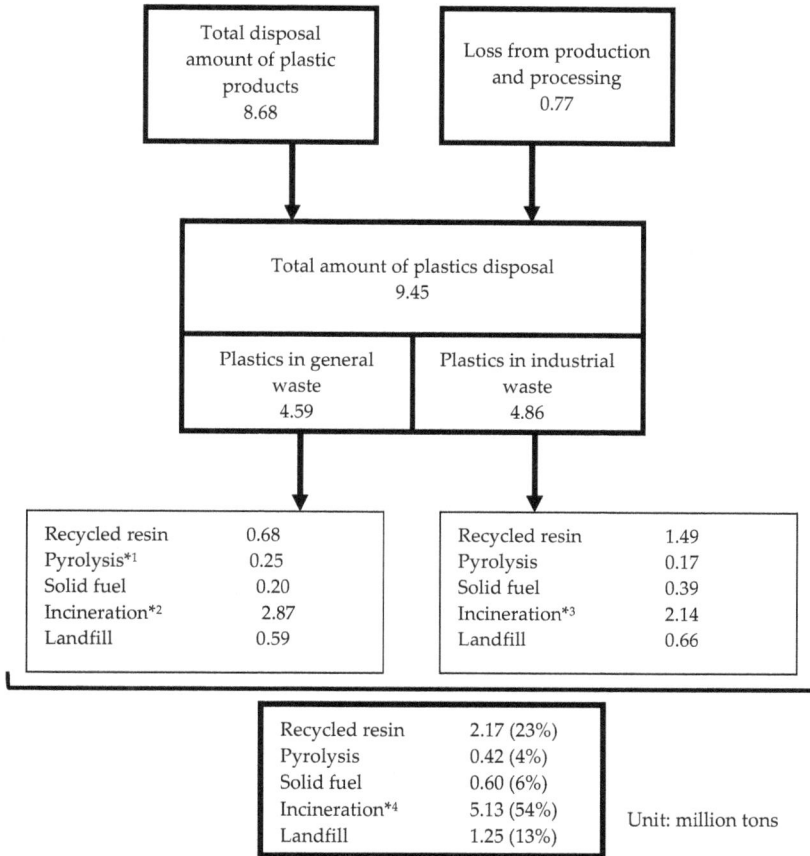

Figure 3. Material flow of waste plastics in Japan (2010)

*1 Pyrolysis includes cokes-oven treatment, blast furnace treatment and gasification.
*2 Incineration with power generation 1.84, incineration with heat recovery 0.36 and incineration without use of combustion heat 0.67 million tons.
*3 Incineration with power generation 1.18, incineration with heat recovery 0.66 and incineration without use of combustion heat 0.30 million tons.
*4 Incineration with power generation 3.03 (32%), incineration with heat recovery 1.03 (11%) and incineration without use of combustion heat 0.97million tons (10%).

Table 2 shows waste-plastic generation from various sources. Some figures are based on actual collection amounts, and some are estimated amounts based on the references cited in the latter. The target wastes of the containers and packaging recycling law is the containers and packaging that made of paper, glass, metal and plastics in general wastes. For waste plastics of containers and packaging in general wastes, PET bottles, food tray made of expanded polystyrene sheet and mixed plastics of the other plastics are separately collected by municipalities. The containers and packaging recycling law allows a municipal government to decide on the separation rules for the collection of municipal wastes.

In Entry 1 of Table 2, annual PET generation is estimated as 594,689 tons (fiscal year 2010), which is given by the total sales of PET bottles including the sales of domestic products (579,782 tons) and imported products (14,907 tons) [8]. The total sales of PET bottles are the weights of PET resin that does not include screw caps and labels of non-PET plastics. The breakdown of PET in general wastes is based on the current status of implementation of PET-bottles collection by municipalities under the Containers and Packaging Recycling Law in FY 2010 (Table 3) [9]. The recycled amount of PET bottles, 286,009 tons, corresponds to PET-resin production from PET bottles in general wastes. If all municipalities implements PET-bottles collection, the expected recovery from general wastes by municipalities is 290,364 tons (=286,009 tons collection / 0.985 population coverage). The difference between the total sales of PET bottles and expected total recovery from general wastes is 305,000 tons, which would correspond to PET in industrial wastes.

The generation amount of mixed plastics is 1,155,000 tons (=705,707 tons collection / 0.611 population coverage), which is plastic containers and packaging other than PET bottles and plastic trays. In a waste sorting facility of a municipality, mixed plastics are compacted to prepare a cubic bale for the cost-effective transportation. The bale often contains serious amounts of contaminants such as metals and moisture. Table 4 shows an example of the composition of a bale that used for the life cycle assessment of waste-plastics recycling [10]. If the plastic content of mixed plastics that are collected under the containers and packaging law is 90.1%, the total plastic amount is 1,040,658 tons (= 1,155,000 × 0.901).

There are 1,742 municipalities in Japan (April 1, 2012). As shown in Table 4, 61.6 % of all municipalities adopted separate collection and recycling rules for mixed plastics under the Containers and Packaging Recycling Law. The other municipalities adopt incineration of mixed wastes or landfills avoiding the high cost of collection and handling because the recycling law in Japan still allows a municipality to choose mixed collection of waste plastics with the other general wastes for incineration or separate collection of waste plastics for recycling. Due to the economic benefits of PET recycling for municipalities, the implementation rate of separate collection and recycling of PET bottles reaches to 99.1% of the whole municipality, population coverage 99.5% under the same scheme of the recycling law.

About 4.16 million end-of-life-vehicles (ELV) generates in 2011[11]. When mean contents of plastic parts is 8 wt% [12,13] and averaged weights of ELVs are assumed as 1,200 kg, waste plastics generation is 399 thousand tons per year, which includes plastics with commercial values as replacement parts and automobile shredder residues without any commercial value. Waste plastics generation from automobiles are estimated as 399 thousand tons a year (= 4,160,000 × 1.2 ton × 0.08). Disassembly of ELV and crushing process give automobile shredder residue (ASR) of 584,305 ton in FY 2007 [14]. Typical composition of ASR is as follows: Plastics of non-polyurethane 33, polyurethane 16, textile 15, rubber 7, wood 3, paper 2, iron 8, non-ferrous metals 4, wire harness 5, glass 7 (wt%) [15]. Thus, the generation amounts of plastics of non-polyurethane and polyurethane are estimated as 193 and 93 thousand tons, respectively.

The Home Appliance Recycling law provides safe treatment and effective recovery of valuable materials from the five home appliances, air conditioners, television sets, refrigerators/freezers, washing machine and clothes dryers [16]. Personal computers are also encouraged to collect for recycling, but total weights of plastics are not clear.

In the fiscal year 2006, the total recovery of plastics was 102,257 ton was recycled with commercial values. The recovered plastics are further processed in mechanical recycling (60,020 ton, 59%), energy recovery (as crushed plastics 1,400 ton, 1%; as RPF 4,800 ton, 5%), and the rest (36,037 ton, 35%) was disposed [17]. In the fiscal year 2010, the total amount of recycled plastics was 181,884 ton [18]. The details of recycling are not announced. Polyurethane is widely used as heat insulation, and this is a major material that is not suitable for any type of recycling application but incineration with special attention to the complete combustion of fluorocarbons in polyurethane.

Aizawa et al. estimated the annual generation of the wastes of small electrical and electronic equipment, and estimated the total plastics as 20,000 ton [19]. The estimate is based on the equation of a typical content times shipment of video cassette recorder, DVD player, video camera, digital camera, flash memory player, HDD-equipped audio equipment and gaming equipment in 2007. The paper gave the estimated amounts of various metals and plastics of the potential resources.

There are the other sources of waste plastics. Recycling laws targeting various types of waste plastics would be considered based on the lifestyles of each country. Following to the recycling of waste plastics of containers & packaging, the more efforts will be considered to expand recycling activities. Substantial amounts of plastics are also used as daily commodities such as kitchen utensils and clothes cabinets. Plastics are also one of the main components of toys and E-wastes (wastes of electric and electronic equipment), which have been treated by landfill. Considerable amounts of plastic products are imported to the Japanese market. The amount of those imported plastic products is not clear. As an agriculture material, plastic film is a widely used product. Mulching, tunnel and green house are the typical uses in agriculture. As shown in Table 5, the current methods for the treatment of these plastic films are incineration, landfill and recycling including mechanical recycling and production of solid fuel [20]. The details of recycling application are not clear in the report. Waste generation data from industry has been collected through questionnaire survey by the plastic waste management institute [21,22].

Waste item	Collection, tons	Amount of recycled plastics, tons	Number of municipalities under implementation	Implementation rate to the total number of municipalities, %	Population coverage, %
PET bottles	296,815	286,009	1,711	97.8	98.5
Food tray	3,242	2,959	505	28.9	35.8
Other plastics (mixed plastics)	705,707	668,775	1,078	61.6	61.1

Table 4. Current status of the implementation of waste plastics under the containers and packaging law in the fiscal year 2010

Plastics	Percent weight
Polyethylene	30.2
Polypropylene	21.1
Polystyrene	17.7
PET	13.8
PVC	4.9
EVA	2.4
Metals	2.6
Moisture	7.3

Table 5. Typical composition of composition of mixed plastics

Treatment method	Polyethylene	PVC	Others
Recycling	43,128	30,373	2,831
Incineration	4,831	2,278	1,460
Landfill	10,416	8,404	4,846
Other	4,403	1,797	451

Table 6. Treatment of used agriculture films in the fiscal year 2008 (unit: ton)

Table 6 shows the typical gate fees of waste plastics by various treatment methods. Gate fee is a payment from a waste generator to a waste management company for waste treatment often including transportation cost. When the waste has a commercial value, a waste management company buys and sells it after a suitable processing. When a waste has a commercial value of higher than transportation cost, it is usually considered to be a commercial article rather than waste. Business sectors can handle them without any license or permission of business or facility installation for transportation, treatment or the other commercial dealing.

Waste plastics from general wastes such as containers and packaging contain various types of plastics such as a sheet, film, bag and bottle of polyethylene, polypropylene, polystyrene, polyamide and PET. And many items are laminates of two or more plastics, paper or aluminum. The recycling cost increase due to these complex compositions of mixed plastics as a feedstock of recycled resins. There are many business sectors buying thermoplastics of good quality for recycled resin production. Waste plastics, especially polyethylene, propylene, polystyrene, PET and PVC, are exported to China over some 1.5 million tons per year. In 2011, 1.6 million ton of waste plastics of commercial values were exported to mainly China for mechanical recycling (Table 7) [25]. China's country share as the exporting destination is 1.48 million ton (90.5 % of total exported amounts from Japan), which includes the mainland 890 thousand ton and Hong Kong 586 thousand ton. Polypropylene is considered as the major component of the waste plastics defined as "the other plastics."The mean price of the waste plastics is 46 yen/kg. It is varied depending on the conditions of wastes plastics. Generally, shredded, clean and colorless plastics are of the higher commercial value.

Entry	Treatment method or products	Type of waste plastics*	Gate fee** yen/kg	Remarks
1	Mechanical recycling	Mixed plastics, C&P recycling law (household)	70	Ref.[23]
		PET Bottles, C&P recycling law (household)	▲49	Ref.[23]
		Industrial wastes of good quality	▲20-30	Hearing
2	Solid fuel (RPF)	Industrial wastes of moderate quality	20-25	Hearing
3	Blast furnace treatment	Mixed plastics, C&P recycling law (household)	34	Ref.[23]
		Mixed plastics from industry	35	Ref.[22]
4	Cokes oven treatment	Mixed plastics, C&P recycling law (household)	45	Ref.[23]
		Waste plastics from industry	35	Ref.[22]
5	Liquid fuel production	Mixed plastics from household	80-100	Hearing
		Mixed plastics from industry	30-40	Hearing
6	Gasification	Mixed plastics, C&P recycling law (household)	31	Ref.[23]
7	Cement kiln treatment	Industrial wastes of moderate quality	10	Ref.[22]
8	Incineration with power generation	Mixed wastes from household	13-17	Ref.[24]
		Mixed wastes from industry	35-50	Ref.[22]
9	Incineration (no energy recovery)	Mixed wastes from household	19	Ref.[24]
		Mixed wastes from industry	30	Ref.[22]
10	Landfill	Industrial plastic wastes	8 yen/m³	Ref.[22]

*C&P designates containers and packaging.
**The symbol "▲" designates payment from a waste business sector to a waste generator.

Table 7. Examples of the gate fees of waste plastics by various treatment methods

Waste plastics	Amount, ton	FOB price*, thousand yen	Mean price, yen/kg
Polyethylene	360,779	15,383,077	42.6
Polystyrene	252,337	11,415,978	45.2
PVC	90,233	2,730,090	30.3
PET	392,291	22,375,368	57.0
Others	535,815	23,339,954	43.6
Total	1,631,455	75,244,467	46.1

*Free On Board. The seller of goods pays for transportation of the goods to an exporting port and the loading cost. The buyer covers the transportation cost after loading on a ship.

Table 8. Export amounts and prices of exported waste plastics from Japan (2011)

The apparent treatment cost of incineration is very low (entries 8 and 9 of Table 6). Substantial subsidies from the national government make the construction cost of an incineration plant lower than the other recycling methods such as mechanical recycling and liquid fuel production. Different from a recycler as a private company, municipalities use the different accounting system, in which fixed costs are not involved in the treatment cost. The actual cost would be 35 - 50 yen/kg by accounting construction costs of incineration plants. The lower apparent cost of incineration derives some municipalities to incinerate mixed wastes rather than recycling.

2.2. Legislation system for recycling waste plastics

To establish a sustainable society throughout Japan, the Basic Law for Establishing the Recycling-based Society as the basic framework. It is also called as the Sound Material-Cycle Society. Table 8 lists the laws for waste management and recycling [26]. Under the individual recycling laws, each targeted wastes have been recycled for several years.

For waste plastics recycling, there are some differences of the preferred recycling and business system among several recycling laws. For example, mechanical recycling is preferred in the Containers & Packaging Recycling Law, but heat recovery through incineration is allowed in the ELV recycling law. The contract system is different between the plastic mixed wastes under the Containers & Packaging Law and the other plastic wastes such as ELV and home appliances. Recyclers receive the mixed plastic wastes based on the competitive bidding among recyclers of mechanical recycling. When the additional amounts of wastes are left, the second bidding will be held among the recyclers of feedstock recycling. The contractor is fixed by each stock yard. The contract is for one year. In the treatment of ELV and wastes of home appliances, limited numbers of waste management companies constantly receive the wastes in connections with automobile manufacturers or home appliance manufacturers. There are strong arguments on the C&P recycling from many stakeholders in the points of recycling cost, bidding system with the preference of mechanical recycling to the other methods and participation of the recyclers of solid fuel production. We also started a discussion for widening the coverage of waste plastics to plastic articles such as dairy necessaries, toys and electronic equipment.

3. Technologies and businesses of waste plastics recycling

3.1. Recycled resin production

Recycled resin or recycled plastic goods are produced from 2.17 million ton of waste plastics (Figure 3). Some 1.6 million ton of waste plastics are exported (Table 7). These plastics are considered to go to mechanical recycling. The difference between 2.17 and 1.6 million is about 600,000 ton, which is the feedstock for recycled resin in the domestic market. The domestic demands for recycled resin are quite weak than ones in China.

Many Japanese manufacturers tend to avoid the use of recycled resin, especially from mixed waste plastics of containers and packaging because of the low quality such as strength, color

Law	Content
Basic law for establishing the recycling-based society	Basic framework determining the role of stakeholders for establishing the sound material-cycle society.
Waste management and public cleansing law	Defines municipal wastes and industrial wastes. The roles and duties of a municipality, waste generator, waste management company, and other stakeholders are strictly provided. The related regulations and rules define both technical and social conditions and guidelines to keep the sound business in addition to construction of a facility, installation and operation of equipment.
Law for promotion of effective utilization of resources	Promotion of waste reduction through recycling. The roles and duties of the stakeholders are mentioned. Promoting reduction of wastes through recycling and suitable disposal in several fields of industries and products such as steel production, paper production, construction, automobile, electric and electronic equipment, batteries, metal cans and PET bottles.
Containers and packaging recycling law	Promotion of recycling containers and packaging through separate collection of those wastes made from paper, metal, glass, PET and the other plastics by municipalities with cooperation of citizens. Producers of the material, manufacturers of the commercial products with containers and packaging and retail stores cover recycling costs. Recycling methods are provided in the related regulations.
Electric household appliance recycling law (Home appliance recycling law)	Forcing consumers to give wastes of home appliances to retailers with paying recycling fees. Air conditioner, refrigerator/freezer, television set, washing machine and cloth dryer are recycled with suitable treatment of fluorocarbons and other potential hazardous substances.
End-of-life vehicle recycling law	Forcing car owners to cover the cost for suitable disposal of hazardous wastes and wastes of no commercial value with recovering valuable resources from end-of-life vehicles.
Construction material recycling act	Reducing the amounts of construction and demolition wastes through recycling.
Food recycling law	Reducing the amounts of food residues from restaurants, food processing industry and supermarkets through recycling waste foods.
Law on promoting green purchasing	Promoting the national and local governments to buy products that made from recycled materials.

Table 9. Major laws for waste management and recycling

and smell. As a result, the selling price of recycled resin pellets (typically, a mixture of polyethylene and polypropylene) from the mixed waste plastics is generally very low, 20 to 40,000 yen/ton, whereas the recycling cost of mixed waste plastics is 72,000 yen/ton in average due to the contamination of various components that are not suitable for the production of recycled pellets. The recyclers convert the separated portion of mixed waste plastics into recycled pellets or recycled products such as transportation pallets and imitation wood at about 45 wt%, and the rest goes to incineration with heat recovery or solid fuel production with paying gate fees.

Some recyclers for plastic containers and packaging from general wastes make the efforts to raise the market value of recycled resin. There are three countermeasures for it. One is to change separation categories of waste collection by municipalities, for example, a separate collection of hard plastics like HDPE bottles and laminated soft plastics. The second is to introduce a sophisticated material sorting facility with optical sorting equipment under the cooperation with municipalities. Stable supply and constant production of recycled resin will be possible by the more precise selection of suitable plastics for mechanical recycling at the larger scale. The third is to develop a new application in cooperation with many companies in the wider business fields across many countries.

To improve the quality of recycled resin, collection of hard plastic wastes and recycled resin production with hard plastic wastes were conducted as a research by Akita Eco Plash Co., Ltd. in cooperation with Akita prefecture and Noshiro city authorities based on funding by the New Energy and Industrial Technology Development Organization (NEDO) [27]. Recycled resin has low melt-flow rate (MFR) because the original form of the polyethylene and polypropylene that recovered from C&P wastes is film and sheet. But the major products from recycled resin are not film or bag but hard plastic products. When hard plastic wastes of non C&P wastes were added to recycled resin of C&P at 10 %, the MFR was improved from 3.2 to 3.8. This result suggests the improvement of the qualities of recycled resin and products with reducing an additive. Minato-ward authority in Tokyo has a total collection system of plastic C&P wastes with the wastes of hard plastic products in the criteria of "Resource Plastics."The collection of the wastes of plastic products increased the collection amount of polypropylene in hard plastic wastes [28], and it will help the MFR, which leads to the reduction of additives and cost reduction.

Home appliance manufacturers and automobile manufacturers are actively seeking the idea and technologies to raise the recycling rate of the waste plastics recovered from their wastes. Electric appliance manufacturers have been made efforts the cost reduction of their products. They took actions in recycling waste plastics in the products to reduce the waste amounts. Additionally, some companies started commercial operation of the precision separation system of some plastics by applying the difference of electrostatic properties of plastics [29]. In 2010, Green Cycle Systems Corporationl aunched Japan's first large-scale, high-purity plastic recycling center under the technical and business support by Mitsubishi Electric Corporation. The announcement states that Green Cycle Systems takes the shredded mixed plastic chips recovered by Hyper Cycle Systems and separates them into reusable plastic on a scale of unprecedented magnitude. And it also tells that the combined output of these two enterprises have increased Mitsubishi Electric's rate of recycled, industrial-grade plastic from 6% to a paradigm-shifting 70%.

3.2. Refuse-derived solid fuel

There is a variety of solid fuel that has been prepared from wastes, which include wood, straw, rice husk, garbage from households, plastics and so on. To control moisture contents, some processes such as drying and carbonization are often performed. Any non-hazardous combustible wastes can be used as the raw material for solid fuel with or without preparation of pellets and briquettes. Preparation of pellets and briquettes contributes to constant quality of heating values, easy transportation and smooth feeding to a combustor such as a boiler.

Table 9 shows the heating values of various combustible wastes and fuels [30-33]. Waste plastics have high heating values, and coal substitutes can be prepared by mixing them with the wastes of low heating values. Thermoplastics act as a suitable binder, and paper, textile and thermosetting plastics can form pellets and briquettes despite to their properties that they are not solidified each other.

Waste	Higher heating value MJ/kg	Reference
Polyethylene	47.7	[30]
Polypropylene	45.8	[30]
Polystyrene	43.7	[30]
Poly(methyl methacrylate)	26.9	[30]
Poly(ethylene terephthalate)	24.1	[30]
RPF	>25	[31]
RDF	>12.5	[32]
Diesel oil	37.7	[33]
Heavy oil	39.1	[33]
Coal	25.7	[33]

Table 10. Heating values of combustible wastes and fuels

From municipal wastes includes kitchen wastes, refuse-derived fuel (RDF) are used in various countries over US [34], Europe [34,35] and Japan [36]. It is also called as solid recovered fuel (SRF). In 2005, RDF was produced from general wastes in 58 facilities [37]. There are 27 facilities are cooperated with power generation plants and the feedstocks are collected from 92 municipalities. Table 10 summarizes the compositions of municipal wastes in Japan and the properties of RDF [36]. A specification guideline of RDF is shown in a technical specification document of the Japanese Industrial Standard as shown in Table 11 [38]. Typical dimension of RDF briquettes is 10 - 50 mm diameter and 10 - 100 mm length. A simply-densified brick and crashed product are not considered as RDF in this guideline.

Prefecture	Ishikawa	Mie	Fukuoka	Ibaraki
Area	Hakui	Kuwana	Omuta-arao	Kashima
Municipal waste composition				
Paper, textile, %	60.6	62.7	42.0	58.7
Wood, %	3.4	6.1	5.3	4.4
Plastics, rubber, %	23.7	6.5	27.3	17.4
Kitchen wastes, %	9.8	24.1	16.3	5.2
Incombustibles, %	0.0	0.0	2.5	10.3
Others, %	2.6	0.6	6.6	4.0
RDF properties				
Moisture, %	2.3	3.3	6.0	4.4
Combustibles, %	85.4	83.0	81.0	84.8
Ash, %	12.3	13.7	13.0	13.7
Higher heating value MJ/kg	17.5	17.3	19.6	19.0
Lower heating value MJ/kg	17.3	15.8	18.0	17.5
Total chlorine, %	0.56	0.70	0.10	0.78

Table 11. Typical compositions of municipal wastes in Japan and the properties of RDF

Category	Specification
HHV	>12.5 MJ/kg
Moisture	<10 %
Ash	<20%

Table 12. Specification of RDF in TS Z 0011:2005.

We need technical countermeasures to the formation of hydrogen chloride and dioxins upon combustion of RDF due to chlorine-containing plastics and salt. The specification in Table 9 does not mention anything about chlorine content because of the difficulty of the removal of chlorine from municipal wastes. Due to the low heating value and high contents of ash and moisture, RDF is considered as low-quality fuel. There are not so many users of RDF except the power generation plants in the cooperation with municipalities.

Different from RDF, densified solid fuel called as RPF (Refuse derived paper and plastics densified fuel) are popular as coal substitutes. Recently, the specification was defined in the Japanese Industrial Standards (Table 12) [32]. It is produced by using paper, wood, plastics, textile and the other wastes. The raw material is dry and non-hazardous combustibles. It does not include any putrefactive wastes. About 1.6 million ton of RPF was shipped to, mainly, paper and steel manufacturers as coal-substitute for coal-combustion boilers in 2011 [39].

Category	Specification in JIS*	Analytical value of RPF	Analytical values of coal
HHV, MJ/kg	>25 or 33	27.6	26.9
Moisture, weight %	<3 or 5	2.75	7.9
Ash, weight%	<5 or 10	5.54	8.5
Total chlorine, weight%	<0.3 - 2.0	0.25	-
Nitrogen, weight%	-	0.54	1.35

*Four classes are defined depending on the values in the category. JIS Z7311:2010.

Table 13. Specification guideline of RPF in JIS and typical analytical values of RPF and coal

It is possible to vary the calorific value of solid fuel in the range of 5,000 - 10,000 kcal/kg by controlling the ratio of input paper and plastic. A 50:50 mix, for example, provides a calorific value of 6,190 kcal/kg (measured LCV), which is about the same as that of coal. Use of RPF as an alternative to fossil fuel helps to reduce CO_2 emissions and is considered environmentally friendly.

The most serious trouble in the use of RPF is the formation of hydrogen chloride upon combustion and the resulting corrosion of equipment in the RPF users. Some users often stop purchasing RPF because of the high chlorine content.

3.3. Liquid fuel

Some thermoplastics, for example polyethylene, polypropylene and polystyrene, are thermally decomposed under an inert gas to yield liquid hydrocarbons at about 450 °C or above [40]. The resulting liquid hydrocarbons have the similar heating values to those of fuels from petroleum.

Decomposition products and fuel quality depends on the types of plastics and decomposition conditions. Polyamides and polyurethane give oily products of high nitrogen content at low yields. Poly(ethylene terephthalate) known as PET does not give liquid hydrocarbon upon pyrolysis but solid products including terephthalic acid.

Many types of reactors of tank, screw, externally-heating rotary kiln and fluidized-bed are developed. Some plants were used in demonstration, and some are commercially operated in Japan [41]. Under the containers and packaging law, mixed plastics were converted into fuel oil through pyrolysis using a 20-ton/day tank reactor in Niigata and four 10-ton/day rotary kilns in Sapporo until recently. Those commercial operations were shut down due to the higher cost (about 80 yen/kg) than the other treatment costs like that of cokes oven treatment (about 40 yen/kg).

However some recyclers bearing pyrolysis plants commercially produce liquid fuel from plastics of industrial wastes. Most recyclers have a tank reactor with a simple distillation system (Figure 4). In 2011, a recycler in Fukuoka prefecture started fuel oil production using a pyrolysis reactor with a paddle mixer (Figure 5). This reactor is commercially operated for mixed plastics from a separate collection of municipal wastes. The product fuel mixed with commercial fuel at 1:1 is used for the boilers in public facilities.

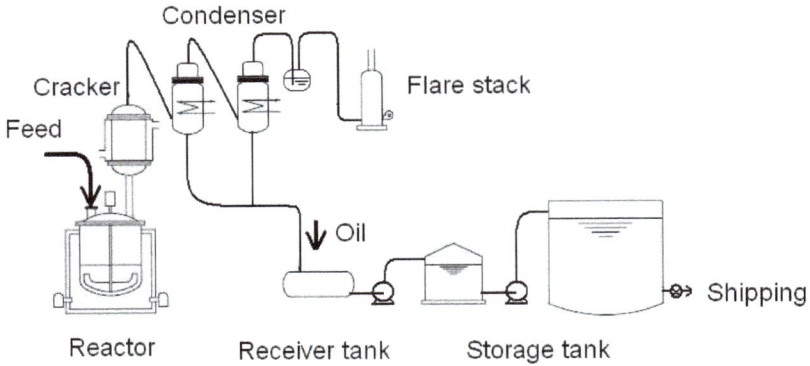

Figure 4. Typical layout of a liquid fuel production plant equipped with a tank reactor.

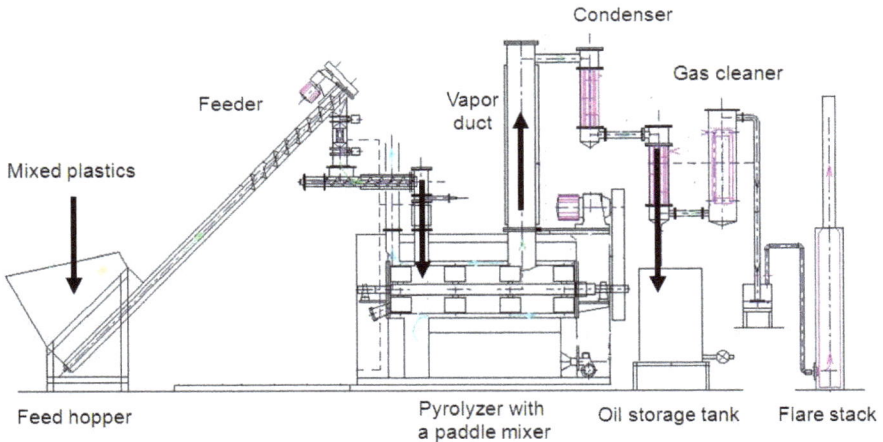

Figure 5. Fuel production system of a pyrolyer with a paddle mixer. Copyright ECR Co., Ltd.

Currently, there is no specification standard of liquid fuel from waste plastics because it is not widely produced. Table 13 summarizes the technical specification that were announced by the Japanese Industrial Standards Committee and expired in 2010 [42]. Table 14 shows typical specifications of plastics-pyrolysis oil, commercial diesel fuel and heavy oil.

Category	Value
Ash	<0.05 %
Total chlorine	<100 ppm
Sulfur	<0.2 %
Nitrogen	<0.2%
Water	No water layer separated from oil
Viscosity, pour point and residual carbon	Depend on consent agreement

Table 14. Specification of pyrolysis oil from waste plastics in TS Z0025:2004

Category	Whole distillate of pyrolysis oil	Middle distillate of pyrolysis oil	Kerosene	Heavy oil Grade A
density (15 °C), g/cm³	0.8306	0.8430	0.8284	0.8511
flash point, °C	-18 (PM)	68.0 (Tag)	69.0 (Tag)	64 (PM)
kinematic viscosity, 30 °C/50 °C, mm²/s	1.041/-	-/1.73	3.822/-	-/2.29
Residual carbon 90% distilled, wt%	-	0.85	0.01	0.46
Ash, wt%	0.00	<0.001	-	0.006
HHV, MJ/kg	47.2	44.9	45.5	44.8
Total Cl, wt ppm	47	10	<1	1.6
Nitrogen, wt%	0.14	0.033	-	0.015
Sulfur, wt ppm	100	910	310	0.41 %
Cetane value	27.0	42.9	58.4	46.3
Distillation range, °C				
Start point	47.0	180.0		164
10%	69.0	199.0		195
50%	148.0	233.0		276
90%	294.5	323.5	344.0	347
End point	374.0	351.5		>370

Table 15. Typical specifications of pyrolysis oil from plastics and commercial fuels

There are two major methods of thermal decomposition of plastics. One is pyrolysis and another is catalytic decomposition. Acidic catalyst such as silica-alumina and zeolite are used in the latter case. A decomposition temperature range is lowered to around 400 °C comparing pyrolysis temperature 450 to 550 °C. Advantage of catalytic decomposition is in the lower energy consumption for decomposing polymers. Disadvantage is the economic expense of purchasing catalysts, regeneration or disposal of waste catalyst. Catalytic decomposition often yields a large amount of gasoline fraction in certain reaction conditions. In this case, the product oil is not suitable for the use of diesel engines or heavy oil burners if the product oil is used without distillation. It is noteworthy that plastics pyrolysis oil does not contain lubricant portion. There is little knowledge on mechanical durability of an engine cylinder and a fuel injection pump of a burner in the use of plastics-derived oil. Mixing with a lubricant or commercial fuel is highly recommended by practitioners and.

Under the containers and packaging law, some recyclers constructed pyrolysis plants to convert mixed plastics into fuel oil. About thirty companies introduced such plants since 2000, when collection and recycling of mixed plastics started under the recycling law. There were three large-sized plants (20 to 40 ton/day) and the others were small-sized plants (1 to 1.5 ton/day). After ten years practice, all plants were shut down. Fuel oil production in the small scale costs high, and the large scale production also had a difficulty to collect such a large amount of waste plastics under the gate fee competition [43]. On the other hand, middle-sized plants of about 3 to 6 ton/day capacity are commercially operated for pyrolysis of industrial wastes. Recently, a waste management company in Fukuoka started new middle-sized plant for fuel oil production from separated waste plastics from the municipalities nearby in order to supply the fuel to public facilities.

3.4. Gaseous fuel

Gaseous products from pyrolysis of waste plastics are categorized into two major types. One is syngas, a mixture of hydrogen and carbon monoxide [44]. Another is gaseous hydrocarbon such as methane and ethylene [45]. Depending on gasification conditions, a mixture of hydrogen and methane will be obtained [46].

Gas composition depends on temperature of a reactor, residence time of decomposing species during gasification of plastics, and other reaction conditions, which are often governed by a reactor structure. Syngas is originally for the production of methanol from hydrogen and carbon monoxide, or ammonia from hydrogen. However, syngas, a mixture of hydrogen and carbon monoxide, is also used as gaseous fuel in some facilities for generating electricity.

Table 15 summarizes the compositions of pyrolysis gas from waste plastics and the gases generating in steel production facilities. Gasification plants in Ciba city produces syngas at the scale of two series of 150 ton/day, and it is used for power generation in the adjacent steel manufacturer through combustion. The two-stage pressurized process in Ube city for mixed plastic wastes under the container and packaging law was already shut down due to the higher operation cost comparing with that of the other methods such as cokes oven treatment.

Gasification technology covers not only waste plastics and the other combustibles in municipal wastes but also biomass and automobile shredder residue (ASR).

There is another type of gasification for the production of gaseous hydrocarbon. Thermal treatment of polyethylene and polypropylene at about 600 °C or above with the residence time around 20 min gives gaseous hydrocarbons mainly. The initial decomposition products

	Heating value High/Low (MJ/Nm^3)	Composition (Volume%)							
		CO	CO$_2$	H$_2$	CH$_4$	C$_2$H$_2$	C$_2$H$_6$	O$_2$	N$_2$
Plant / waste source	Thermoselect method*, Chiba city / Combustibles from municipal waste								
	8.0 / 7.4	32.5	33.8	30.7					2.3
Plant / waste source	Thermoselect method*, Chiba city / Combustibles from industrial wastes								
	9.6 / 8.9	43.1	18.8	32.4					
Plant / waste source	Two-stage pressurized process of EUP**, Ube city / Mixed plastics separated from municipal wastes								
	8.1 / 7.7	30-35	20-25	40-45	0				4-7
Plant / waste source	By-product gases in steel production facilities / Coal								
Cokes oven gas	21.1 / 18.7	6.9	2.4	56.1					3.6
Blast furnace gas	3.1 / 3.1	24.1	20.5	2.7	27.6	2.8	0.4	0.2	52.7
Converter gas	8.2 / 8.2	64.4	15.0	1.8					18.8

*Ref.[47]. **Ref.[48,49].

Table 16. Compositions of pyrolysis gas from waste plastics and the gases in steel production facilities

of plastics are liquid hydrocarbon, and the further heat transfer to the vaporized portions results in the conversion of vaporized hydrocarbons into gaseous hydrocarbons of methane, ethylene, ethane, propylene, propane, and the other gaseous hydrocarbons with the formation of liquid hydrocarbons at about 10 to 20 %. This is still in research stage to demonstration stage [45].

Applying the effective heat transfer of a heating medium in an external-heating rotary kiln, waste polyethylene was gasified in the coprocessing with asbestos-containing waste building material. The resulting flammable gas was used as fuel for a heating gas to melt asbestos(Figure 6) [46].In asbestos removal works, waste polyethylene generates as protective clothing and shielding curtain Addition of a flux to asbestos-containing demolition wastes makes the melting range of asbestos at 800 to 900 °C. At the similar temperature range, polyethylene was readily converted into a mixture of hydrogen, methane, ethylene and other hydrocarbons. The gaseous products were supplied to a furnace to generate heating gas for the pyrolyzer of the asbestos melting system.

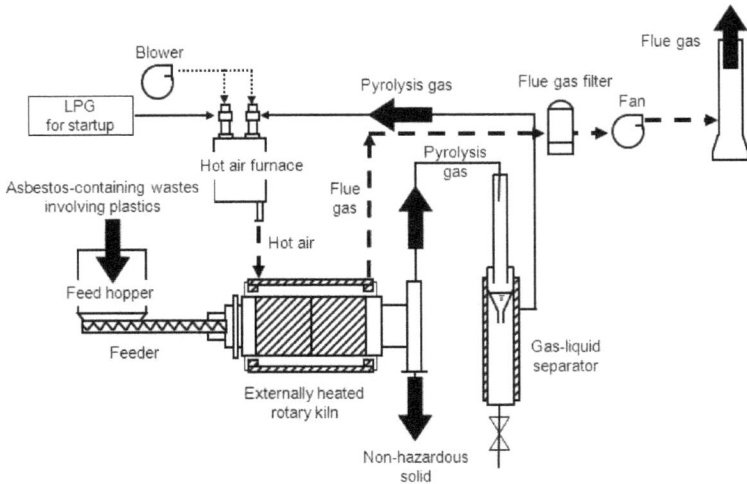

Figure 6. Gaseous fuel production from waste polyethylene in the coprocessing with asbestos-containing demolition wastes

4. Tasks for the future

Dissemination of recycling in the local societies and technology transfer to developing countries are of importance with respect to the promotion of the sustainability of the world. Life cycle assessment has been recognized as an important tool for judging how a certain product or a manufacturing process is green. For the transfer of waste plastics recycling technology, the technology should be socially and economically accepted by the stakeholders in developing countries.

Technology providers should develop the technologies of suitable specifications and easy operation with being aware of the local conditions. For example, a plant size should meet average collection amounts of target components of wastes. Table 16 shows an average generation amounts of waste plastics in each factory.

The average amounts 2.8 and 6.8 ton/day suggest a general idea of the suitable capacity of the equipment for waste treatment. In Japan, strict laws and regulations, a waste management company has to have a special permission in case that a treatment facility or equipment has a capacity larger than 5 ton/day. The permission is given by a prefectural government after strict check of planning, inspection of the entire facility including equipment, buildings and yard. In addition, consensus building with local residents is required as the one of conditions of the permission by the local government. A long history of pollution problems and conflicts between a polluter and the local residents resulted in the severe conditions to both private companies and municipalities including waste management and recycling works.

Category	average amount of generation or treatment of waste plastics ton/day
Chemical industry	5.0
Manufacturer of plastic products	3.1
Manufacturer of rubber products	2.6
Manufacturer of electric machine	1.1
Manufacturer of transportation machine	1.8
Manufacturer of paper and pulp	3.2
Mean value of generation in one factory	2.8
Incineration facility for industrial wastes	3.3
Shredding facility for industrial wastes	6.9
Mean value of treatment in one facility	6.8

Table 17. Waste plastics generation in factories of manufacturing industry and waste management facilities

For waste plastic recycling, there is a limited pattern of successful business. It is a narrow pathway to connect waste plastics with a product of commercial value. The compositions, quantities and qualities of plastics determines a recycling method, system configurations and the business scale, which also leads to the type of the product with a certain commercial value. The product price, number of users and consumption are the important factors for establishing a sound flow of waste plastics recycling.

Technology assessment methodology has been discussed in UNEP [50,51], and its application to recycling technologies for waste plastics will be developed through the discussion among the stakeholders and experts. The promotion of technology transfer is fulfilled through the cooperative efforts among the experts of policies, economics, technologies and the people in local communities. The Global Partnership on Waste Management, an open-ended partnership for everyone was launched in November 2010 [52]. The more discussions and experiences are required to find the effective solutions for managing and recycling wastes to make our society sustainable.

Author details

Yoichi Kodera

Research Institute for Environmental Management Technology, National Institute of Advanced Industrial Science & Technology (AIST) , Tsukuba, Japan

5. References

[1] For a general scope of wastes and policies, see Annual report on the environment and the sound material-cycle society in Japan 2010, Ministry of Environment, 2011. Available: http://www.env.go.jp/en/recycle/smcs/a-rep/2010gs_full.pdf. Accessed 2012 Apr 12.

[2] Waste treatment in Japan, fiscal year 2009 (Heisei 21), Ministry of Environment, March 2011. Available: http://www.env.go.jp/recycle/waste_tech/ippan/h21/data/disposal.pdf (in Japanese). Accessed 2012 Apr 12.

[3] Survey report of the composition of burnable wastes in Chuo incineration plant (FY 2010), Waste Management Federation of Tokyo's 23 Cities. Available:
http://www.union.tokyo23-
seisou.lg.jp/gijutsu/kankyo/toke/nakami/documents/chuo.pdf (in Japanese). Accessed 2012 Apr 12.

[4] Generation and treatment of industrial wastes, fiscal year 2009 (Heisei 21), Ministry of Environment, February 23, 2012. Available:
http://www.env.go.jp/recycle/waste/sangyo/sangyo_h21.pdf (in Japanese). Accessed 2012 Apr 12.

[5] For English information of the macro flow of plastics in 2009, See Plastic products, plastic wastes and resource recovery [2009], Plastic Waste Management Institute, 2011. Available: http://www2.pwmi.or.jp/siryo/ei/ei_pdf/ei40.pdf. Accessed 2012 Apr 12.

[6] Plastic products, plastic wastes and resource recovery [2009], Plastic Waste Management Institute 2012. Available:
http://www2.pwmi.or.jp/siryo/flow/flow_pdf/flow2010.pdf (in Japanese). Accessed 2012 Apr 12.

[7] Statistic data of chemical industry, Ministry of Economy, Trade and Industry. Available:
http://www.meti.go.jp/statistics/tyo/seidou/result/ichiran/02_kagaku.html#menu2.
Accessed 2012 Apr 12. See also, Production of synthetic resin, Japan Petrochemical Industry Association. Available: http://www.jpca.or.jp/4stat/02stat/y7gousei.htm (in Japanese). Accessed 2012 Apr 12.

[8] Record of sorted collection and re-merchandizing based on the Act for Promotion of Sorted Collection and Re-Merchandizing of Containers and Packaging," Chapter 26-6 Environment, Disaster and Accidents, Japan Statistical Yearbook, Ministry of Internal Affairs and Communications, 2012. Available:
www.stat.go.jp/data/nenkan/zuhyou/y2606000.xls. Accessed 2012 Apr 12. See also, Calculation of recycling rate, the Council for PET Bottle Recycling. Available:
http://www.petbottle-rec.gr.jp/data/calculate.html (in Japanese). Accessed 2012 Apr 12.

[9] Yearbook of Paper, Printing, Plastics Products and Rubber Products Statistics, pp. 94, Ministry of Economy, Trade and Industry, 2010. Available:

http://www.meti.go.jp/statistics/tyo/seidou/result/gaiyo/resourceData/06_kami/nenpo/h
2dgg2010k.pdf (in English and Japanese). Accessed 2012 Apr 12.

[10] Management report on the Automobile recycling law in fiscal year 2011, Japan
Automobile Recycling Promotion Center. Available:
http://www.jarc.or.jp/automobile/manage/pdf/12_02.pdf (in Japanese). Accessed 2012
Apr 12.

[11] Committee material, Ministry of Economy, Trade and Industry.
http://www.meti.go.jp/report/downloadfiles/g30121b031j.pdf (in Japanese, Accessed
2012 Apr 12.)

[12] For a general review in English, Environmentally conscious design for electric home
appliances in Japan, Association for Electric Home Appliances, March 2012. Available:
http://www.aeha.or.jp/assessment/en/ECD_Mainly_Recycling.pdf. Accessed 2012 Apr
12.

[13] Committee material, the current conditions of recycling home appliances in fiscal year
2006, pp.11, Ministry of Environment. Available:
http://www.env.go.jp/council/03haiki/y0319-05/mat02.pdf (in Japanese). Accessed 2012
Apr 12.

[14] Annual report of the Recycling of Home Appliances Heisei 22 (2010), p.27, Association
for Electric Home Appliances, July 2011. Available:
http://www.aeha.or.jp/recycling_report/pdf/kadennenji22.pdf (in Japanese). Accessed
2012 Apr 12. Accessed 2012 Apr 12.

[15] Aizawa H, Hirai Y, Sakai S. (2009) Recycling of Small Electrical and Electronic
Equipment Waste, J. japan soc. material cycles and waste management, 20: 371-382.
Available:
http://repository.kulib.kyoto-
u.ac.jp/dspace/bitstream/2433/152209/1/haikibutsushigen20_371.pdf   (in   Japanese).
Accessed 2012 Apr 12.

[16] Survey on the horticultural facilities and agricultural waste plastics, Japan Greenhouse
Horticulture Association, April 2011 (in Japanese).

[17] Survey report (Fiscal year of Heisei 16) on disposal and treatment of waste plastics from
industry, Plastic Waste Management Institute, March 2005 (in Japanese).

[18] Survey report (Fiscal year of Heisei 18) on disposal and treatment of waste plastics from
industry, Plastic Waste Management Institute, March 2007 (in Japanese).

[19] Detailed data of contract prices (weighted average) in 2012 (Heisei 24), Japan Containers
and Packaging Recycling Association. Available:
http://www.jcpra.or.jp/recycle/recycling/recycling04/h24a.html (in Japanese). Accessed
2012 Apr 12.

[20] Cost comparison of biogas facility and incineration facility, committee material,
Ministry of Environment, 2006. Available:

http://www.env.go.jp/recycle/waste/conf_raw_g/06/ref01.pdf (in Japanese). Accessed 2012 Apr 12.

[21] Trade Statistics of Japan. Available: http://www.customs.go.jp/toukei/info/tsdl_e.htm. Accessed 2012 Apr 12. The code numbers of waste plastics in the statistic tables are polyethylene 3915.10 000, polystyrene 3915.20 000, PVC 3915.30 000, PET 3915.90 100 and the other plastics 3915.90 900.

[22] State of implementation of legal systems for establishment of a sound material-cycle society, section 3, pp.58-67, Annual report on the environment and the sound material-cycle society in Japan 2010, Ministry of Environment, 2011. Available: http://www.env.go.jp/en/recycle/smcs/a-rep/2010gs_full.pdf. Accessed 2012 Apr 12.

[23] NEDO report, Demonstration research on the collection of hard plastics from waste plastic products in Noshiro city and experimental research on the production of sophisticated recycled products, Recycle One, Inc. and Akita Eco Plash Co., Ltd., February 2011. NEDO management number 20110000000544.

[24] Survey on the environmental impacts of the collection and recycling of resource plastics, pp.61, Minato ward, March, 2010. Available: http://www.city.minato.tokyo.jp/jigyoukeikaku/kurashi/gomi/seso/chosa/documents/ka nkyouhukatyousahonpen.pdf (in Japanese). Accessed 2012 Apr 12.

[25] For example, electrostatic plastics separator by Hitachi Zosen Corporation. See http://nett21.gec.jp/jsim_data/waste/waste_2/html/doc_374.html. Accessed 2012 Apr 12.
See also, Dodbiba G, Shibayama A, Miyazaki T, Fujita T (2002) Electrostatic Separation of the Shredded Plastic Mixtures Using a Tribo-Cyclone. Magnetic and electrical separation. 11:63-92. Available: http://downloads.hindawi.com/archive/2002/096372.pdf. Accessed 2012 Apr 12.

[26] Walters R.N., Hackett S.M., Lyon R.E. (2000) Heat of Combustion of High Temperature Polymers, Fire and materials. 24:245-252. Available: http://www.fire.tc.faa.gov/pdf/chemlab/hoc.pdf. Accessed 2012 Apr 12.

[27] Refuse derived paper and plastics densified fuel, Japanese industrial standard JIS Z7311:2010 (in Japanese).

[28] Densified refuse derived fuel, Japanese industrial standard committee, Technical specification TS Z 0025:2004 (in Japanese).

[29] Kaino K (2007) Guide for annual energy report, On the revision of standard calorific values of various energy sources, pp. 366. Available: http://www.rieti.go.jp/users/kainou-kazunari/download/pdf/2007EBXAN1000.pdf (in Japanese). Accessed 2012 Apr 12.

[30] Annual sales and expected consumption, Japan RPF Association, April 2011 (in Japanese).
See http://www.jrpf.gr.jp/rpf-6.html

[31] Grassie N (1989) Products of thermal degradation of polymers. In: Brandrup J, Immergut E.H., editors. Polymer Handbook, Third Edition. New York: Wiley Interscience. pp. II/365-II/367.

[32] Converting waste plastics into a resource, compendium of technologies, United Nations Environment Programme, 2009. Available: http://www.unep.or.jp/Ietc/Publications/spc/WastePlasticsEST_Compendium.pdf. Accessed 2012 Apr 12.

[33] Pyrolytic oil from waste plastics - boiler fuel and diesel generator fuel, Japanese industrial standard committee, Technical specification TS Z 0025:2004 (in Japanese).

[34] Kodera Y, Ishihara Y, Muto D, Kuroki T (2008) Technical and Economic Studies for the Promotion of Fuel Production through Waste Plastic Recycling. J. jpn soc. of waste management experts. 19:35-43 (in Japanese).

[35] Yamada S, Shimizu M, Miyashi F (2004) Thermoselect Waste Gasification and Reforming Process. JFE technical report No.3: 21-26. Available: http://www.thermoselect.com/news/2004-07%20JFE%20Technical%20Report%20Thermoselect%20Process.pdf. Accessed 2012 Apr 12.

[36] Steiner C, Kameda O, Oshita T, Sato T (2002) EBARA's Fluidized Bed Gasification: Atmospheric 2 x 225 t/d for Shredding Residus Rcycling and Two-stage Pressureized 30 t/d for Ammonia Synthesis from Waste Plastics. The 2nd international symposium on feedstock recycling of plastics. Oostende. September 9. Available: http://www.ebara.ch/downloads/ebara_ISFR02_EBARA_manuscript_Oostend.pdf. Accessed 2012 Apr 12.

[37] Sugiyama H, Imaizumi T, Fukuoka D, Chiba S, Oku S, Sato T (2001) Pressureized Two-Stage Gasification System for Plastics. The Ebara Ube Process (EUP). Ebara engineering review. 193: 45-48.
See also, http://nett21.gec.jp/jsim_data/waste/waste_3/html/doc_453.html (in English). Accessed 2012 Apr 12.

[38] Kodera Y, Ishihara Y, Kuroki T (2006) Novel Process for Recycling Waste Plastics To Fuel Gas Using a Moving-Bed Reactor. Energy and fuels. 20:155-165. Available: http://www.aseanenvironment.info/Abstract/41012780.pdf. Accessed 2012 Apr 12.

[39] Kodera Y, Sakamoto K (2011) Fuel Gas Production and Utilization in Thermal Decomposition of Asbestos and Waste Plastics. pp. P-E-17. Proceedings of ISWA 2011.

[40] Anticipating the environmental effects of technology, UNEP. Available: http://www.unep.or.jp/ietc/publications/integrative/enta/aeet/index.asp. Accessed 2012 Apr 12.

[41] Environmentally sound technologies for sustainable development (draft version), UNEP. Available:

http://www.unep.or.jp/ietc/techtran/focus/sustdev_est_background.pdf. Accessed 2012 Apr 12.

[42] Global Partnership on Waste Management, UNEP. See http://www.unep.org/gpwm/. Accessed 2012 Apr 12.%

Management of Special and Hazardous Wastes

Chapter 12

Agricultural Waste Management Systems and Software Tools

John J. Classen and Harbans Lal

Additional information is available at the end of the chapter

1. Introduction

As the demand for animal products such a milk, meat, etc. has increased, producers have found ways to increase productivity and decrease the unit cost of production. Fossil fuels, inorganic fertilizer, pesticides, improved genetics of production species, better management techniques, and mechanization allowed productivity to increase to meet these demands. This has also meant concentration of more animals at each location. Confining some types of animals to houses or barns through all or most of their life cycle protects them from the weather and from predators and facilitates feeding, animal movement, and materials handling. Producers have benefited from economies of scale and product uniformity to provide the consumer with low-cost, high-quality meat and animal products.

These housing and confinement facilities employ specialized systems for materials handling, feed distribution, and, in the case of dairy, product collection and processing. Because of the large scale of these facilities, specialized waste collection and management systems are required. The manure, litter and process wastewater contains nitrogen, phosphorus, and potassium that are useful to plants if managed properly but, along with other pollutants such as pathogens, metals, and pharmaceuticals, could pollute the environment or harm human health if not handled properly. When properly applied to crop land as fertilizer, nutrients are used by crops, and other materials are generally rendered harmless in the soil. The purpose of waste management is to protect the environment and the public by keeping manure and contaminated waters out of surface and ground water and controlling application of manure nutrients to crop land such that nutrients are available in the right quantity, at the right time and at the right place.

This chapter describes the purpose and design of manure management systems and demonstrates how two software tools (AWM and SPAW) can be used to assist with the design and evaluation process.

2. Manure management systems

Manure is the collection of feces, urine, spilled feed and water from animal production and is collected in different forms, depending on the animal species. Swine and cattle produce thick liquid manure called slurry while manure from broilers and laying hens is much dryer. Storage systems depend on the animal species, how manure is collected, and local practices. Swine and dairy production in some areas collect and store slurry manure in storage ponds or tanks while some systems use liquid from anaerobic treatment lagoons to flush manure from collection pits or alleys. Although beef cattle excrete very wet manure, many of the concentrated feedlots are in dry regions where excessive rainfall and runoff do not create large storage requirements.

Selection of a manure management system is largely up to the producer based on the needs and goals of the individual operation. Slurry can be stored in earthen storage ponds or in above-ground glass-lined steel tanks. Earthen storage ponds can be less expensive to construct than steel tanks but will use more space than a steel tank of the same capacity. Storage tanks are more expensive and require installation by specifically trained teams. Storage tanks are installed with a central drain pipe through which manure can be loaded to slurry wagons or pumped back into the top of the tank for mixing. Such mixing prior to loading gives the applied manure a more consistent nutrient concentration and makes complete emptying and cleaning of the tank easier. A major concern with both ponds and tanks is odor emission, especially during mixing and land application.

Lagoons combine the storage capacity of ponds and tanks with a functional anaerobic treatment capacity [1]. By having a larger structure with more dilute contents, naturally occurring organisms convert manure organic matter to methane and carbon dioxide and transform organic nutrients into plant-available mineral forms [2]. This dilute liquid can be applied to crop land by irrigation, a less expensive and less labor-intensive operation than applying slurry. Drawbacks to lagoons include significant amounts of ammonia volatilization from lagoon surfaces and during spray irrigation, higher construction costs due to the larger size, and the need to irrigate lagoon effluent frequently during the growing season. Effluent irrigation requires a careful balance of preventing ponding on and runoff from the application fields with the proper timing and rate of nutrient application. Although odor reduction is a consequence of anaerobic treatment and odor from lagoon effluent is not as intense as that of slurry manure, odor emissions and neighbor complaints are still problems for producers using lagoon treatment systems [3,4].

Manure management facility design includes consideration of the amount and type of manure, requirements of any treatment system that will be used, any wash water or bedding that is added to the manure, and any limitations on land application such as applying when the ground is not frozen to allow infiltration and only when a crop is actively growing (or within 30 days prior to emergence in some cases). Expected manure mass and volume production of different species are available from the Manure

Production and Characteristics Standard produced by the American Society of Agricultural and Biological Engineers (ASABE) [5]. The data and procedures needed to calculate these volume components are available from the Animal Waste Management Field Handbook published by the United States Department of Agriculture (USDA) Natural Resource Conservation Service (NRCS) [6]. The Animal Waste Management Field Handbook also provides additional information needed for system design such as bedding characteristics and typical use rates, lagoon loading rates, sludge accumulation ratios for different species, expected depth of storms of different intensities, typical wash water requirements, and other guidelines regarding accepted practices of animal waste management. Typical rainfall and evaporation numbers must be obtained from local sources or national databases such as the National Climatic Data Center at the National Oceanic and Atmospheric Administration in the United States. National and state or provincial regulations must also be considered for requirements or limitations on facility siting or design.

Any design calculation or model depends on accurate data and sound procedures for reliable results; the critical data needs for manure management system design are weather data and manure production characteristics. The United States government took on the task of collecting and organizing weather data, assured the quality of these data, and has made these available to the general public. Animal manure production and characteristic data have been collected and verified primarily through university research efforts and reported in peer reviewed journals. These data have been checked for quality and organized by both the ASABE and the NRCS and are proven through frequent user designs and verification.

Collecting and manipulating these data are time consuming and require a great deal of care to obtain an acceptable facility design. Changing that design or comparing two or more designs takes almost as much time as creating the first design. In addition, it is highly recommended that liquid animal waste storage structures – especially ponds and lagoons should be adequately designed and frequently evaluated for the changing operational scenarios and for long term weather conditions. Changes in animal population and management style of a CAFO coupled with unexpected weather event(s) can significantly impact the waste generation and the need for the storage volume even for a well-defined storage period. Design software permits users to carry out such tasks in a fraction of the time.

AWM and SPAW are two such tools that have been developed to facilitate the design process. The capabilities of these tools have been recognized nationally by regulatory and non-regulatory agencies and private entrepreneurs. The United States Environmental Protection Agency (USEPA) CAFO Rules suggest using AWM and SPAW for evaluating liquid animal waste storage structures with respect to adequate storage for various storm events. This chapter describes these two tools and how they can be used for evaluating liquid waste storage structures using a hypothetical example scenario.

3. Animal Waste Management (AWM)

AWM is the NRCS national software tool for designing storage and treatment facilities for liquid and solid animal waste using site-specific characteristics and monthly weather data [7-9]. Key AWM components include: 1) Manure production; 2) Adjustments: bedding, wash water, runoff; 3) Withdrawals; 4) Storage Design/Analysis (Figure 1).

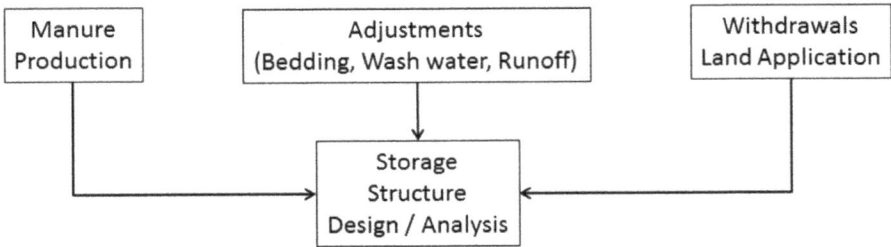

Figure 1. Key components of the AWM (Animal Waste Management) Software

The AWM production component estimates the quantity of manure based upon the type and size of the animal herd. The adjustments component adjusts the quantity and consistency of manure due to wash water, flush water, bedding etc. It also accounts for the amount of contaminated runoff from the pervious and impervious areas from the normal rainfall and design storm events (25 year-24 hour return period) contributing to the waste stream from the operation. It uses a database of monthly rainfall averages for the specific location in estimating runoff from the contributing area(s) but the rainfall amounts can be modified if more accurate values are available to the user.

The storage facility design is the key AWM output based upon the waste produced and withdrawal schedule for on-farm uses or off-farm disposal to meet regulatory requirements which are generally defined by months of waste removal or by the storage period.

The AWM software has the capability to evaluate existing waste storage structures in addition to designing new facilities. This feature permits evaluating dimensional adequacy for storing the waste flowing into the structure for the defined storage period or for the withdrawal schedule and can be used to determine if planned expansion of an animal operation will require additional storage capacity. The waste flowing into the structure includes: animal manure with additions, normal runoff, and the runoff from the 25 year-24 hour storm events, if applicable. Figure 2 shows the screen shot of the Evaluate function of AWM for a dairy operation used as an example in this chapter.

The AWM generates standard and customized reports that document the information furnished by the user, design and evaluation features of storage facilities and an estimate of land area needed to effectively utilize the available nutrients (N, P, and K) for the current or proposed cropping systems.

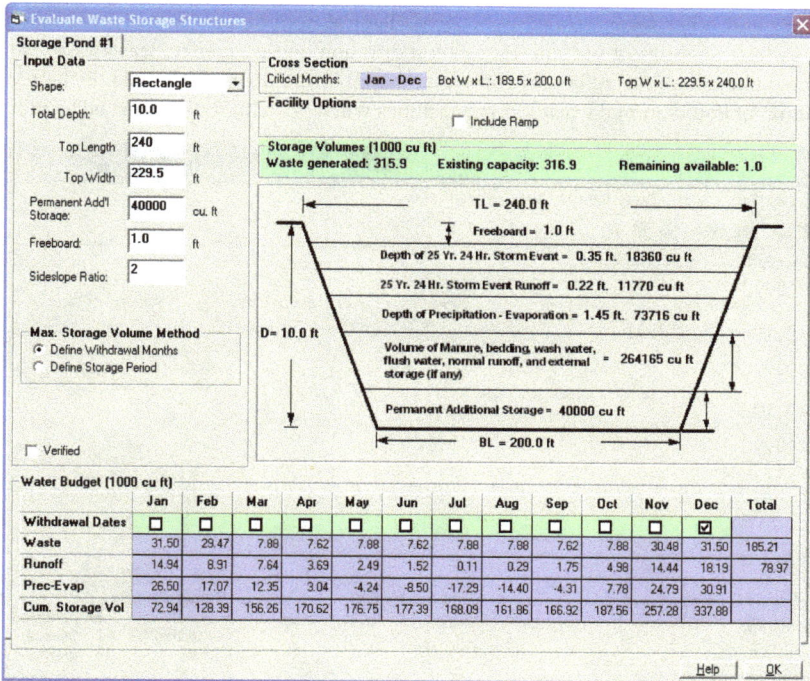

Figure 2. The "Eval" Screen of AWM for the example pond. This function evaluates an existing storage pond design against information about the farm operation. Note that in this example only a small storage volume is considered available.

4. Soil-Plant-Air-Water (SPAW)

SPAW (**S**oil-**P**lant-**A**ir-**W**ater) is a hydrologic water budget model that consists of two main connected routines: one for farm fields and the second for impoundments such as wetland ponds, lagoons or reservoirs [10, 11, 12].

The "Field" module of the SPAW simulates daily vertical, one-dimensional water budget depth of all major hydrologic processes such as runoff, infiltration, evapo-transpiration, and percolation occurring on a field. Input to this budget include: 1) daily rainfall, temperature and evaporation; 2) a soil profile of interacting layers, each with unique water holding characteristics; 3) annual crop growth and target yields with management options for rotations, irrigation and fertilization. The volumes for different water budget components are estimated by multiplying the component depth by the associated field area.

The "Pond" module simulates the hydrology of impoundments such as wetland ponds, lagoons or reservoirs (Figure 3). These simulations are based upon multiple input sources and depletion processes affecting the impoundment such as runoff from agricultural fields, irrigation and other water related operations. Typical applications of a SPAW simulation could include analyses of wetland inundation duration and frequency, wastewater storage design evaluation, and reliability of water supply reservoirs [11]. This is helpful in evaluation of liquid animal waste storage facilities when open lot runoff must be included.

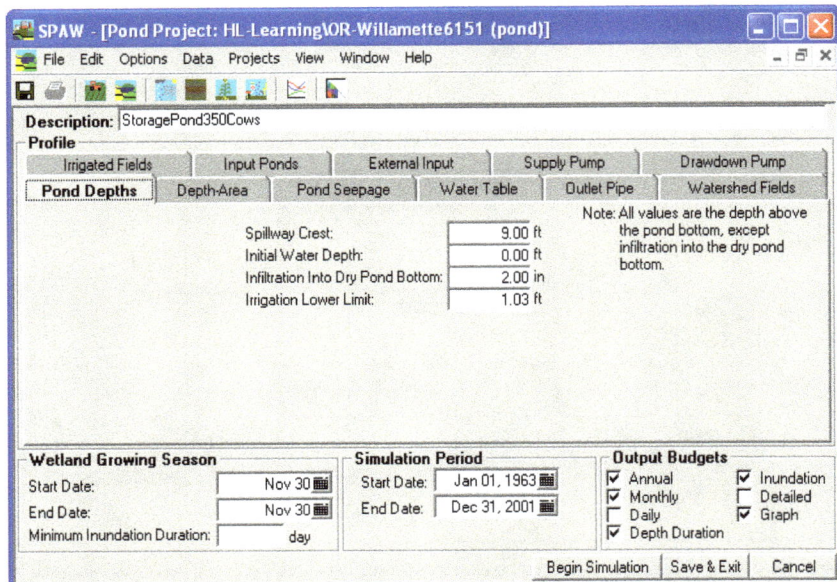

Figure 3. SPAW Screen for pond evaluation project

In this chapter AWM and SPAW have been used to evaluate an existing waste storage structure on a dairy farm (referred to as An Example Dairy). The AWM verifies the design parameters of the structure and the SPAW evaluates the operational characteristics of the structure.

5. An example dairy operation

The example is a hypothetical dairy operation located in Clackamas County, Oregon close to the N. Willamette Ext STN OR6151 weather station. The operation has a capacity to milk 350 animals and includes dry cows, heifers and calves as shown in Table 1. Animal barns have a total of 3,730 ft² of roof area and the animals have access to a 1-acre un-surfaced open exercise lot except during the coldest part of the winter. This farm follows typical practices of housing the calves in the barn at all times but the larger animals have access to the pasture during the warmer months of the year but are kept in the freestall barn during the months of November to February (Table 2). In this example all the waste collected from the operation is directed to a storage pond which is designed for a storage period of 12 months and the pond is emptied in the month of December.

Animal Type	Animal	Quantity	Body Weight (lbs)
Dairy	Calf	90	150
Dairy	Dry Cow	50	1000
Dairy	Heifer	12	900
Dairy	Milker 60 Lbs	350	1000

Table 1. Animal herd size and characteristics of the example dairy.

Location	Calf	Dry Cows	Heifer	Milker (60 lb Milk)
1ˢᵗ *Operating Period (November – February)*				
Freestall Barn	100	100	100	80
Milking Parlor	0	0	0	20
Pasture	0	0	0	0
2ⁿᵈ *Operating Period (March – October)*				
Freestall Barn	100	0	0	0
Milking Parlor	0	0	0	20
Pasture	0	100	100	80

Table 2. Percentage of manure deposited by animals at different locations during two operating periods.

The as-built dimensions of the storage pond are shown in Table 3. It stores the waste generated from the operation and also the runoff from a pervious watershed of 1 acre and an impervious area of 3,730 (0.09 acres) sq. ft of roofs, slabs and walkways.

Key Parameter (s)	Units	Value(s)
Top Length × Width	ft × ft	240 × 229.5
Side-slope ratio	ft/ft	2
Bottom Length × Width	ft × ft	200 × 189.5
Total Depth	ft	10
Freeboard	ft	1
Permanent Storage Volume	ft³	40,000
Storage Period	months	12
Withdrawal Month		December

Table 3. Key characteristics of the existing storage pond for the example farm

6. Evaluating the design with AWM

To evaluate the storage structure on the example farm, we entered and/or selected all the operational characteristics of the farm in the AWM as described by Lal [9]. With the complete set of AWM inputs, the evaluation function becomes active and shows the existing capacity is 1,000 ft³ larger than the waste generated (see the green strip just above the pond graphic, Figure 2). This indicates that the storage pond is appropriately designed and has enough capacity to store the waste based on the waste generated and monthly averages of rainfall and evaporation. The user can also produce a separate report (Figure 4) with these parameters for inclusion in the overall pond report or as a component of a Comprehensive Nutrient Management Plan (CNMP) report.

AWM uses long term monthly rainfall averages and provides estimate of the storage volume required for the critical months [7]. AWM cannot evaluate the impact of what is known as "chronic" rainfall events, or a series of rainfall events that in total exceed the depth of a 25 year – 24 hour rainfall event. For such an evaluation, the SPAW model, which is based upon a daily time step of historical rainfall, is considered more appropriate and is discussed in the next section.

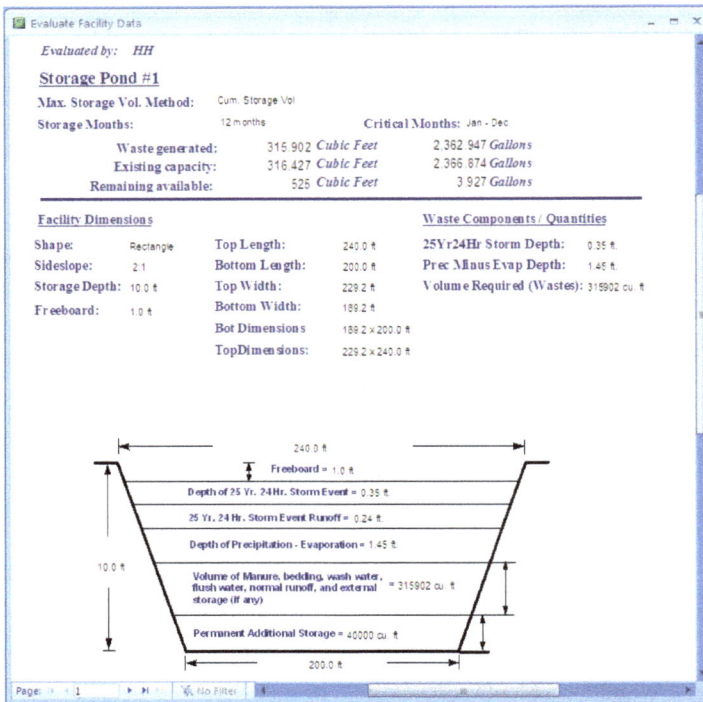

Figure 4. The pond evaluation report by AWM

7. Evaluating the design with SPAW

The hydrological analysis of a pond using SPAW is carried out in two linked simulations: the "Field" simulation and the "Pond" simulation. The Field simulation is performed to estimate the runoff that will be generated from the field and is routed to the pond and/or for the application of irrigation of the pond water to the field. A field represents any surface receiving rainfall that can be modeled by the SCS Runoff Curve Number hydrology method. Fields can be vegetated areas (defined as pervious watershed in AWM) or impervious areas such as rock outcrops, roofs, parking lots, and feedlots. Fields are further characterized by their soil type, crop, and management.

The "Pond" hydrology is simulated by accounting for the water entering and exiting the pond. Input sources include runoff, rainfall, evaporation, pumped inflows, and outflows such as irrigation and seepage. The user defines one or more fields that supply runoff to the "pond" based upon the site-specific climatic data for the simulation period. The user also defines the rules for wastewater withdrawals with an irrigation schedule and/or pumping rates and durations.

In the example scenario, there are three sources of wastewater to the storage pond: 1) runoff from the pervious area of 1 acre, 2) runoff from the impervious area of 3730 sq. ft (0.09 acre), and the waste from the waste producing locations such as free stall barn and milking parlor. Waste flow includes manure, flush water, wash water, etc. which are accounted as an "External Input" for SPAW pond simulation. A depth/volume curve for the example pond is created and entered in SPAW. Other information such as spillway crest height and lower limit for withdrawal are defined which were set respectively at 9 ft and 1.03 ft in the example. The spillway crest limit is found by subtracting the freeboard (1 ft.) from the total depth of 10 ft. The lower limit 1.03 ft the depth of the minimum treatment volume.

SPAW simulates removal of water/waste from the storage pond at scheduled irrigations and/or withdrawal events. The withdrawal is specified by the dates and rates of withdrawal. SPAW removes wastewater from the pond on the scheduled date regardless of the current climate conditions. The withdrawal (drawdown) removes liquids from the pond defined by the upper and lower limits and creates storage space for the future rainfall events. The two limits define the total volume of water to be removed during the drawdown period. If the upper limit is set to "zero", the SPAW starts pumping on the start date irrespective of depth of water in the pond. Otherwise, it starts at the next scheduled irrigation event or between the start date and end date whenever the pond depth exceeds the upper limit and it runs until either the end date is reached or the pond depth drops below the lower limit.

Table 4 shows the mapping of AWM parameters to SPAW inputs and also sources of additional data for SPAW simulations. These include: the location file (daily climatic data, monthly evaporation values), soils, pond volume/surface area relationship, external input of waste flow, and the pond depths including spillway and irrigation lower limit depth.

Parameter	AWM Value	SPAW Value
Location file Precipitation Evaporation	Built-in Monthly Averages Built-in Monthly Averages	Daily precipitation file from either of the following two source USDA/NRCS High Resolution Climate Extractor website (http://199.133.175.81/HCEWebT/) USDA/NRCS Water & Climate Center fttp site (ftp://ftp.wcc.nrcs.usda.gov/support/climate/daily-data/) Transferred from AWM monthly averages
Soil	Not Required	Layered information on Soil Texture (percentage Sand, Silt, Clay) % Organic Matter (OM) % Gravel Bulk Density USDA/NRCS Web Soil Survey Website (http://websoilsurvey.nrcs.usda.gov/app/WebSoilSurvey.aspx) (Also special SQL query for accessing attribute data at http://sdmdataaccess.nrcs.usda.gov/Query.aspx)
Runoff 1.Pervious Surface 2. Impervious Surface	Area (1 acre) and SCS curve number of 90 Area (3730 sq. ft) and SCS curve number of 98	Management files with appropriate selection of cropping, irrigation and fertilizer files both for pervious and impervious layer
Pond Depth	Total Depth = 10 ft Free Board = 1 ft Permanent Additional Volume (40,000 cu. ft) = 1.03 ft Waste Volume (264, 165 cu. ft) = 5.95 ft Preci-Evap (73,703 cu. ft) = 1.45 ft On-site and runoff from 25 yr-24 Hr Storm (30,126 cu. ft) = 0.56 ft	Spillway Crest = 9 ft Irrigation Lower Limit = 1.03 ft

Pond Dimensions	Top Length * Width (240*229.5 ft) Bottom Length *Width (200*1189.5) Side Slope Ratio = 2:1			
		Depth (ft)	Area (Ac)	Volume (ac-ft)
		0	0.87	0.00
		1	0.91	0.89
		3	0.99	2.78
		5	1.06	4.82
		7	1.14	7.02
		9	1.22	9.38
		10	1.26	10.62
Pond Withdrawal	Emptying once every year in the month of December	Drawdown to the Lower limit (1.03 ft) every December irrespective the upper level (set upper Level = 0)		
Waste Flow	Internally estimated based upon the herd size, bedding, waste water and flush water, etc. (values from the AWM Eval Screen)	Calculated based upon the monthly waste flow estimated in AWM pumping rate of 15 gal/min (equal to a Nelson Model 70 sprinkler with 5/16 inch nozzle at 30 psi)		
	Mon / Waste (1,000 cu. ft)	Start Date	End Date	Duration (hrs/day)
	Jan — 31.50	Jan 1	Jan 31	9
	Feb — 29.47	Feb 1	Feb 29	8
	Mar — 7.88	Mar 1	Mar 31	2
	Apr — 7.62	Apr 1	Apr 30	2
	May — 7.88	May 1	May 31	2
	Jun — 7.62	Jun 1	Jun 30	2
	Jul — 7.88	Jul 1	Jul 31	3
	Aug — 7.88	Aug1	Aug 31	2
	Sep — 7.62	Sep 1	Sep 30	2
	Oct — 7.88	Oct 1	Oct 31	2
	Nov — 30.48	Nov1	Nov 30	9
	Dec — 31.50	Dec 1	Dec 31	9

Table 4. Mapping of AWM data value to SPAW input parameters and also source of additional information needed for running SPAW model (Numerical values shown are for the example dairy)

Once all the input data files have been created, simulations in SPAW are run initially for the two fields (pervious surface and impervious surface) and then for the pond by selecting the appropriate project files. The field simulations are made prior to the pond simulation because they provide runoff and climate data for the pond simulation. The simulation period (Jan 01, 1963 to Dec. 31, 2001) is selected within the dates of available climatic data with appropriate selection of output files. Each simulation generates the specified output files which can be viewed in a tabular or graphic form for analysis. Budget summaries for annual, monthly and daily time periods are provided. Average data for each time period (annual, monthly or daily) are shown at the end of each summary table. The graphical routine provides a visual representation of the daily hydrologic values within the field and pond budgets. Daily and cumulative values for most variables are selectable. The pond graph is similar to that of the field of both daily and cumulative variables over each calendar year. The flexibility in selecting the time-period for displaying graphs is a unique SPAW feature. It enables analyzing SPAW results over a variable time period by months (1-24) and years. The graphs can be saved using the "File/Save As" option.

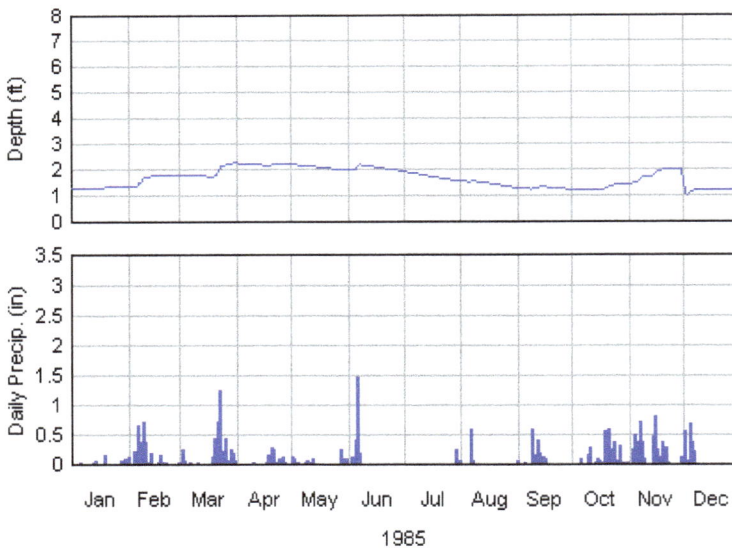

Figure 5. Pond storage depth and daily precipitation for the example dairy for 1985 -- the driest year of the simulation period. Annual precipitation was 26.45 inches with impervious layer runoff from 3730 ft².

The primary objective of the SPAW example simulation is to evaluate the daily variation of the pond's depth during the simulation period. Figures 5 and 6 present the daily precipitation and pond depth for the two extreme weather years (1985 and 1996), respectively. The year of 1985 was the driest year with the total annual precipitation of 26.4

inches. On the other hand, 1996 was the wettest year with the precipitation of 72.94 inches. During the dry year of 1985 the maximum depth of wastewater in the pond was 2.26 ft on Mar. 30th as compared to 7.20 ft during the wet year of 1996. It was satisfying to note that the storage pond designed and evaluated as being "satisfactory" using AWM was confirmed through SPAW simulation to be able to withstand the waste flow during one of the wettest years on record.

To validate the sensitivity of the SPAW model to input changes, another simulation was made by changing the area of impervious layer considerably from 0.09 acre to 1 acre. Figures 7 and 8 show the daily storage depth and spillway volume from the smaller impervious area for the same two extreme years as shown above. The maximum depth with the increased impervious area increased to 3.37 ft compared to 2.26 ft for the dry year (1985) and to 9 ft for wet year 1996. As the spillway crest was set at 9 ft (Figure 3), the pond started flowing soon after the pond depth reached 9 ft on Nov 17 and continued until Nov. 30th with the total spill amount of 1.16 acre-ft. This shows that the SPAW model is sensitive to the runoff generating areas which are clearly reflected in the variation of the waste storage depths in the pond. In the event a producer wanted to implement a modification that resulted in such an increase in impervious area, additional pond capacity would be required; the calculation and design of that capacity would easily be completed with the AWM software.

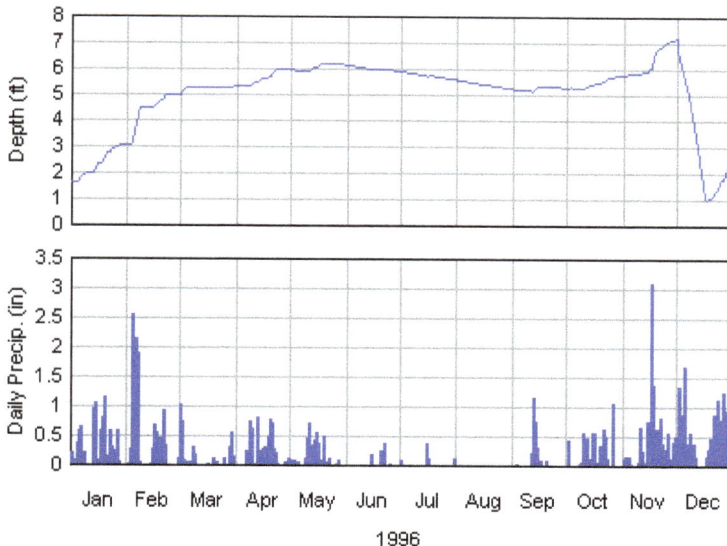

Figure 6. Pond storage depth and daily precipitation for the example dairy for 1996 -- the wettest year of the simulation period. Annual precipitation was 74.94 inches with impervious surface runoff area of 3730 ft².

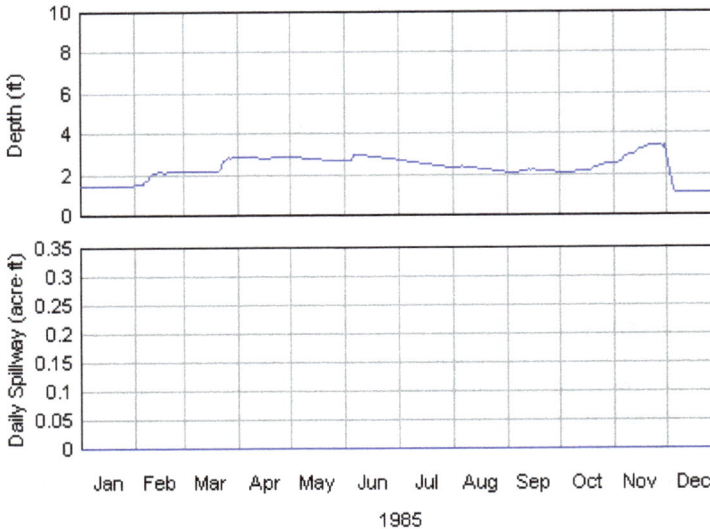

Figure 7. Storage pond depth and spillway dicharge during 1985 – the driest year with total precipitation of 26.45 inches from the enlarged impervious surface area of 1 acre. Please note there was no spillway discharge during the year.

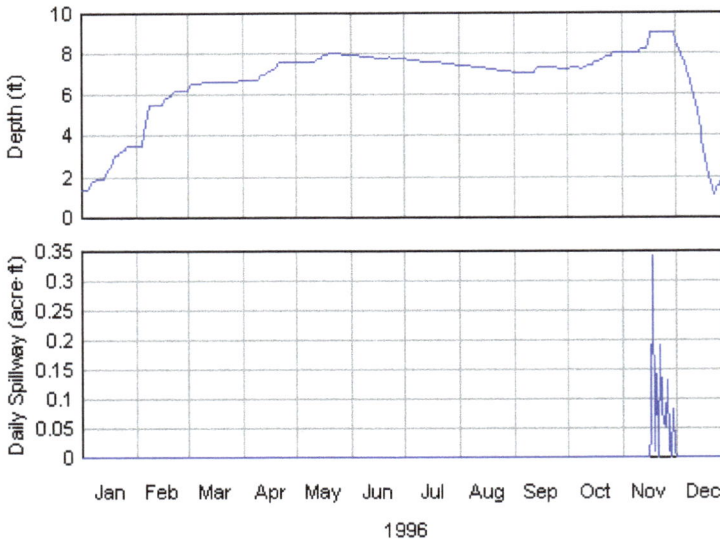

Figure 8. Storage pond depth and spillway discharge volume during 1996 – the wettest year with total precipitation of 72.94 inches from the enlarged impervious surface area of 1 acre.

8. Concluding remarks

This chapter described the basics of animal production systems as related to liquid animal waste storage. Design of storage and treatment facilities on the farm requires accurate data about the manure production characteristics, the size and operation of the farm, local weather conditions, and regulatory requirements and procedures. Liquid storage structures must provide sufficient capacity for manure, wash water, bedding, contaminated runoff and any additional inputs for that period of time during which land application operations do not normally occur. For some operations, that period may be a year and for others, it may be only several months. Local conditions and practices are critical to a successful design.

It was successfully demonstrated how two engineering software packages, namely AWM (Animal Waste Management) and SPAW (Soil-Plant-Air-Water), supported by the NRCS can be used to design and evaluate waste storage ponds and treatment lagoons. They serve evaluation processes that are complementary to each other. AWM evaluates the design of the storage pond while SPAW evaluates the operation of the pond, identifying how the pond will behave for extreme events. The test example demonstrated the pond designed and successfully evaluated by AWM was also able to withstand the wettest year (1996) when evaluated using SPAW model. However, when the impervious surface area contributing runoff to the pond was increased to 1 acre from 0.09 acre, the pond reached the maximum storage height of 9 ft on Nov. 17 and spilled continuously till Nov. 30th.

Author details

John J. Classen
Biological and Agricultural Engineering Department, North Carolina State University, Raleigh, NC, USA

Harbans Lal
National Water Quality and Quantity Team, NRCS/USDA, Portland, OR, USA

9. References

[1] Bicudo JR, Westerman PW, Safley LMJ. (1999) Nutrient content and sludge volumes in single-cell recycle anaerobic swine lagoons in North Carolina. Trans ASAE ;42: 1087-1093.

[2] Burns JC, King LD, Westerman PW. (1990) Long term swine lagoon effluent applications on to coastal bermudagrass: I. yield, quality, and elemental removal. Journal of environmental quality ;19: 749-756.

[3] Lim T, Heber, A J, Ni, J, Sutton, A L, Shao, P. (2003) Odor and Gas Release from Anaerobic Treatment Lagoons for Swine Manure. J Environ Qual /3;32(2):406-416.

[4] Mukhtar S, Borhan MS, Rahman S, Zhu J. (2010) Evaluation of a field-scale surface aeration system in an anaerobic poultry lagoon. Appl eng agric ;26: 307-318.

[5] ASAE. (MAR2005) D384.2 Manure Production and Characteristics. ASAE: St. Joseph, MI.

[6] USDA NRCS. National Engineering Handbook - Part 651 - Agricultural Waste Management Field Handbook. United States Department of Agriculture: Washington, D.C.

[7] Moffitt, DC, Wilson, B, and Wiley, P. (2003) Evaluating the design and management of waste storage ponds receiving lot runoff. ; ASAE Paper 034129; St. Joseph, Mich.: ASAE.

[8] Moffitt, DC, and Wilson, B. (2004) Evaluating the design and management of waste storage ponds - Part II. ; ASAE Paper 044072; St. Joseph, Mich.: ASAE.

[9] Lal H. (2011) Animal Waste Management (AWM) System New and Improved - A tool for evaluating and designing waste storage facilities. Resource May/June 2011;18(3):16-19.

[10] Saxton K.E. (2002) SPAW Soil-Plant-Atmosphere-Water Field and Pond Hydrology Model. United States Department of Agriculture-Agricultural Research Service: .

[11] Saxton, KE, Wiley, P, and Rawls, WJ. (2006) Field and pond hydrologic analysis with the SPAW model. ; ASAE Paper 062108; St. Joseph, Mich.: ASAE.

[12] Lal, H, Wiley, P, and Khanal, P. 2011. Using AWM and SPAW for Evaluating Animal Waste Storage Structures, presented at the Southern Regional Water Conference, USDA, Athens, GA, September 13-16, 2011.

Chapter 13

Greenhouse Gas Emissions from Housing and Manure Management Systems at Confined Livestock Operations

Md Saidul Borhan, Saqib Mukhtar, Sergio Capareda and Shafiqur Rahman

Additional information is available at the end of the chapter

1. Introduction

As the name implies, the gases that assist in capturing heat in the atmosphere are termed as greenhouse gases (GHGs). The continuously rising concentrations of these gasses are believed to work against nature's natural process, trapping more heat than what is needed leading to an increase of earth's climate temperature. Livestock production operations contribute both directly and indirectly to climate change through the emissions of greenhouse gases such as carbon dioxide (CO_2), methane (CH_4) and nitrous oxide (N_2O). Generally, swine and ruminant livestock operations, especially dairy cows and beef cattle, contribute to the production of GHGs mainly CH_4, N_2O, and CO_2 in the environment. The CH_4, CO_2, and N_2O are considered as direct greenhouse gases. The indirect GHGs include carbon monoxide (CO), oxides of nitrogen (NOx), and non-methane volatile organic compound (NMVOCs). Characterization and quantification of N_2O and CH_4 emitted from livestock operations are important because these gases are believed to play a major role in the increase of Earth's temperature. During the last two hundred and fifty years, anthropogenic activities, including demanding agricultural production, have increased the global atmospheric concentration of GHG, namely CO_2, CH_4, and N_2O by 36, 148, and 18%, respectively [1]. Total greenhouse gas (GHG) emissions in the US increased by 14.7% from 1990 to 2006. All agricultural sources combined were estimated to have generated 454 Tg (10^{12}g) of CO_2 equivalents in the U.S. during 2006 [2]. The CH_4 emissions from enteric fermentation and manure management represent about 25 and 8% of the total CH_4 emissions from anthropogenic activities. The US Environmental Protection Agency (USEPA), Inventory of U.S. Greenhouse Gas Emissions, and Sinks identified manure management as generating 24 and 5% of CH_4 and N_2O emissions, respectively, from agricultural sources [2-3]. The USEPA has begun to consider regulating GHGs emitted by the stationary sources, including manure management from CLOs. Thus, it is essential to obtain accurate estimates of GHG emissions from various ground level area sources (barns/housings, lagoons, pens, settling basins, silage piles, pasturelands, etc.) within CLOs to improve emissions inventories and to devise source-specific abatement strategies. In this chapter, GHG emission sources, emissions process, measurement methods and gas sampling protocol, and migration strategies including air scrubbing technology, biofilters, and best manure management practices in the context of livestock waste management were reviewed and discussed.

1.1. Sources of GHG in CLOs

Main sources of pollutant gases are broadly classified as natural (geogenic and biogenic) and anthropogenic. The anthropogenic sources again can be divided into mobile (vehicle, ships, trains, etc.) and stationary (power plants, chemical industries, refineries, intensive land uses, confined animal operations, etc.). Biogenic sources of GHGs, such as those contained in grass, hay, silage, and grains are a major part of bovine diets and are emitted from these biogenic sources during fermentation of starches, lipids, and proteins in the digestive system of cattle (enteric fermentation) and later in the feces and urine. Tables 1 and 2 describe the salient features of the characteristics of manure voided by the animal at CLOs. Ruminant livestock is the principal source of enteric methane emissions to the atmosphere, while manure management such as manure storage and treatment are the most important sources of CH_4 and N_2O emissions [4]. Globally, CH_4 is contributing 22%, and N_2O is contributing 6% of the total GHG. Enteric CH_4 is produced as a waste product of this fermentation process in the rumen. Figure 1 describes the number of factors affecting CH_4 production from rumen.

GHG emissions from livestock vary by animal type and growth stage due to different diets, feed conversion mechanisms, and the manure management [5]. Methane is produced by the microbes in the stomach of ruminants due to enteric fermentation, from freshly deposited manure due to bacterial degradation of organic matter, and from storage lagoons and settling basins due to anaerobic degradation of volatile solids by bacteria. Methane, with a global warming potential (GWP) of 21, can affect climate directly through its interaction with long-wave infrared energy and indirectly through atmospheric oxidation reactions. Methane is second in rank to CO_2 in importance and contributes around 18% of the overall greenhouse effect [6]. Table 3 describes the salient features of the three major GHGs. In addition to the anaerobic degradation of the organic materials, CO_2 is released from the use of fertilizers in crop/pasture production, fossil fuel used to run farm machinery (tractors, loaders, and irrigation pumps) and feed processing operations, the loss of tree for crop production on land adjacent to CLOs, and carbon loss from the soil for feed production.

Nitrous oxide is a GHG that contributes to stratospheric ozone depletion and is 310 times more potent as a GHG than CO_2. Nitrous oxide emissions are associated with manure management and the application and deposition of manure in crop/ pasture land. Indirect N_2O emissions from livestock production include emissions from fertilizer use for feed production, emissions from leguminous feed-crops, and emissions from aquatic sources following fertilizer application. Nitrous oxide is produced in soils through microbial processes of nitrification and denitrification and is released from manure and urine excreta, fertilizer and manure slurry applied for feed-crop production, dry manure piles and aerobic and anaerobic degradation of livestock manure/wastewater in lagoons. The amount of these gaseous emissions from livestock vary by animal type and growth stage due to different diets, daily feed intake, and quality of diet feed conversion mechanism, while GHG emissions from storage and treatment of manure depend on the type of storage, duration of storage, ambient temperature, and manure management practices.

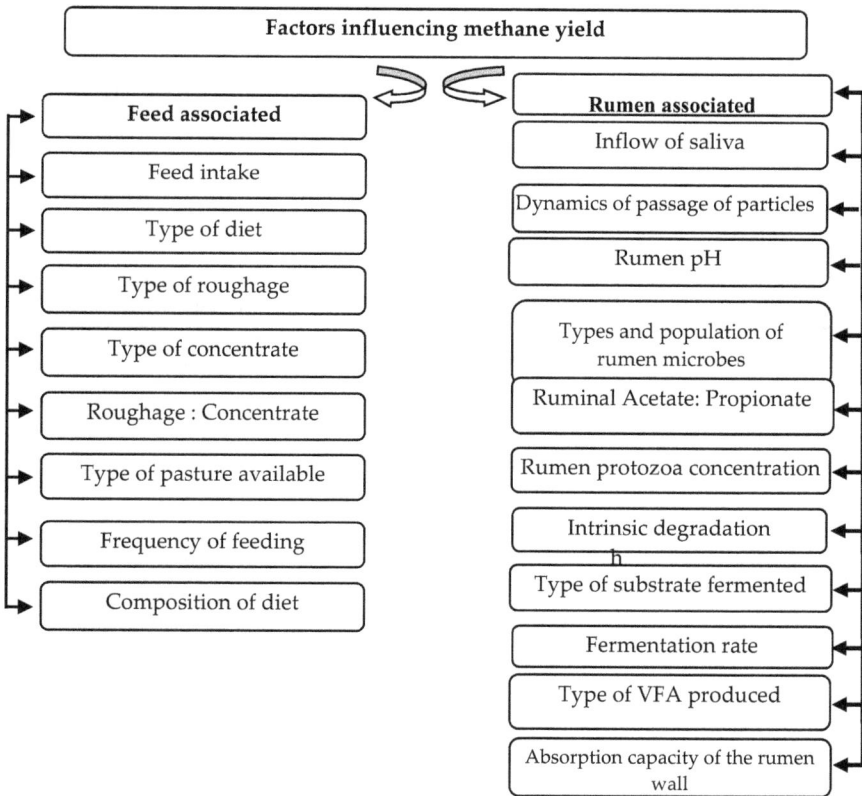

Figure 1. Factors affecting methane production [7].

Animal Type	Average weight (pound)	Days on feed	Total solids (TS)	Volatile solids (VS)	N	P	P_2O_5	K	K_2O	Manure	Moisture
			Pounds per day per animal on an "as excreted" basis								%, wb[1]
Cattle											
Cows/ Heifers **	1000	365	9.5	8.1	0.36	0.048	0.11	0.23	0.28	82	88
Finishing	1200	153	5.1	4.2	0.36	0.048	0.11	0.248	0.30	64	92
Bulls **	1100	365	6.2	5.7	0.54	0.092	0.21	0.267	0.32	80	92
Calves**	450	210	3.4	2.9	0.14	0.044	0.10	0.092	0.11	26	92
Dairy-Milk cows[1]	1300	365	18	15.3	0.92	0.16	0.36	0.44	0.53	141	87
Swine											
Nursery	27.5	36	0.3	0.2	0.025	0.0042	0.01	0.01	0.01	2.4	90
Finishing	154	120	1.0	0.8	0.083	0.0142	0.03	0.037	0.04	10	90
Gestating	440	365	1.1	1.0	0.071	0.02	0.05	0.048	0.06	11	90
Lactating	423	365	2.5	2.3	0.19	0.055	0.13	0.12	0.14	25	90
Sheep **	100	365	1.1	0.9	0.04	0.009	0.02	0.03	0.04	4	75
Poultry											
Layers	3	365	0.05	0.04	0.0035	0.0011	0.003	0.0013	0.002	0.19	75
Broilers	2.8	48	0.06	0.04	0.0025	0.00073	0.002	0.0014	0.002	0.23	74
Turkeys[2]	25	140	0.12	0.1	0.0072	0.00212	0.005	0.0033	0.004	0.47	74
Litter[3]			2	1.6	0.089	0.038	0.086	0.049	0.059	2.5	21
Horse[4]	1100	365	8.5	6.7	0.27	0.05	0.12	0.14	0.16	56.5	85

*ASAE standard D384.2 2005. Manure Production and Characteristics. ASABE, St. Joseph, MI 49085-9659
**Manure Characteristics. 2000. Mid West Plan Service, Ames, IA 50011-3080.MWPS-18,Section I.
[n]Milk vow data on TS,N,P,K and manure provided by Dr. Tamilee Nenich, TCE Dairy Specialist. Volatile solids (VS) estimated to be 85% of TS.
[1]%, wb = percent wt basis.
[2]Days on feed data from "economic Impact of the Texas Poultry Industry," 2004, TCE publication, L-5214.Average weight, TS,VS,N,P,K and total manure averaged from data for female and male turkeys.
[3]Poultry Waste Management Handbook," 1999. Natural Resource, Agriculture, and Engineering Service. Ithaca, NY 14853-5701. NARES-132. Pounds of whole poultry litter (as removed from production houses) per broiler sold. N,P and K values in pounds per 2.5 pounds of litter.
[4]Average weight, TS,VS,N,P,K and total manure averaged from data for sedentary and intense exercise horses.

Table 1. Animal manure production and characteristics [8].*

Animal type	Year*	Animal numbers** (thousands)	Total manure	TS	VS	N	P	P₂O₅	K	K₂O	Total energy***
			Thousands of tons per year on an as excreted" basis								BTU x 10¹² Tera BTUs
Cattle											
Cows/Heifers		5780	86498	10021	8544	380	51	116	243	291	145
Finishing		5520	27026	2149	1764	152	20	46	105	126	30
Bulls	2006	370	5402	419	385	36	6	14	18	22	6.43
Calves		2430	6634	868	740	36	11	26	23	28	12.6
Milk Cows		334	8581	1096	936	56	9.7	22.3	32	32	15.8
Swine											
Nursery		270	12	1.36	1.16	0.12	0.02	0.05	0.05	0.06	0.02
Finishing	2006	565	339	34	28	2.81	0.48	1.10	1.24	1.49	0.48
Other¹		95	312	31	29	2.26	0.65	1.49	1.46	1.75	0.48
Sleep & Goats²	**2006**	**2140**	**1759**	**484**	**400**	**18**	**4**	**9**	**13**	**16**	**6.8**
Poultry											
Layers³	2005	18688	648	167	123	12	4	9	4	5	2.1
Broilers	2005	627900	3451	874	659	38	11	25	21	26	11.2
Turkeys⁴	2004	14100	468	120	96	7	2	5	3	4	1.63
Litter			785	622	494	28	12	27	15.4	18	8.4
Horses⁵	**1998**	**1067**	**11000**	**1655**	**1304**	**53**	**10**	**23**	**26**	**32**	**22.2**

*Year of estimated total population or production data from National Agricultural Statistical Services
** Animals finished or on feed per year.

*** Dry and ash free basis
¹Includes all hogs other than nursery and grow-finish. Estimates based on average nutrient data from gestating lactating sows in Table I.
²Includes sheep and goats. Manure and nutrient totals calculated using sheep data only.
³Include hens and pullets of egg-laying age.
⁴animal numbers for turkey estimated from difference between total turkey and broiler population in Texas (615.6 million from TCE publication L-5214) and National Agricultural Statistical Service estimated number of broilers (601.5 million in 2004).
⁵ Animal numbers for horses adopted form Texas Horse Industry Report, 1998, and from the Texas Horse Industry Quality Audit initiative, TCE, January 1998.

Table 2. Animal manure production and characteristics [8].*

GHG	MW (g mol⁻¹)	Typical ambient concentration (ppm)	Life time (Yr)	Radiative efficiency (W m⁻² ppb⁻¹)	Global Worming Potential
CO₂	44.01	380	Up to 100	1.4×10^{-5}	1
CH₄	16.04	1.7	12	3.7×10^{-4}	21
N₂O	44.01	320	114	3.03×10^{-3}	310

Table 3. Global warming potential of the GHGs [1]

1.2. Greenhouse gas inventory

The livestock industry is a significant contributor to the economy of any country. More than one billion ton of manure is produced annually by livestock in the United States. Animal manure is a valuable source of nutrients and renewable energy. However, most of the manure is collected in storage/treatment structures or left to decompose in the open, which poses a significant environmental hazard. Tables 4a, 4b, and 4c summarized the studies on GHG emission rates (ERs) from dairy, feed yard and swine operations. Based on a literature review using limited data for free-stall and naturally ventilated dairy operations, emissions of CO_2 from the dairy slurry manure storage facilities averaged 72 kg CO_2 m^{-3} yr^{-1} (data ranged from 8.6 to 117 kg CO_2 m^{-3} yr^{-1}) [9-11]. Emissions of CO_2 from dairy housing averaged 1,989 kg CO_2 hd^{-1} yr^{-1} (data ranged from 1,697 to 2,281 kg CO_2 hd^{-1} yr^{-1}, where hd^{-1} is per head) [11]. Kinsman et al. [12] reported that the mean daily CH_4 emission per dairy cow (602 kg mean bodyweight) in a tie-stall barn ranged from 373 to 617 g CH_4 $AU^{-1}d^{-1}$(436 to 721L) ,while the mean daily CO_2 emission per cow ranged from11,900 to 17,500 g CO_2 $AU^{-1}d^{-1}$(5,032 to 7,427 L). In a study by Amon et al. [13], CH_4 and N_2O emissions per livestock unit (LU=600 kg of body weight) or animal unit (AU=500 kg of body weight) from tie stalls for dairy cows were measured several times in the course of a year. Average emissions were 619.2 mg N_2O $LU^{-1}d^{-1}$ (516 mg N_2O $AU^{-1}d^{-1}$), and 194.4 g CH_4 $LU^{-1}d^{-1}$(162 g CH_4 $AU^{-1}d^{-1}$). Emissions of CH_4 and N_2O from animal housing averaged 54 (1.0-100) kg CH_4 hd^{-1} yr^{-1} and 0.3 (0.0-0.6) kg N_2O hd^{-1} yr^{-1}, respectively [11,13]. Ngwabie et al. [14] reported CH_4 emissions ranging from 25 to 312 g $hd^{-1}d^{-1}$ (9 to 114 kg hd^{-1} yr^{-1}) in a naturally ventilated dairy barn. Methane emission estimates for dairy cows have been reported to range from 230 g/cow/day [15] to 323 g $cow^{-1}d^{-1}$ [16]. Fiedler and Muller [17] show that CH4 emissions from naturally ventilated dairy barns ranged from 672 to 528 g $cow^{-1}d^{-1}$. A study conducted in California, USA, indicated that CH_4 emissions of 296 and 438 g $cow^{-1}d^{-1}$ for dry and lactating cows, respectively, were mainly due to enteric fermentation and fresh manure produced negligible amount of CH_4 [18]. Most dairy facilities are naturally ventilated and do not have controlled air exchange. Therefore, in addition to the large variations in methane emission rates causing a range of methane concentrations in dairy facilities, CH_4 concentrations in dairy barns will also vary with geographical locations, weather conditions, and ventilation management practices. Only limited studies have been conducted on indoor air methane concentrations in dairy barns.

Most of the published literature reporting CH_4 emissions from feedlot manure systems used atmospheric dispersion modeling (inverse dispersion, backward Lagrangian stochastic model, IPCC tiers I and II algorithm, and Blaxter and Clapperton algorithm) to estimate emissions from a whole farm [19-21]. Zoe et al. [19] estimated summer CH_4 ER data for two Australian feedyards using an open-path tunable near infrared diode laser coupled with backward Lagrangian stochastic model of atmospheric dispersion. Methane ERs reported were 146 and 166 g hd^{-1} d^{-1} for Victoria and Queensland, respectively. Using the same techniques, the average CH_4 emissions were 166 and 214 g CH_4 hd^{-1} d^{-1} for feedlots in Queensland and Alberta, respectively [20]. Average daily CH_4 emissions were estimated to be 323 g hd^{-1} d^{-1} for a large beef feedlot in western Canada using the inverse dispersion model [21]. Phetteplace et al. [22] determined GHG emissions from simulated beef and dairy livestock systems in the United States using a computer spreadsheet program. The methane N_2O and CO_2 ERs reported were 1.56, 11.4 and 3411 g hd^{-1} d^{-1} from manure management systems of a feedlot. Direct measurements using micrometeorological mass difference technique reported 70 g CH_4 hd^{-1} d^{-1} emissions from a confined beef feedyard in Australia where animals were fed a highly digestible high grain diet [23].

Emission rates of CO_2, CH_4 and N_2O from different pig housing systems are presented in Table 4c. CH_4 was observed to be emitted from all swine housing systems showing a large variation because of the different animal types, housing systems, and manure handling methods [24-27]. Methane emissions from fattening pigs range between 0.5 to 135 g pig^{-1} d^{-1}, whereas emissions of 0.77 and 5.8 g pig^{-1} d^{-1} (Table 4c) were reported for sows and weaners, respectively. Similarly, CO_2, CH_4, and N_2O ERs in gestation pigs in North Dakota (USA) ranged from 5,350-15,830, 116-572, and 0.06-7.3 g d^{-1} pig^{-1}., respectively [28]. Similarly, CO_2 ERs for different growing stages from swine operation ranged from 5,920 to 30,000 g pig^{-1} d^{-1}. The highest N_2O ER was estimated from swine nursery in China [29]. Animal feces temporarily stored indoors deep pits are the principal source of CH_4 emissions in swine housings. The quantity of CH_4 emitted by the animal itself and the amount emitted from barns of fattening pigs is influenced by the diet and digestibility, daily weight gain of the pigs, and the temperature and type of housing system. Methane emissions are lower in summer when compared with autumn and winter due to higher air exchange rates. Also, the CH_4 generation might be influenced by the availability of oxygen over the emitting surfaces [30]. Significant amount N_2O emits from pig manure handling system is exclusively originated from deep litter and compost systems. The variation in the N_2O emissions mainly depends on the kind of housing system. Fattening pigs raised on partly or fully slatted floor emit very little N_2O while higher emissions reported for fatteners in deep litter and compost systems [30].

Facility type and ground sources	Animal Number	CH_4 g/LU/d	CO_2 kg/LU/d	N_2O g/LU/d	Technology used	Remarks	References
Free-stall dairy: Barn, settling basins, loafing pen, primary and secondary lagoon, walkway, silage pile.	500	181	6.6	6	Dynamic flux chamber coupled with GC and chromatograms acquired directly at the field	Five consecutive days in summer and winter; Reported values are annualized	[31]
Free-stall dairy: Barn, settling basins, loafing pen, primary and secondary lagoon, open-lot, compost piles.	3500	836	5.5	3.4	Dynamic flux chamber coupled with GC and chromatograms acquired directly at the field	Five consecutive days in summer	[32]
Open-lot dairy: Open-lot pen,	700	0.20 to			OP-PATH FTIR	One to two days	[33]

Facility type and ground sources	Animal Number	CH₄	CO₂	N₂O	Technology used	Remarks	References
storage lagoon, and composting areas.		0.55			(MDA Atlanta) /BLS (Wind-Trax 2.0)	in January, March and June, and September.	
Tied-stall dairy: Housing, Storage, and spreading	12 cows	194.4		0.619	Mobile dynamic chamber coupled with FTIR and GC. Sampled alternately incoming-outgoing air of the chamber.	Several times in course of a year, and 24 hours a day during each experiment. LU =600 kg and AU=500kg	[13]
Tied-stall dairy: Housing (enteric), Manure storage,	118	373 24	14 – 20		DTs: 0-5000 ppm for CO_2 and 0-1000 ppm for CH_4.	112 days BW 602±62	[12]
Naturally ventilated dairy building with daily manure collection	720	305	8.9	NR	Ventilation by CO_2 balance and concentrations measured by INNOVA 1312	5 consecutive days in spring, summer, autumn, and winter	[34]
Free-stall dairy: Dry cows (Holstein) Lactating cows	9 9	296 438	NR NR	ND ND	PAS Multi-gas Monitor (INNOVA 1412)	Manure and Enteric CH_4 in environmental chambers. BWs 770 and 565 kg	[18]
Open-lot Dairy: Total emissions rates from open-lot, runoff pond, and compost area	10,000	1390	NR	20	INNOVA 1412 OP-FTIR coupled with BLS Wind-Trax2.0	Four months (January, March, June, September) and 2-3 days in each month	[35]
Free-stall dairy: Open lot Run-off pond Compost area	10,800	490 952 125	28 6 12	10 4.5 8.3	PAS Multi-gas Monitor (INNOVA 1412) coupled with BLS (Wind-trax 2.0)	2 or 3 days in each month over a year; BW 635 kg	[36]
Dairy with Conventional bedding Straw bedding		700 ± 400 1400 ± 200			TDLAS (DT: 100 ppb) coupled with Gaussian plume model	7 conventional farms 3 straw farms	[37]
ND = not detected by sensor; NR= not reported							
				a.			
Facility type and ground sources	Animal Number	CH₄	CO₂	N₂O	Technology used	Remarks	References
Feedyard: Feedlot	16,995	323			OP-TDLAS/ BLS (Wind-Trax 2.0)	12 days, High grain diets and	[21]

Runoff pond		6.2				BW 185-635 kg 280-700 kg	
Feedyard: Queensland Victoria	13,800 16,500	146-166			OP-TDLAS/ BLS (Wind-Trax 2.0)	High grain diets and animal weight 265-620 kg 280-700 kg	[19]
Feedyard: Pen surface Run-off pond Composting areas	42,000	1.71 2.05 0.04	1.31 0.035 0.055	0.57 0.01 0.10	Dynamic flux chamber coupled with gas chromatograph and chromatograms acquired directly at the field	Five consecutive days in summer	[32]
Open Feedlot: Queensland (Australia) Alberta (Canada)	13,800 22,500	166 214			OP-TDLAS/ BLS (Wind-Trax 2.0)	High grain diets and animal weight 350-600 265-620 kg	[20]
Feedlot and Grazing: Grazing Feedlot	A group of cattle	230 70			Microclimatological mass difference; same group of cattle tested at each the source.	High grain diets Grazed Animal weight 436 ± 21 kg	[23]
Simulated beef system	100	1.56	3.4	11	Computer spreadsheet Program (Gibbs and Johnson, 1994)	Data collected from nine beef and dairies in US	[22]

b.

Country	Animal Stage	Housing and or manure Handing type	GHG emission factor				Reference
			Unit	CO₂	CH₄	N₂O	
Germany	Fattening	Fully slatted floor Kennel housing	g d⁻¹ AU⁻¹ g d⁻¹ AU⁻¹	$17000 – 23000$ $11000 – 13000$	69-135 18-36	N/A N/A	[26]
Germany	Fattening	N/A	g d⁻¹ AU⁻¹	N/A	0.5-1	N/A	[5]
Holland	Sows Weaner Finisher	N/A N/A N/A	mg h⁻¹ pig⁻¹ mg h⁻¹ pig⁻¹ mg h⁻¹ pig⁻¹	N/A N/A N/A	2406 445 1269	N/A N/A N/A	[24]
Italy	Fattening	Fully slatted floor Vacuum system	g d⁻¹ AU⁻¹ g d⁻¹ AU⁻¹	N/A N/A	7.9±1.6 6.4±2.0	0.02±0.15 0.05±0.03	[38]
Belgium	Weaned Pigs	Straw litters Sawdust litters	g d⁻¹ pig⁻¹ g d⁻¹ pig⁻¹	463 481	1.58 0.77	0.35 1.4	[39]
Denmark	Finishing	Partly slatted floor	g fattening period⁻¹	5540	302	9.1	[40]

USA	Farrow-to-finish	N/A N/A	$g\ d^{-1}\ pig^{-1}$	N/A	6.9-29.2	N/A	[27]
	Farrow-to-weaner		$g\ d^{-1}\ pig^{-1}$	N/A	46.2	N/A	
USA	Gestation	Deep pit (3m) manure storage emptied twice a year	$g\ d^{-1}\ AU^{-1}$	5350-15830	116-572	0.06-7.3	[28]
United Kingdom	Fattening	Slurry	$g\ d^{-1}\ AU^{-1}$	N/A	85	0.4	[41]
China	Fattening	Manure scrapped twice a day	$g\ d^{-1}\ AU^{-1}$	16730 ±1060	32.1± 11.7	0.86±0.75	[29]
	Nursery	twice a day	$g\ d^{-1}\ AU^{-1}$	29670±1090	58.4±21.8	1.29±0.37	
	Gestation	Flushed twice a	$g\ d^{-1}\ AU^{-1}$	5920±440	9.6±1.9	0.75±0.56	
	Farrowing	day Manure removed twice a day Removed as produced	$g\ d^{-1}\ AU^{-1}$	7490±110	9.6±3.6	0.54±0.15	
Taiwan	NA	3 stages (solid separation, anaerobic and aerobic treatment) plug-flow type wastewater treatment plant.	$g\ hd\text{-}1\ d^{-1}$	11.5	13.3	0.03	[42]
				c.			

Table 4. a. Summary of GHG emission rates (ERs) estimated from dairy operations at different ground level area sources (GLAS) as reported by previous researchers (LU = live weight, BW= body weight). b. Summary of GHG emission rates (ERs) estimated from feedyard operations at different ground level area sources (GLAS) as reported by previous researchers. c. Summary of GHG ERs of swine operations with different housing and management schemes as reported in the literature (updated after Dong et al. [29]). AU = animal weight = 500 kg live weight.

2. Emission process

2.1. Methane emission process

During anaerobic fermentation, organic wastes are biologically degraded in the absence of oxygen to CH_4, CO_2, N_2, and H_2S. Methanogenic fermentation of organic materials occurs under strictly anaerobic and low redox potential (Eh < -200 mV) conditions where sulphate and nitrate concentrations are low [43]. Methanogens produce methane by breaking down organic matter in the absence of oxygen (anaerobically), releasing CO_2 and CH_4 according to the following equation:

$$\text{C}_6\text{H}_{12}\text{O}_6 \xrightarrow{\text{Microbial Action}} 3\text{CO}_2 + 3\text{CH}_4 \qquad (1)$$

This transformation requires the successive action of four different types of micro-organisms as shown in Figure 2 that degrade complex molecules to simpler compounds [44]:

a. **Hydrolytic microflora**: *hydrolysis* of longer chain carbohydrates, fats and proteins are broken into shorter chain molecules. This can be aerobic, facultative, or strictly anaerobic.

b. **Fermentative microflora**: *acidogenesis* of shorter chain molecules produce carbon dioxide, hydrogen sulfide, alcohol, and more volatile fatty acids. This can be facultative or strictly anaerobic.

c. **Homoacetogenic or syntrophic microflora**: *acetogenesis* from previous metabolites. Simple molecules created through the first two steps are digested by specific bacteria to produce acetic acids as well as hydrogen and carbon dioxide.

d. **Methanogens**: *methanogenesis* of simple compounds such as $H_2 + CO_2$ and acetate. Products developed in stages 1-3 and convert ($H_2 + CO_2$ and acetate) them into methane, carbon dioxide and trace amount of other gases).

Numerous physical, chemical, and biological factors influence the physiology of methanogenic archaea and characteristics manure (organic waste) and the micro-environment of the anaerobic systems. One of the important obvious factors is temperature. Methane production increases with increasing temperature if other parameters are kept constant. The main factor determining the degree of CH_4 production is the amount of degradable organic matter contained in the effluent and organic animal waste. This fraction is commonly expressed in terms of biochemical or chemical oxygen demand (BOD or COD). The higher the BOD/COD value, the more CH_4 is produced. The potential amount of CH_4 formation from animal feces will depend on the amount of fecal matter excreted, the physical form of the deposit (shape, size), excretal form (solid, slurry, and effluent), climatic and soil conditions, and the length of time these deposits remain intact before being decomposed [43].

Methane production from manure when managed in a controlled setting will depend on the type of waste, temperature, and duration of storage, and the manner in which the waste is handled. On the other hand, emissions during composting of dung depend on aeration rate, water content, thermal insulation, weather conditions, and manure composition. Methane production during composting is related to the lack of oxygen in the decomposing biomass. A study reported that anaerobic digestion of the slurry reduced CH_4 emissions after field application, because the easily degradable organic compounds were already converted to CO_2 and CH_4 during digestion in the biogas plant [45]. The factors affecting CH_4 emission by soils are summarized as follows [43, 44]:

a. Gas diffusion in relation to oxydo-reduction level and CH_4 transfer, in particular the water content, the nature of clays, and the type of vegetation.

b. Microbial activities in general temperature, pH, Eh, substrate availability, physicochemical properties of soils.
c. Methanogenesis and, in particular, the competition with denitrification and sulphate reduction.
d. Methane-mono-oxygenazse activity—concentrations of H_2, CH_4, NH_4 +, NO_3 -, Cu.

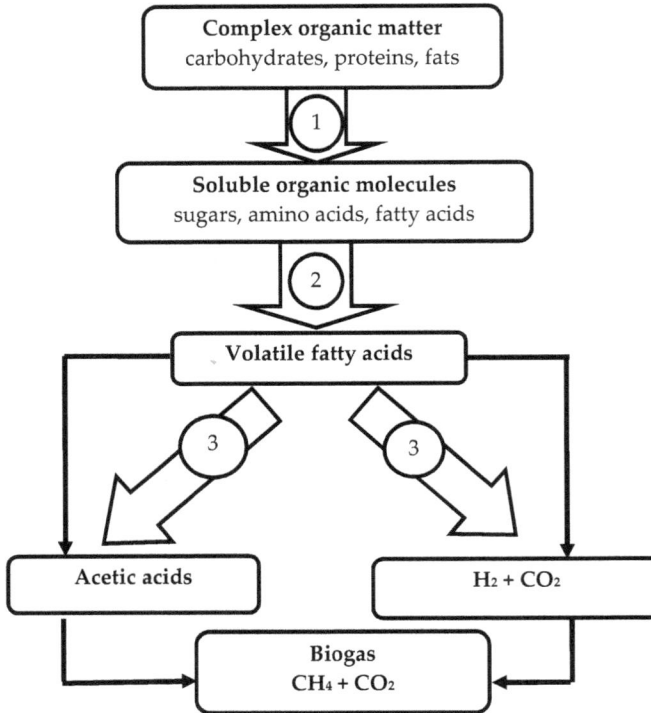

Figure 2. Schematic diagram showing anaerobic fermentation process.

2.2. Nitrous oxide emission process

Nitrous oxide (N_2O), nitrogen monoxide (NO), and nitrogen dioxide (NO_2) are the most plentiful nitrogen oxides in the atmosphere and being produced lavishly by biogenic sources such as plants and yeasts. Nitrous oxide is an ozone (O_3) depleting substance which reacts with O_3 in both the troposphere and in the stratosphere and has a long half-life (100 - 150 years). In livestock agriculture, N_2O emissions are associated with manure management and the application and deposition of manure in crop/ pasture land. Indirect N_2O emissions from livestock production include emissions from fertilizer use for feed production, leguminous feed crops, and emissions from aquatic sources following fertilizer application. Thus, fertilized soils are important sources of N_2O. Soils contribute about 65% of the total

N_2O produced by terrestrial ecosystems [46]. Nitrous oxide gas is formed in soils during the microbiological processes of nitrification and denitrification as shown in equations 2 and 3. Nitrous oxide production by nitrifying bacteria may arise either during NH_4^+ oxidation to NO_2^- or during dissimilatory NO_2^- reduction when O_2 supply is limited. During denitrification, N_2O is an intermediate product in the dissimilatory reduction of NO_3^- and NO_2^- to N_2 under anaerobic conditions and may, therefore, be produced and consumed by denitrifying bacteria in soil [47].

Fertilizer and manure type may affect N_2O emission in several ways [48] such as: (1) the type of N (NO_3^-, NH_4^+, and organic N) which affects N_2O production during nitrification and denitrification, (2) the presence of freely available C, which stimulates denitrification activity and O_2 consumption in the soil following its application, and (3) effects on biological, chemical and physical soil processes because of changes in pH and the addition of other compounds (salt, water). The availability of N (NH_4^+ and NO_3^-) [43, 49, 50], and the factors that alter the redox potential of the soil, such as changes in soil moisture [51-53], soil texture, and organic C, have major effects on the production of N_2O in soils. In addition, several soil management practices such as tillage, soil compaction [54-55], irrigation, and drainage affect the production and transport of N_2O release by influencing the physical condition of the soils such as aeration and soil water content.

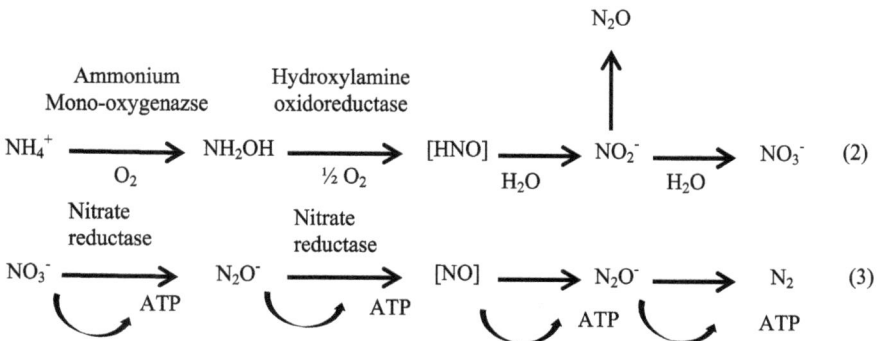

$$
\begin{array}{c}
\qquad\qquad\qquad\qquad\qquad\qquad\qquad\qquad\qquad N_2O \\
\text{Ammonium} \qquad \text{Hydroxylamine} \qquad\qquad\qquad \uparrow \\
\text{Mono-oxygenazse} \quad \text{oxidoreductase} \\
NH_4^+ \xrightarrow{\ \ O_2\ \ } NH_2OH \xrightarrow{\ \ \frac{1}{2}O_2\ \ } [HNO] \xrightarrow{\ \ H_2O\ \ } NO_2^- \xrightarrow{\ \ H_2O\ \ } NO_3^- \quad (2) \\[4pt]
\text{Nitrate} \qquad\qquad\qquad \text{Nitrate} \\
\text{reductase} \qquad\qquad\qquad \text{reductase} \\
NO_3^- \xrightarrow{\ \ ATP\ \ } N_2O^- \xrightarrow{\ \ ATP\ \ } [NO] \xrightarrow{\ \ ATP\ \ } N_2O^- \xrightarrow{\ \ ATP\ \ } N_2 \quad (3)
\end{array}
$$

3. GHG measurement methods

3.1. Measurements of Greenhouse gas concentrations

Uncertainties in the accurate emission estimate are mainly dependent on the errors associated with sampling protocol and devices and gas analyzers. Thus, the validation of emission inventories using emission measurements is extremely important as well as source-related emission measurements that are feeding emission inventories. To develop GHG mitigating strategies, it is required to quantify GHG emissions from livestock operations under a wide range of productions and management circumstances. Gases at trace levels can be measured using different techniques. GHGs can be measured using infrared spectroscopy (IR), photoacoustic spectroscopy (PAS), gas chromatography (GC), mass spectroscopy (MS), tunable diode laser absorption spectroscopy (TDLAS) technology, open path Fourier Transform Infrared Radiation (OP-FTIR) technologies, and solid-state electro-chemical technology. Instruments with mass spectrophotometers have very rapid response, can detect many gases at one time, and exhibit linear responses over a wide range of concentrations, while behaving very accurately and with stability. However, mass spectrophotometers, TDLs, and OP-FTIRs are expensive. Solid state electrochemical sensors are relatively inexpensive but they are unstable and require frequent calibration. The shelf lives of those sensors also vary from 12-18 months.

3.2. Flux measurement process at the ground level of the livestock facilities

The GHG emission estimation from different ground level area sources (GLAS) of manure managements in livestock operations such as lagoons (primary and secondary), barns, settling basins, silage piles, loafing pens, feedlots pens, compost windrows, and crop/pasture land is a very complicated process. Generally, two basic processes such as device independent and sampling device are widely used to estimate emission from emitting surfaces. In the device-independent techniques, the emission rate (amount (g or kg) of compound emitted per head per day or year is estimated from the concentrations of the measured emission across the plume of emitted material using local micrometrical data, especially wind velocity profile data [56-57]. When using a sampling device, a chamber and winds tunnel is deployed on an emitting surface under some recommended operating conditions. Those devices may be static (sealed or vented) or flushed with zero grade air (containment free) at a known velocity or flow rate (known as dynamic). Generally, the emission rate is estimated as the product of concentration and air flow through the device [57-59]. There is debate about the suitability and accurateness to quantify pollutant emissions at CLOs and other area sources due to the creation of microenvironments in the chamber and the small measurement footprint relative to the size of the source [57-60]. Hudson et al. [61] compared and reported that odor emissions from a wind tunnel rates were 60 to 240 times higher than those in a flux chamber [62]. Parker et al [60,71] also demonstrated that water evaporation, wind speed, and temperature would be useful to standardize and compare emission rates from flux chamber and wind tunnels. They also suggested developing correction factors for each device, which depend on the geometry of the wind tunnel and chamber. Instruments and devices commonly used to measure gas emissions from CLOs were presented in the Table 1.

4. GHG scrubbing technology

In the livestock industries, **End-of-Pipe** technologies such as biofilters and wet scrubbers are commonly used in process-air applications i.e. potentially harmful particulates matter (PM) and pollutants in exhaust air of the housings/barns are treated. Generally, water with added active chemicals such acids and oxidizing agents (H_2O_2, H_2SO_4, O_3, $kMnO_4$, HOCl, etc) are tailored with the process to spray into the air stream coming out of the exhaust. This approach for reducing emission is basically a treatment of the exhaust air released from mechanically ventilated animal housings. The main advantages for this approach are air can be treated without affecting the routine management operations and structural design inside the barn. Broadly, two types of air scrubbers are presently available: acid scrubbers and bio-trickling filters. The main purpose of these scrubbers is ammonia abatement; the scrubber systems are commercially available and considered as off-shelf techniques in such as the Netherlands, Germany and Denmark [64].

4.1. Scrubber descriptions

4.1.1. Spray type wet-scrubber

In an acid scrubber for ammonia, diluted acid mainly sulfuric acid (H_2SO_4) with pH of 2-4 is used to scrub ammonia from air and the ammonium salt is removed from the system with the discharge water. Spray scrubbers consist of empty cylindrical or rectangular chambers in which the gas stream is contacted with liquid droplets generated by spray nozzles. The spray nozzles (hydraulically or air or steam atomized), are used to extend the surface area of the scrubbing liquid and produce target droplets size that facilitates mass transfer of the contaminants gas(es) into liquid. They are mainly used for gas absorption. Particulate matters (PMs) and gaseous pollutants in the air stream are removed by either absorption or chemical reactions with the water solution. PM and pollutions from the scrubber process are removed periodically through the drain. Schematic of a typical spray nozzle scrubber configuration along with system components is shown in the Figure 3.

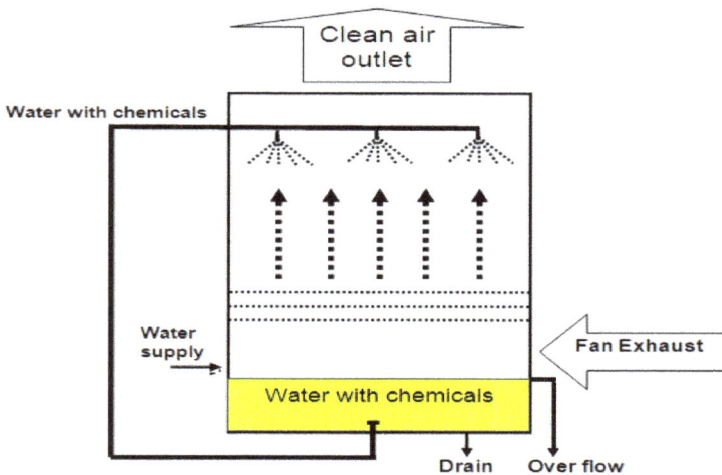

Figure 3. Schematic diagram spray type wet-scrubber system

4.1.2. Packed bed wet-scrubber

A packed tower air scrubber or bio-trickling filter is a reactor that is filled with an inert or inorganic packing material. The packing material usually has a large porosity, or void volume, and a large specific area [64]. Water with added a chemicals is sprayed either continuously or intermittently from the top of the packed bed to keep it wet. The contact between the air and water, facilitates a mass transfer from soluble gases to a liquid phase when exhaust air is introduced wither horizontally (cross-current) or upwards (counter-current). A fraction of the trickling water is continuously recirculated, while another fraction is discharged and replaced by fresh water [65]. Schematics of typical spray nozzle scrubber and packed bed acid scrubbers configuration along with system components is shown in Figure 4.

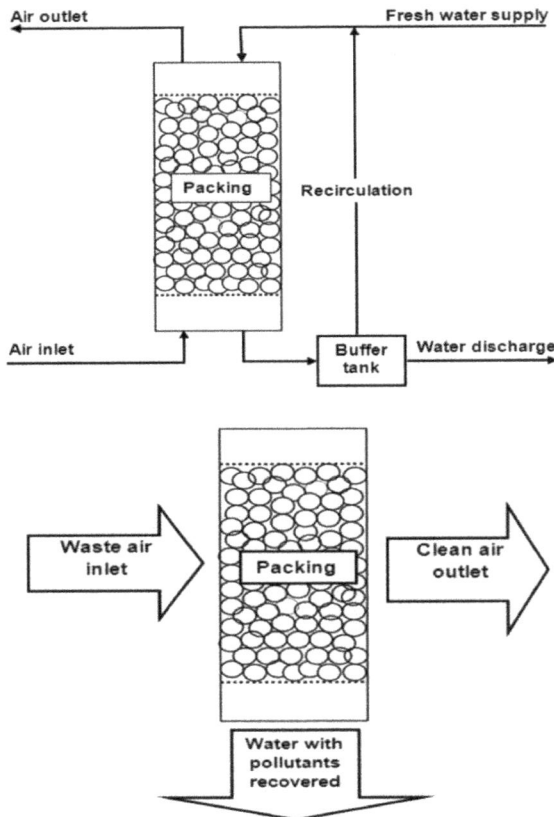

Figure 4. Schematic diagrams of packed bed trickling filters: top) counter current packed wet-scrubber, and bottom) cross current (adopted from [64-65).

4.1.3. Pollutants removal efficiency calculation

Gaseous pollutants removal efficiency is generally used as the criteria for determining the spray type wet scrubber performance can be defined as [66]:

$$\gamma_{total} = \frac{C_{in} - C_{out}}{C_{in}} \times 100 \qquad (4)$$

Where:

γ_{total} = Pollutant collection efficiency (%)
C_{in} = Airborne Pollutant concentration before the scrubber (ppm)
C_{out} = Airborne Pollutant concentration after the scrubber (ppm)

Similarly, the difference in the weight of the PM filters before and after scrubber sampled during 24 hours and the standardized airflow were used to calculate the average PM_{10} concentration. The details on the used method for PM_{10} determination can be found in [67]. For wet-scrubbers, the air flow rate is used to calculate average Empty Bed air Residence Time (EBRT). The air flow rate through the scrubbers can be determined either by measuring fans or by means of a CO_2 balance method [68]. The EBRT can be defined as follows:

$$EBRT = \frac{Packing\ volume\ (m^3)}{Air\ flow\ rate\ (m^3 s^{-1})} \times 100 \qquad (5)$$

4.2. Wet scrubber applications in AFOs

The development of wet air-scrubbers for mitigating air emissions from CLOs has started a longtime ago. To begin with, wet scrubbers were employed to reduce odor and particulates being discharged from livestock facilities [69-70]. Later on, scrubbing other airborne contaminants such as NH_3 [66,71,65], H_2S [72], and pathogens [73] were also investigated to test their scrubbing efficacy. The collection efficiencies reported ranging minus to 100% for odor, 23% to 96% for NH3, and 36% to 96% for particulate matter. An acid scrubber and a bio-trickling filter (BTF) were developed to reduce ammonia and odor from swine and poultry houses in the Netherland. Melse and Ogink [71] reported an average ammonia removal efficiency of 96% in acid scrubbers (ranging from 40% to 100%). The average efficiency estimated using the air balance method was 71% (±4%). At least 24 measurement days are recommended to keep the relative error below 5% when using the air balance method in determining the NH_3 removal efficiency of an acid packed bed scrubber [65].

Chemical scrubbers and bio-scrubbers have shown tremendous potentials in reducing high particulates and ammonia, however, are not very effective in removing typical odors [74-75]. The major limitations encountered in the development of wet scrubber technology for CLOs are low collection efficiency of the odorous compounds, high pressure drop, and high operating costs. An acid spray wet scrubber has the greatest potential for adaptation to existing swine facility ventilation fans because they do not cause excessive backpressure to the fans and do not significantly reduce building ventilation airflow [66].

Recent literature showed that the majority of scrubbers designed for CLO applications were employed for removing ammonia. In Europe, the most common commercial scrubbers are packed-type, which can be bought off the shelf are for application at CLOs [76] and have been proven to effectively remove NH_3 by up to 96%, but their packing material resulted in large pressure drop [71]. Other types of wet scrubbers such as impingement plate, fiber bed 71], and rotating beds [77] have been used for NH_3 collection. These wet scrubbers also resulted high pressure drop and did not work well with the existing ventilation systems of AFOs because axial fans are typically used for movement of large volume of airflow under small differential static pressure conditions. Recently, a new generation multi-pollutants scrubbers have also been developed to address ammonia, odor, and particulates abatement released from livestock operations. This scrubber mainly consist of two or more scrubbing stages (combining the concepts of acid scrubbing, bio-scrubbing, and bio-filtration), each stage aims for the removal of one type of compound [78]. Three multi-stage scrubbers, one double-stage scrubber (acid stage+ bio-filter), one double-stage scrubber (acid stage + bio-scrubber), and one triple-stage scrubber (water stage + acid stage + bio-filter) were evaluated to test their effectiveness in reducing airborne dust, total bacteria, ammonia, and CO_2 emissions from swine houses in Netherlands. Those scrubbers reduced PM_{10}, $PM_{2.5}$, total airborne bacteria, and ammonia emissions from 61 to 93%, 47 to 90%, 46 to 85%, and 70 to 100%, respectively [79]. Concentrations of CO_2 were not affected.

Most scrubbing technologies for CLOs are still in the developmental stages in the US. The major limitations encountered in the development of wet scrubber technology for CLOs are low efficiency, high pressure drop, and high operating costs. The spray-type wet scrubbers usually are generally shown low collection efficiency for NH_3 gas [80,70]. A spray scrubber with water has shown to be a collection efficacy of approximately 20%, although, ammonia is fairly water soluble 20% [81]. Higher NH_3 absorption can be achieved by spraying diluted acidic solution as a scrubbing liquid [73]. In addition, use of acidic substances for collecting ammonia is highly preferred because of its great potential to get the NH_3 recycled into liquid fertilizer. Ohio state university has developed spray wet scrubber and three of those were installed on a commercial deep-pit swine building, a poultry manure composting facility, and a covered swine manure storage, respectively. The field tests of the wet scrubber showed an ammonia collection efficiency up to 98% and 80% for exhaust air with low (5 ppmv) and high ammonia concentrations (200 ppm), respectively. However, it is not tested for GHG mitigation.

4.3. Biofilters

Biofiltration is an air-cleaning process which absorbs pollutant gases and particulates into a biofilm on the filter media. Microorganisms in the filter media degrade and break the volatile organic compounds (VOC) and oxidizable inorganic gases. Selection of a proper biofilter media is a critical factor for developing an efficient biofilter. These factors such as optimum environment for microorganisms (moisture, temperature, porosity, etc.), large surface area to maximize attachment area and sorption capacity, stable compaction properties, high moisture holding capacity, high pore space to maximize empty bed residence time (EBRT), and minimize pressure drop [72]. Recently, Chen et al. [72] evaluated a pilot-scale wood chip-based (e.g., western cedar and hardwood) biofilter to reduce odor, H_2S, and NH_3 from swine barn ventilation air for 13 weeks. They found that hardwood and western cedar biofilters can remove odor by 70.1 and 82.3%, respectively, and H_2S by 81.8 and 88.6 %, respectively. Biofilters saturate easily [82] and large pressure drops across them making it difficult for them to be adopted by animal facilities. Difficulty also rests with the stringent moisture and temperature requirements for the process and more frequent media replacement [83]. It reported that biofilters may not be suitable to reduce high odor concentrations due to nitrogen accumulation in the biofilter material that causes the release of other pollutants including nitrous oxide (N_2O), a highly potent greenhouse gas [74]. The biofiltration is a simple technology but requires careful monitoring of operating parameters for treating contaminated air effectively. However, microbial process taking place in the filter beds are very complicated, which depends on few environmental and physical factors as summarized below [82,84]:

High ammonia loads generally trigger excessive nitrite/nitrate concentrations that inhibit a proper functioning of micro-organisms in the filter bed and leads to acidification which forms nitrite/nitrate salts. This in turn, declines the removal efficiency of the filter. Thus, by replacing saturated biofilter packing at regular interval this issue can be addressed.

Maintaining adequate moisture in the filter bed is the critical factor for proper functioning by the filter because of drying out inlet side of the filter bed when relatively dry air is coming out of the exhaust housings. Thus, the biofilter bed has to be kept moist to ensure proper microbial functioning.

Biofilters are inherently prone to dust loads, thus, clogging the packing bed and increasing the pressure drop. The total pressure drop over the filter bed can be very high and in practice it is >200-300 Pa. This clogging when coupled with inadequate moisture in the filter bed may lead to reduced air flow which will decrease overall scrubbing performance of the filter bed. These in turn need an increase in energy input unit of air volume handled. The functional lifespan of the biofilter can be enhanced by pre-treating incoming air, routing it first through an acid scrubber and mist eliminator before entering the biofilters.

Thus, design and operational parameters such selection of packing material, maintaining optimum moisture content, weed control and assessing pressure drop are very critical for efficient operation of the biofilters. The functional lifespan of the biofilter can be enhanced by pre-treating incoming air, routing it first through an acid scrubber and mist eliminator before entering the biofilters.

4.4. GHG scrubbing technology

Existing GHG mitigation options related to manure management are focused on feed manipulation, animal management, and processes to treat and manage animal manure. GHG emitted from animal buildings, especially swine buildings with deep pits and dairy buildings, accounted for a large portion of GHG emission. However, mitigation technologies for GHG emissions from these animal buildings are lacking. Recent adaptation of mechanical ventilation systems for these buildings made it possible for better maneuvering the air stream in routing it through a suitable scrubbing system. Gaseous emissions including methane must be reduced before it escapes into the atmosphere. Housings or sources with mechanical exhaust ventilation systems are advantageous for capturing methane emission before it enters the atmosphere. However, this process inherently needs to handle large quantities of exhaust/ ventilation air at low cost. Thus, higher energy requirement to process huge amount of exhaust may not be economically feasible. Therefore, it is urgently needed to develop effective and economically feasible GHG mitigation technologies for the reduction of these emissions from animal barns to ensure sustainable and viable swine and dairy industries.

Nitrous oxide (N_2O) and NH_3 are fairly soluble in a wide variety of solvents including water, alcohols, sulfuric acid, etc.). Unlike ammonia and N_2O, CH_4 is highly insoluble in water, and thus, scrubbers developed for NH_3 cannot be used for mitigating CH_4. Methane is a very stable molecule at ambient temperatures so it cannot be removed by many of the scrubbing techniques that are used for other gases. It has been previously shown in rendering facilities that oxidants like chlorine dioxide are effective for the removal of VOCs and other organic compounds using exhaust wet scrubbers. However, these may be expensive when applied as scrubbing liquid for swine facilities. In swine pits and dairy buildings, CH_4 concentrations are too lean to burn; oxidation by scrubbing can be an effective and safe alternative for reducing CH_4 emissions. This situation warrant researching alternative scrubbing liquids to make the CH_4 scrubbing process economically feasible. Other possible oxidants such as hydrogen peroxides (H_2O_2), Ozone (O_3), potassium permanganate ($KMnO_4$), and hypochlorus acid (HOCl), etc) can be tested to reduce CH_4 in a suitable scrubber. Ozone, a strong oxidant, has been used extensively to improve air and water quality. It has also been used by agricultural engineers to control air quality in animal buildings [85,86]. Ozonating water has been proven to effectively oxidize organic compounds dissolved in water. Transfer of organic compounds from air to water by spray scrubber absorption is very promising for the capture of GHG, and deodorizing and sanitizing air without affecting the CLOs ventilation system. It is well known that ozone reacts with methane either in presence of ultraviolet (UV) light or high temperature. Therefore, to explore the methane reduction in a cost effective manner, a spray type wet scrubber can be tested with ozonated water (or other effective oxidants) and UV light. The scrubbing system can be consisted of a spray-wet-scrubber column made from a material (either glass or acrylic) with high UV light transmittance, flow control meters, an ozone generator, and oxidant liquid supply and collection systems. A dust cleaning system needs to be incorporated to allow UV light to always pass through the scrubber.

5. Management practices to reduce GHG emissions

Methane from enteric fermentation, manure storage and spreading, and nitrous oxide mainly from application of manure on land are the major sources of agricultural GHG emission. Prior to being applied to crop or pasture field, manure from CLOs is generally

stored in a liquid or solid form. For most cases, in order to produce CH_4 gas to use as bio-
fuel, manure is anaerobically digested or composted before land application. By adapting
manure management and treatment practices to enable methane collection, methane
emissions from anaerobic digestion can be recovered and used as energy. Anaerobic
digestion provides an appropriate environment for the complete degradation of organic
matter to low-odor end products [[87]. It also produces methane (biogas), which can be used
for the production of electricity and heat [10]. However, due to high content of ammonia,
the digestion of only swine manure was not favorable [85,88]. Generally, CH_4 mitigation
approaches can be broadly divided into four categories as follows:

1. Preventative or feed management
2. Manure managements including treatment process and land application methods
3. Adaptation of housing system design including inside manure storage
4. End of pipe air treatment.

5.1. Preventive method (feed management)

Preventative measures are the reduction in carbon/nitrogen inputs into the system of animal
husbandry through a dietary manipulation to achieve reduced CH_4 production.   An
effective tool to reduce nutrient/mineral pollution and GHG emissions is proven to be
dietary manipulation. Numerous studies have revealed that reducing crude protein in the
diet could substantially lessen nitrogen excretion and ammonia volatilization without
compromising productivity [89-93]. Hao et al. [94] studied the effects of DDGS on feces and
manure composition in feedlot cattle and revealed that as the ratios of wheat DDGS (e.g., 0,
20, 40, and 60%) in animal diet increased (40 and 60% wheat DDGS), the likelihood of
volatile fatty acids (VFAs) also increased. This led to the growth of odors produced from the
breakdown of fiber and protein [89]. They suggested that it might be a practicable option to
obtain 20% or less DDGS in animal diet to limit VFAs produced from the breakdown of fiber
and protein [89]. Enteric fermentation produced approximately 80% of the CH_4 produced
from ruminants. The chemical composition of diet is a vital feature, which affects rumen
fermentation and methane emission by the animals. Furthermore, dietary manipulation also
impacts the amount of GHG emissions, particularly from enteric fermentation. For example,
feeding cattle with a high starch and low fiber diet reduces creation of acetate in the rumen
and leads to lower methane production [93]. As a proportion of energy intake, a higher
proportion of concentrate in the diet leads to a reduction in CH_4 emissions. Stored manure
and high fiber fed animals tend to have higher emissions. Diet affecting emissions from
manure applied soil has significant evidence. Replacing fibrous diets by starchy feedstuff
has been shown to reduce methane from enteric fermentation and manure storage [95-96].
As the level of production is increased to meet global demand for ruminant meat and milk
products, dietary manipulation will be useful in addressing environmental concerns. Sejian
and Naqvi [7] described enteric methane reduction strategies under four categories as
shown in Figure 5. Likewise, a detailed evaluation of mitigation options of methane
emissions from enteric fermentation is presented in Table 5 [97]. Abatement of GHG
emissions from ruminant animals has been focused on diet, rumen and animal

manipulations, such as improving forage quality, adding dietary supplement, reducing unproductive animals, and supplementing probiotics to change microbial population in rumen [96]. Dietary fat seems a promising alimentary alternative to depress ruminal methanogenesis without lessening ruminal pH as opposed to concentrates [98]. Beauchemin et al. [99] recently reviewed the effect of level of dietary lipid on CH_4 emissions over 17 studies and reported that with beef cattle, dairy cows and lambs, for every 1% (DMI basis) increase in fat in the diet, CH_4 (g/kg DMI) was reduced by 5.6 %.

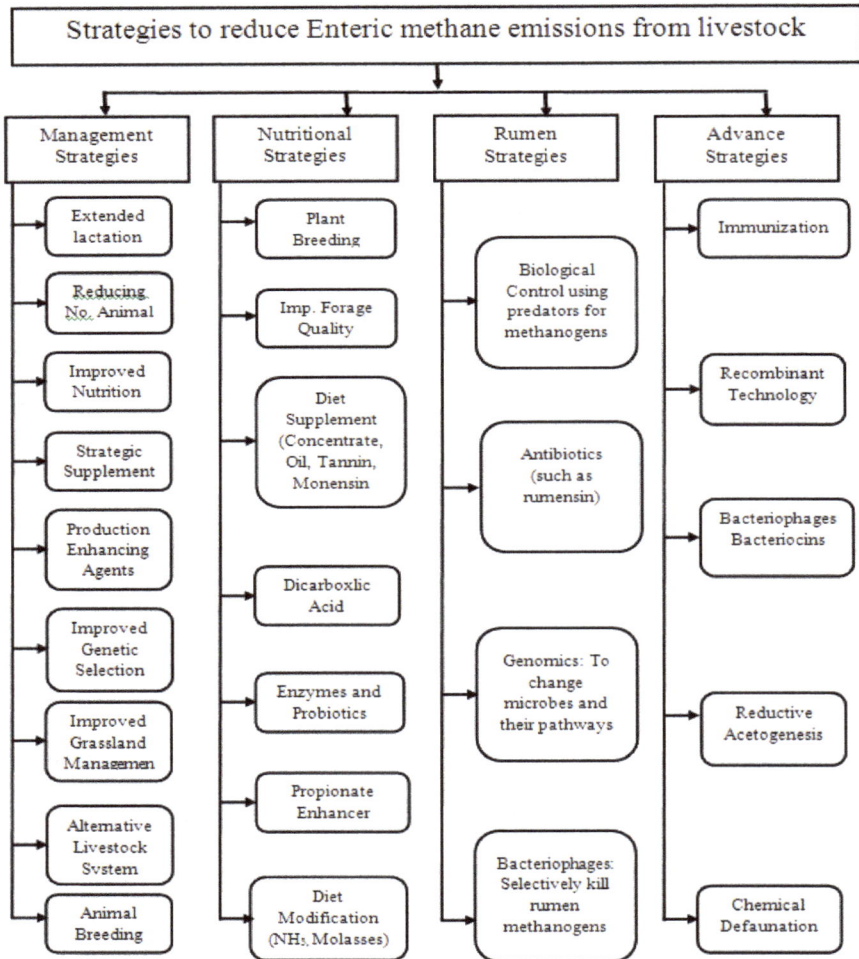

Figure 5. Different enteric methane mitigation strategies [7]

Strategy	Potential CH4 reduction	Technology/feasibility	Cost/production benefits
Improving animal productivity	20-30%	Feasible and practical	Increased feed cost increased milk production use of fewer animals less feed per kg of milk
Increasing concentrate levels at high levels of intake	25% and more	Feasible, for high producing cows, but may increase N_2O and CO_2 emissions	Increased feed cost increased milk production use of fewer animals less feed per kg of milk production
Processing of forages, grinding/pelleting	20-40%	Feasible	Increased cost of processing improved feed efficiency increased milk production
Forage species and maturity	20-25%	Feasible	Increased feed efficiency increased milk production
Rotational grazing of animals/early grazing	9% or more	Feasible	Increased cost of fencing increased management of animals increased feed intake increased milk production
Managed intensive grazing vs. confined feeding		Feasible need more investigation	Cheaper feed cost may need supplements reduced milk fat/protein content higher net return
Use of high quality forages/pastures	25% or more	Feasible	Increased feed intake increased milk production
Preservation of forage as silage vs. hay/additives	up to 33% (model prediction)	Feasible	Limited studies
Addition of fats	up to 33%	Feasible and practical, but usage limited to 5-6% in diet	Increased cost of diet increased or no effect on milk production may or may not affect milk fat

Strategy	Potential CH4 reduction	Technology/feasibility	Cost/production benefits
Use of ionophores, e.g., monensin, lasolocid	11-30%	Feasible but long lasting public concern	Increased feed efficiency decreased feed intake increased milk production
Use probiotics	10-50% (*in vitro*)	Feasible, needs more investigation	May increase feed intake may increase milk production or no change
Use of essential oils	8-14%	Feasible, needs more investigation	Not quantified
Use of bovine somatotropin (bST)	9-16%	Not approved for use in Canada	Reduced feed cost
Protozoa inhibitors	20-50%	Not available for practical use	Practically and cost to be assed
Propionate enhancers (fumarate, malate)	5-11% (in vitro) up to 23% (in vivo)	Possible microbial adaptation to fumaric acid	Economic feasibility ruminal adaptation and level of inclusion need to be evaluated
Use of acetogens	not quantified	Not available, needs more investigation	Needs further investigation
Use of bacteriocins, e.g., Nisin, bovicin HC5	up to 50% (in vitro)	May provide alternatives to ionophores needs more investigation	Production effects are to be evaluated
Use of methane inhibitors, e.g., BES, 9,10-anthraquinone	up to 71% (in vitro)	No compounds registered for use No long lasting effects identified	Increased cost of chemicals production effects not established
Immunization	11-23%	Not available, needs more investigation	May increase cost of production increased gain
Genetic selection (use of high Net Feed Efficiency animals)	21%	Long term feasibility	May increase cost of production increased gain

Table 5. Summary of methane mitigation strategies for dairy cows [97]

5.2. Manure management

It is well known that GHG emissions (mainly CH_4 and N_2O) from manure differ significantly depending on the management system employed to process them. Therefore, strategies for mitigating net GHG emissions should be aimed to manipulate manure properties or the conditions under which CH_4 and N_2O are produced and utilized during manure storage and treatment. However, GHG mitigation options are critical and depend on several factors. These factors are economic, technical and material resources, climatic conditions, existing manure management practices, bio-energy sources, and a source of high quality fertilizer and soil amendments. One such approach is to manipulate livestock diet composition and/or include feed additives to alter manure pH, concentration and solubility of carbon and nitrogen, and other properties that are pertinent to CH_4 and N_2O emissions [7]. Nitrogen excreted in urine is predominant in the form of urea that can easily be converted into ammonia and carbon dioxide by the enzyme urease (which is present in feces), thus resulting in emission of ammonia. Nitrogen excreted in feces is mainly present as protein, which is less susceptible to decomposition into ammonia [64]. Therefore, feed management aims at either reducing the nitrogen excretion in feces and urine by matching the amount and composition of feed more closely to animal requirements at various production stages, or shifting nitrogen excretion from urine to feces by increasing fibrous feedstuffs in the diet [64]. The use of these strategies can reduce the ammonia emission both for pigs [100-101], poultry [102-104] and dairy cattle [105-106]. About 50% of ammonia emissions to the environment were reduced through feed management for pigs and poultry when compared to standard feed composition. However, feed manipulation for ammonia abatement may negatively affect the emission of methane and nitrous oxide during storage and after land application of the manure [107].

Another manure management option is to change the material used for bedding the animals, which could also affect manure pH and soluble C and N levels and thus, the emissions during manure storage and treatment. Composting technology, control of aeration, use of amendments, or co-composting livestock manure with other organic waste could also potentially modify conditions for GHG production and emission. The use of covers may also help retain N nutrients during storage. Floating covers of natural and synthetic, origin or composites of both have shown substantial reduction in NH_3 and H_2S emissions when compared with uncover liquid manure. However, little is known about the effect of covers on GHG emissions. In a two week study, covers generally increased CO_2 and CH_4 emissions [108].

5.2.1. Animal population and low N grass

A large part of N from animal waste and farm effluents is lost to the environment as excess NH_3 or N_2O from urine spots and animal manures instead not being recovered in livestock-production systems. There are various options for reducing NH_3 and N_2O emissions from livestock facilities, but the most significant option is to improve overall N efficiency. By reducing the livestock numbers, the amount of excreta would be reduced, hence the amount of emissions. Another mitigation option would be to manipulate the N economy of the animal to reduce N excretion. A lower N content of pasture or silage would reduce N excretion by animals and NH_3 volatilization loss [43]. Excretal N could be reduced by using grass grown with moderate fertilizer application [93].

5.2.2. Anaerobic digestion and gas capture

With the use of liquid-based livestock facilities, the primary method for reducing emissions is to recover the methane before it is emitted into the air. Methane recovery involves capturing and collecting the methane produced in the manure management system. This recovered methane can be flared or used to produce heat or electricity. Because most of the manure facility methane emissions occur at large confined animal operations (primarily dairies and hog farms), the most promising options for reducing these emissions involve recovering the methane at these facilities and using it for energy. Additionally, in the effluent management systems, where the animal waste is gathered and/or stored in a covered digester lagoon and permitted to decay anaerobically, farmers are allowed to collect the generated CH_4 for heating and bio-energy use. Once CH_4 is removed from the waste, the remaining digestate can be used as a fertilizer and soil conditioner [43]. This option saves farmers money for energy costs and reduces CH_4 emissions to the atmosphere [109]. Additionally, during anaerobic digestion of the waste/manure, N_2O emission is negligible since N_2O is formed during aerobic nitrification and anaerobic denitrification [40]. This is an important N_2O mitigation option which reduce N_2O emission in the farming system as follows [43]: (1) reduce the total amount of excreta N returned to pasture; (2) increase the efficiency of excreta and/or fertilizer N; and (3) avoid soil conditions that favor N_2O emissions

5.2.3. Land application

GHG emissions from animal manure and wastewater management systems are influenced by different physicochemical and biological factors. The key factors responsible for CH_4, CO_2, and N_2O emissions are soil moisture, temperature, manure loading rates by the animal, depth of manure in the pen, redox potential, available C, diets, and microbial process. Limited studies on the overall impact of effluent application on the whole suite of gaseous emissions have indicated different effects on the emissions of greenhouse gasses. For example, injecting slurry into soil may reduce NH_3 emissions. In contrast, such slurry incorporation into the soil may trigger N_2O emissions. Similarly, anaerobic digestion of effluents and its subsequent land application can reduce N_2O emission. However, under anaerobic condition with substrate pH may increase higher NH_3 emissions and also higher emissions of CH_4. On the other hand, the direct applications of animal waste either solids or liquid form to pasture and/or crop land can result in CH_4 emissions. Additionally, this method is prone to N losses, and up to 90% of manure N losses occur in various forms, including N_2O. Ammonia (NH_3) also acts as a precursor for N_2O and NO production when emitted from animal excreta [110] and thus, any approach mitigating NH_3 will also reduce N_2O emissions. Brink et al. [111] reported that NH_3 abatement may have a contrasting effect on N_2O emissions, while abatement of N_2O results in a net decrease in NH_3 volatilization.

Application of swine slurry to crop/pasture field resulted the high emissions of gaseous N, which also led to constraints on the amount of slurry N that can be applied per hectare of land. One potential option could be decreasing N content of the slurry by feeding low N diets. However, little work has been done on the dietary manipulation as a means of decreasing N losses without compromising the animal production. It was reported that a greater decrease in N excretion can be achieved by decreasing the crude protein (CP) content of the diet. The denitrification rate was lower from slurry collected from pigs on low CP diet (140 g kg^{-1} CP) than a standard diet (205 g kg^{-1} CP), showed similar N$_2$O emissions from both treatments [112]. Addition of available C to soil was previously found to increase denitrification and also the ratio of N$_2$:N$_2$O produced [113]. Therefore, higher C in the low-CP diet would have favored the production of N$_2$ rather than N$_2$O as the product of denitrification [43]. Therefore, the abatement strategies to reduce gaseous emissions of NH$_3$, N$_2$O, and CH$_4$ from animal waste and farm effluents would therefore require some trade-offs among these three gases.

5.3. Housing system design and management

The structure of a housing system, for example the combination of the floor-system, manure collection, and the manure removal system, largely determines the level of the emission of gaseous compounds, especially the emission of ammonia. Housing systems that reduce gaseous emissions basically comprise of at least one or more of the subsequent abatement principles [64]:

1. Reduction of emitting manure surface.

2. Fast and complete removal of the liquid manure from the pit to external slurry storage.

3. Applying an additional treatment, such as aeration, to obtain flushing liquid.

4. Cooling the manure surface.

5. Changing the chemical/physical properties of the manure, such as decreasing the pH.

The housing systems that have been developed to include the above principles are able to reduce their gaseous (ammonia) emissions to the atmosphere from approximately 30% to 80%. Brink et al [111] in Europe, estimated that while it may increase nitrous oxide emissions significantly the emission of methane was hardly affected by animal housing adaptations for ammonia abatement. Usually limited with mixed results, the effect of animal housing adaptations on odor emission was demonstrated. Furthermore, control of the indoor climate in terms of reducing air velocity at the manure surface, which decreases mass transfer at the manure-air interface [114-115], and has relatively low indoor temperatures, and results in less fouling of floors especially for pigs [116], can reduce ammonia and odor emissions to the atmosphere even further if emitting surface is reduced. The slurry-based manure management system methane emissions increase with the temperature of the stored slurry. The reduction of slurry storage temperature from 20 to 10°C resulted in a reduction in CH$_4$ emissions of 30 -50% [117]. In animal houses the volatilization of NH$_4$$^+$ is linked to the ammonium ($NH_4{}^+$) concentration, the pH and surface area of the manure stored in the house, the area contaminated by the animals, and the temperature and ventilation of the housing system. Decreasing the surface area soiled by manure has the potential of reducing NH_3 emissions. In addition, the cattle housing ammonia and CH_4 emissions can be reduced through a more regular removal of manure to a closed storage system and through the systematic, everyday scraping of the floor.

5.4. End-of-pipe air treatment

This approach for reducing emission is basically a treatment of the exhaust air released from mechanically ventilated animal housings. The main advantages for this approach are that air can be treated without affecting the routine management operations and structural design inside the barn. End-of-pipe air treatment techniques are applied mainly for treating ammonia released from the exhaust air of livestock facilities and are commercially available off-the-shelf in the Netherlands, Germany, and Denmark [64]. The state of art of End-of-Pipe techniques and their scrubbing performance are briefly discussed in the section 4. However, existing scrubber used for ammonia scrubbing does not scrub methane. Thus, an appropriate scrubber needs to be designed to scrub methane released from the exhausts of the animal housing or covered manure storage.

6. Conclusions

The livestock industry is a significant contributor to the economy of any country. More than one billion tons of manure is produced annually by livestock and poultry reared in the United States. Animal manure is a valuable source of nutrients and renewable energy in the country. On the other hand, livestock manure management is extremely challenging and resultant gaseous emissions may contribute to global warming. Livestock manure produces odor and emits GHGs such as carbon dioxide, methane and nitrous oxide that have prompted significant environmental quality degradation concerns. Major sources of agricultural GHG emissions include methane from enteric fermentation, manure storage and spreading, and nitrous oxide mainly from application of manure on land. GHG emissions from animal manure and wastewater management systems are influenced by soil/manure moisture, temperature, manure loading rate by the animal, depth of manure in the pen, redox potential, available carbon, diets, and microbial process. Mitigation options for GHG emissions are source and characteristics dependent. Mitigation of GHG emissions from animal waste must be addressed in the context of integrated waste management. Manure as a biomass goes through different chemical and biological processes for bio-energy recovery and thus, reduced methane emission. Anaerobic bio-digesters, covered lagoons or manure storages with methane flaring systems or small electricity generators are gaining popularity as viable technologies to abate GHG emissions from manure storages. In addition, since methane is generated under anaerobic conditions, switching manure management from liquid to dry manure, such as pack-bedded dairy option and hoop structure swine buildings with bedding, are other possibly effective management strategies to reduce methane emission. Mitigation technologies for GHG emissions released from the animal housings are lacking. Therefore, it is urgently needed that other effective and economically feasible GHG mitigation technologies be developed for the reduction of GHG emissions from CLOs to ensure improved environmental quality and sustainable and livestock agriculture in order to meet the milk and protein demand of an ever increasing world population.

Author details

Md Saidul Borhan* and Shafiqur Rahman
Agriculture and Biosystems Engineering Department, North Dakota State University, Fargo, ND, USA

Saqib Mukhtar and Sergio Capareda
Biological and Agricultural Engineering Department, Texas A&M University, TAMU College station, TX, USA

7. References

[1] IPCC. Chapter 2: Changes in atmospheric constituents and in radiative forcing. In *Climate Change 2007: The Physical Science Basis*; Contribution of Working Group I to the 4th Assessment Report of the Intergovernmental Panel on Climate Change; Cambridge University Press: New York, NY, USA; 2007.

[2] Burns RT, Li H, Xin H, Gates RS, Overhults DG, Earnest J, Moody L. Greenhouse Gas (GHG) Emissions from Broiler Houses in the Southeastern United States. *Presented at the ASABE Annual International Meeting*, St. Joseph, MI, USA, 29 June–2 July 2008; Paper No. 084649.

[3] USEPA. United States Environmental Protection Agency. *Inventory of U.S. Greenhouse Gas Emissions and Sinks*. Report Number EPA 430-R-08-005; U.S. Government Printing Office: Washington, DC, USA, 2008.

[4] Crosson, P., L. Shalloo, D. O'Brien, G. J., Lanigan, P. A. Foley, and T. M. Boland. 2011. A review of whole farm systems models of greenhouse gas emisisons from beef and dairy cattle production systems. Animal Feed Science Technology. 166-167: 29-45.

[5] Chadwick, DR, Pain, BF, Brookman, SKE. Nitrous oxide and methane emissions following application of animal manures to grassland. Journal Environmental Quality 2000; 29, 277–287.

[6] Naqvi SMK, Sejian V. 2010. Global Climate Change: Role of Livestock. Asian Journal of Agricultural Sciences 2010; 3(1) 19-25.

---

* Corresponding Author

[7]   Sejian V, Naqvi SMK. 2012. Livestock and Climate Change: Mitigation Strategies to Reduce Methane Production. Available at http://cdn.intechopen.com/pdfs/30637/InTech-Livestock_and_climate_change_mitigation_strategies_to_reduce_methane_production.pdf [accessed 5.15.12].

[8]   Mukhtar S. *Manure Production and Characteristics: Its Importance to Texas and AnimalFeeding Operation*; University libraries digital Texas A&M: College Station, TX, USA, 2009.Available online: http://tammi.tamu.edu/pubs.html (accessed on 05/02/12).

[9]   Sedorovich, DM, Rotz CA, Richard TL. Greenhouse Gas Emissions from Dairy Farms. In: roceedings ASABE Annual International Meeting, St. Joseph, MI, USA, 17–20 June 2007; Paper o. 074096.

[10] Hansen MN, Kai P, Moller HB. Effects of anaerobic digestion and separation of pig slurry on odor emission. Applied engineering in agriculture 2006; 22(1) 135-139.

[11] Jungbluth T, Hartung E, Brose G. Greenhouse emissions from animal houses and manure stores. Nutrients Cycling Agroecosystem 2001; 60 133–145.

[12] Kinsman R, Saucer FD, Jackson HA, Wolynetz MS. Methane and carbon dioxide emissions from dairy cows in full lactation monitored over a six-month period. Journal Dairy Science 1995; 78 2760–2766.

[13] Amon, B, Amon T, Boxberger J, Alt Ch. Emissions of NH_3, N_2O, and CH_4 from dairy cows housed in a farmyard manure tying stall (housing, manure storage, manure spreading). Nutrient Cycling Agroecosyst. 2001; 60 103–113.

[14] Ngwabie NM, Jeppsson KH, Nimmermark S, Swensson C, Gustafsson G. Multi-location measurements of greenhouse gases and emission rates of methane and ammonia from a naturally-ventilated barn for dairy cows. Biosystems Engineering 2009; 103 68–77.

[15] Crutzen PJ, Aselmann I, Seiler W. Methane production by domestic animals, wild ruminants, other herbivorous fauna and humans. Tellus 1986; 38B 271-284.

[16] Gibbs, M and Johnson, D. E. Methane emissions from digestive process of livestock. International anthropogenic methane emissions: Estimates for 1990. EPA 230 R-93-010, pp. 2-1-2-44, 1994.

[17] Fiedler AM, Muller HJ. Emissions of ammonia and methane from a livestock building natural cross ventilation. Meteorologische Zeitschrift 2011; 20(1) 059-065.

[18] Sun H, Trabue SL, Scoggin K, Jackson WA, Pan Y, Zhao Y, Malkina IL, Koziel JA, Mitloehner FM. Alcohol, volatile fatty acid, phenol, and methane emissions from dairy cows and fresh manure. J. Environ. Qual. 2008; 37 615-622.

[19] Zoe L, Chen D, Mei B, Naylor T, Griffith D, Hill J, Denmead OT, McGinn S, Edis R. Measurement of greenhouse gas emissions from Australian feedlot beef production using open-path spectroscopy and atmospheric dispersion modeling. Aust. J. Exp. Agric. 2008, 48 244–247.

[20] McGinn SM, Chen D, Loh Z, Hill J, Beauchemin KA, Denmead OT. Methane emissions from feedlot cattle in Australia and Canada. Aust. J. Exp. Agric. 2008, *48* 183–185.

[21] Van Haarlem, RP, Desjardins RL, Gao, Z, Flesch TK, Li, X. Methane and ammonia emissions from a beef feedlot in western Canada for a twelve-day period in the fall. Can. J. Anim.Sci. 2008; 88 641–649.

[22] Phetteplace HW, Johnson DE, Seidl AF. Greenhouse gas emissions from simulated beef and dairy livestock systems in the United States. Nutr. Cycling Agroecosyst. 2001; 60 99–102.

[23] Harper LA, Denmead OT, Freney JR, Byers FM. Direct measurements of methane emissions from grazing and feedlot cattle. Journal Animal Science 1999; 77 1392–1401.

[24] Groot Koerkamp PWG, Uenk, GH. Climatic conditions and aerial pollutants in and emissions from commercial animal production systems in the Netherlands. In: Proceedings of International Symposium Ammonia and Odour Control from Animal Facilities. 6-10.10.1997, Vinkeloord, pp. 139–144.

[25] Gallmann E, Hartung E, Jungbluth T. 2000. Assessment of two pig housing and ventilation systems regarding indoor air quality and gas emissions - diurnal and seasonal effects. In: prceedings Ag Eng 2000, Paper 00-FB-002. 2000, Warwick, England, pp. 140–141.

[26] Gallmann E, Hartung E, Jungbluth T. Long-term study regarding the emission rates of ammonia and greenhouse gases from different housing systems for fattening pigs - final results. In: Proceedings of the International Symposium on Gas and Odor Emissions from Animal Production, Horsens, Denmark. CIGR, 2003, pp. 122–130.

[27] Sharpe RR, Harper LA., Simmons JD. Methane emissions from swine houses in North Carolina. Chemosphere Global Change Science 2001; 3 1–6.

[28] Rahman S, Lin D, Zhu J. Greenhouse gas (GHG) emissions from mechanically ventilated deep pit swine gestation operation. Journal Civil Engineering 2012; 2 104, doi:10.4172/2165-784x.1000104.

[29] Dong H, Zhu H, Shang B, Kang H, Zhu H, Xin H. Greenhouse gas emissions from swine barns of various production stages in suburban Beijing, China. Atmospheric Environment 2007; 41 2391–2399.

[30] Hahne J, Hesse D, Vorlop KD. Trace gas emissions from fattening pig housing. Landtechnik 1999; 54 (3) 180–181.

[31] Borhan MS, Capareda S, Mukhtar S, Faulkner WB, McGee R, Parnell Jr CB. Determining Seasonal Greenhouse Gas Emissions from Ground Level Area Sources in a Dairy Operation in Central Texas. J. Air and Waste Management 2011; 61 786–795.

[32] Borhan , MS, Capareda S, Mukhtar S, Faulkner WB, McGee R, Parnell Jr CB. Greenhouse Gas Emissions from Ground Level Area Sources in Dairy and Cattle Feedyard Operations. Atmosphere 2011; 2 303-329.

[33] Bjorneberg DL, Leytem, AB, Westermann T, Griffiths PR, Shao L, Pollard MJ. Measurement of atmospheric ammonia, methane, and nitrous oxide at a concentrated

dairy production facility in southern Idaho using open-path FTIR spectrometry. Transactions of ASABE 2009; 52(5) 1749-1756.

[34] Zhu Z, Dong H, Zhou Z. Ammonia and greenhouse gas in a dairy cattle building with daily manure collection system. In: Proceedings Annual International Meeting, ASABE St. Joseph MI; 2011; Paper no 1110761.

[35] Leytem AB, Bjorneberg DL, Dungan RS. Emissions of ammonia and greenhouse gas from dairy production facilities in southern Idaho. Western Nutrients Management Conference, Reno Nevada 2011; 9 29-34

[36] Leytem AB, Dungann RS, Bjorneberg DL, Koehn AC. Emissions of ammonia, carbon dioxide, and nitrous oxide from dairy cattle housing manure management systems. Journal Environmental Quality 2010 40:1383 - 1394. doi: 10.2134/jeq2009.0515.

[37] Hensen A, Groot TT, van den Bulk WCM, Vermeulen AT, Olesen JE, Schelde K. Dairy farm CH_4 and N_2O emissions from square meter to full farm scale. Agric. Ecosyst. Environ 2006; 112 146–152.

[38] Guarrino, M., Fabbri, C., Navarotto, P., et al., 2003. Ammonia, methane and nitrous oxide emissions and particulate matter concentration in two different buildings for fattening pigs. In: Proceedings of the International Symposium on Gas and Odor Emissions from Animal Production, Horsens, Denmark. CIGR, pp. 140–149.

[39] Baudouin, N, Laitat M, Vandenheede, M. Emissions of ammonia, nitrous oxide, methane, carbon dioxide and water vapor in the raising of weaned pigs on straw-based and sawdust-based deep litters. Animal Research 2003; 52 299–308.

[40] Osada T, Rom HB, Dahl P. Continuous measurement of nitrous oxide and methane emissions in pig units by infrared photoacoustic detection. Transactions of the ASAE 1998; 41 (4) 1109–1114.

[41] Sneath RW, Chadwick DR, Phillips VR, Pain BF, A UK inventory of Methane emissions from farmed livestock. Silsoe Research Institute, Wrest Park, Silsoe, Bedford MK45 4HS, UK Commissions WA0604 and WA0605; 1997.

[42] Su J-J, Liu B-Y, Chang Y-C. Emission of greenhouse gas from livestock waste and wastewater treatment in Taiwan. Agriculture, Ecosystems and Environment 2003; 95 253–263.

[43] Saggar S, Bhandral R, Hedley CB, Luo J. A review of emissions of methane, ammonia, and nitrous oxide from animal excreta deposition and farm effluent application in grazed pastures. New Zealand J. Agric. Res. 2004; 47 513–544.

[44] Le Mer JL, Roger P. Production, oxidation, emission and consumption of methane by soils: A review. European Journal of Soil Biology 2001; 7 25-50.

[45] Wulf S, Maeting M, Bergmann S, Clemens J, 2001: Simultaneous measurement of NH_3, N_2O and CH_4 to assess efficiency of trace gas emission abatement after slurry application. Phyton 2001; 41 131-142.

[46] IPCC 2001: Technical summary. *In:* Houghton, J. T.; Ding, Y.; Grigg, D. J.; Noguer, M.; van der Linden, P. J.; Dai, X.; Maskell, K.; Johnson, C. A. *ed.* Climate change 2001: The

scientific basis. Contributions of Working Group I of the Intergovernmental Panel on
Climate Change. Cambridge, Cambridge University Press.

[47] Robertson G P, Tiedje JM. Nitrous oxide sources in aerobic soils; nitrification,
enitrification and other biological processes. Soil Biology and Biochemistry 1987; 19:
187-193.

[48] Velthof GL, Kuikman PJ, Oenema O. Nitrous oxide emission from animal manures
applied to soil under controlled conditions. Biol Fertil Soils 2003; 37 221-230.

[49] Ball, B. C, Horgan GW, Clayton H, Parker J P. 1997: Spatial variability of nitrous oxide
fluxes and controlling soil and topographic properties. Journal of Environmental
Quality 1997; 26 1399-1409.

[50] Castaldi S, Smith KA. 1998: Effect of cycloheximide on N_2O and NO_3^- production in a
forest and agricultural soil. Biology and Fertility of Soils 1998; 27 27-34.

[51] MacKenzie AF, Fan MX, Cadrin F. Nitrous oxide emission in three years as affected by
tillage, corn-soybean-alfalfa rotations, and nitrogen fertilization. Journal of
Environmental Quality 1998; 27 698-703.

[52] Dobbie KE, McTaggart IP, Smith KA. 1999: Nitrous oxide emissions from intensive
agricultural systems: variations between crops and seasons, key driving variables, and
mean emission factors. Journal of Geophysics Research 1999; 104 26891-26899

[53] Dobbie KE, Smith KA. The effect of temperature, water-filled pore space, and land use
on N2O emissions from an imperfectly drained gleysol. European Journal of Soil
Science 2001; 52 667-673.

[54] Staley TE, Caskey WH, Boyer DG. Soil denitrification and nitrification potentials during
the growing season relative to tillage. Soil Science Society of America Journal 1990; 54
1602 -1608.

[55] Hansen S, Maehlum JE, Bakken LR. N2O and CH_4 fluxes in soil influenced by
fertilization and tractor traffic. Soil Biology and Biochemistry 1993; 25 621-630.

[56] Christensen S, Ambus P, Arah J, Clayton H, Galle B, Griffith D, Hargreaves K,
Klemedtsson, L, Lind A-M., Maag M, Scott A, Skiba U, Smith K, Welling M, Wienhold
F. Nitrous oxide emission from an agricultural field: comparison between
measurements by flux chamber and micrometeorological techniques. Atmospheric
Environment 1996; 30 (24) 4183–4190.

[57] Hudson N, Ayoko GA. Odour sampling 1: Physical chemistry considerations.
Bioresource Technology 2008; 99 3982–3992

[58] Eklund B, Balfour W, Schmidt C. 1985. Measurement of fugitive volatile organic
emission rates. Environmental Progress 1985; 4 (3) 199–202.

[59] Gholson AR, Albriton J.R, Yayanti RKM. *Evaluation for the Flux Chamber Method for
Measuring Volatile Organic Emissions from Surface Impoundments*; Project Summary
EPA/600/S3-89/008; U.S. Environmental Protection Agency: Washington, DC, USA,
1989.

[60] Parker DB, Caraway EA, Rhoades MB, Cole NA, Todd RW, Casey KD. Effect of wind tunnel air velocity on VOC flux from standard solutions and CAFO manure/wastewater. Transactions ASABE 2010; 53 831–845.

[61] Hudson N, Ayoko GA, Dunlop M, Duperouzel D, Burrell D, Bell K, Gallagher E, Nicholas P, and N. Heinrich N. Comparison of odor emission rates measured from various sources using two sampling devices. Bioresource Technology 2009; 100(1) 118-124.

[62] Kienbusch MR. 1986. Measurement of Gaseous Emission Rates from Land Surfaces Using an Emission Isolation Flux Chamber - User's Guide. EPA/600/8-86/008. Washington, DC: U.S. Environmental Protection Agency.

[63] Parker D, Ham J, Woodbury Lingshuang BC, Mindy S, Trabue S, Casey K, Cole A. 2012. Standardization of flux chamber and wind tunnel flux measurements for quantifying volatile organic compound and ammonia emissions from area sources at animal feeding operations. Atmospheric Environment In press.

[64] Melse RW, Ognik NWM, Rulkens WH. Air treatment techniques for abatement of emissions from intensive livestock production. The Open Agriculture Journal 2009; 3 6-12.

[65] Estelles F, Melse RW, Ogink NWM, Calvet S. Evaluation of the NH3 removal efficiency of an acid packed bed scrubber using two methods: a case study. Transaction of ASABE 2011; 54(5) 1905-1912

[66] Manuzon RB, Zhao LY, Keener HM, Darr MJ. A prototype acid spray scrubber for absorbing ammonia emissions from exhaust fans of animal buildings. Transactions of the ASABE 2007; 50(4) 1395-1407.

[67] Zhao Y, Aarnink AJA, Hofschreudera P, Groot Koerkamp PWG. Evaluation of an impaction and a cyclone pre-separator for sampling high PM10 and PM2.5 concentrations in livestock houses. Aerosol Science 2009; 40 868 – 878.

[68] Pedersen S, Blanes-Vidal V, Joergensen H, Chwalibog A, Haeussermann A, Heetkamp MJW, Aarnink AJA. Carbon Dioxide Production in Animal Houses: A literature Review. Agricultural Engineering338 International: the CIGR Ejournal.2008 Manuscript BC 08 008, Vol. X.

[69] Licht LA, Miner JR. 1978. A scrubber to reduce livestock confinement building odors. ASAE Paper No. 78-203, St. Joseph, Mich.: ASAE; 1978.

[70] Zhao LY, Riskowski GL, Stroot P, Robert M, Heber AJ. Development of a wet scrubber to reduce dust and gas emissions from swine buildings. ASAE Paper No. 014075. St. Joseph, Mich.: ASAE; 2001.

[71] Melse RW, Ogink NWM. Air scrubbing techniques for ammonia and odor reduction at livestock operations: review of on-farm research in the Netherlands. Transactions of the ASAE 2005; 48(6) 2303-2313.

[72] Chen L, Hoff S, Cai L, Koziel JA, Zelle B. 2009. Evaluation of wood chip-based biofilters to reduce odor, hydrogen sulfide, and ammonia from swine barn ventilation air. Journal of the Air & Waste Management Association 2009; 59: 520-530.

[73] Aarnink AJ, Le PD, Ognik NWM, Becker PM, Verstegen MWA. Odor for animal production facilities: Its relationship to diets. Nut Res Rev. 2005; 18(1) 3-30.

[74] Hahne J, Krause K-H, Munack A, Vorlop K-D. Environmental engineering-reduction of emissions. In Yearbook Agricultural Engineering, 35-43. Matthies EHHJ, Meier F (eds.) Muenster, Germany: VDMA Landtechnik, VDI-MEG, KTBL; 2003.

[75] Hahne J, Krause K-H, Munack A, Vorlop K-D. Vorlop 2005. Bio-engineering, environmental engineering. In Yearbook Agricultural Engineering, 183-188.Harms H-H, Meier F (eds.) Muenster, Germany: VDMA Landtechnik, VDI-MEG, KTBL; 2005

[76] Milleu Partners. 2007. Home page: http://www.milieupartners.nl/index.htm. Last accessed on May 25, 2012.

[77] Shah SB, Westerman PW, Munilla RD, Adcock ME, Baughman GR. Design and evaluation of a regenerating scrubber for reducing animal house emissions. Transactions ASABE 2008; 51(1) 243-250.

[78] Melse RW, Timmerman M. Sustainable intensive livestock production demands manure and exhaust air treatment technologies. Bioresource Technology 2009; 100 5506–5511

[79] Zhao Y, Aarnink AJA, de Jong MCM, Ognik NWM, Groot Koerkamp PWG. Effectiveness of multi-stage scrubbers in reducing emissions of air pollutant from pig houses. Transactions ASABE 2011; 54(1) 285-293.

[80] Zhang Y, Polakow JA, Wang X, Riskowski GL, Sun Y, Ford SE. 2001. An aerodynamic deduster to reduce dust and gas emissions from ventilated livestock facilities. In: Livestock Environment VI: In: Proceedings of the 6th International Symposium. pp. 596-603. Kentucky; 2001.

[81] Ocfemia K, Zhang Y, Tan Z. Ammonia absorption in a vertical sprayer at low ammonia partial pressures. Transactions of the ASABE 2005; 48(4) 1561-1566.

[82] Feddes, J, Edeogu I, Bloemendaal B, Lemay S, Coleman R. 2001. Odor reduction in a swine barn by isolating the dunging area. In: Stowell, R. R, Bucklin R, Bottcher R.W(eds): Livestock Environment VI: proceedings of the 6th International Symposium. ASAE, Louisville, Kentucky; 2001.

[83] Burgess JE, Parsons SA, Stuetz RM. Developments in odour control and waste gas treatment biotechnology: A review. Biotechnology Advances 2001; 19 35-63.

[84] Schmidt DR, Janni KA, Nicolai RE. 2004. Biofilter Design Information BAE #18, University of Minnesota, Department of Biosystems and Agricultural Engineering, <http://www.manure. umn.edu/assets/baeu18.pdf> accessed 05.15.12

[85] Elenbass A, Zhao LY, Young Y, Wang X, Riskowski GL, Ellis M, Heber AJ. Effects of room ozonation on air quality and pig performance. Transaction ASAE 2005; 48(3) 1167-1173.

[86] Wang L, Oviedo-Rondón EO, Small J, Liu Z, Sheldon BW, Havenstein GB, Williams MC. Farm-scale evaluation of ozonation for mitigating ammonia

concentrations in broiler houses. Journal of Air and Waste Management Association 2010; 60 789-796.

[87] Powers W. Odor control for livestock systems. Journal of Animal Science 1999; 77: 169-176.

[88] Hansen KH, Angelidaki I, Ahring BK. Anaerobic digestion of swine manure: Inhibition by ammonia. Water research 1998; 32(1) 5-12.

[89] Hobbs PJ, Pain BF, Kay RM, Lee PA. Reduction of odorous compounds in fresh pig slurry by dietary control of crude protein. Journal of the Science of Food and Agriculture 1996; 71: 508-514.

[90] Sutton AL, Kephart KB, Verstegen MWA, Canh TT, Hobbs PJ. Potential for reduction of odorous compounds in swine manure through diet modification. Journal of Animal Science 1999; 77 430-439.

[91] Castillo AR, Kebreab E, Beever DE. The effect of protein supplementation on nitrogen utilisation in grass silage diets by lactating dairy cows. Journal of Animal Science 2001; 79 247-253.

[92] Kendall DC, Lemenager KM, Richert BT, Sutton AL, Frank JW, Belstra BA, and Bundy D. (1998). Effects of Intact Protein Diets Versus Reduced Crude Protein Diets Supplemented with Synthetic Amino Acids on Pig Performance and Ammonia Levels in Swine Buildings. Swine Day1998; 141-146.

[93] Kebreab E, Strathe A, Fadel J, Moraes L, France J. 2010. Impact of dietary manipulation on nutrient flows and greenhouse gas emissions in cattle. R. Bras. Zootec 2010; 39 458-464.

[94] Hao X, Benke MB, Gibb DJ, Stronks A, Travis G, McAllister TA. Effects of dried distillers' grains with solubles (wheat-based) in feedlot cattle diets on feces and manure composition. Journal of Environmental Quality 2009; 38: 1709-1718.

[95] Kebreab E, Starthe A, France J, Beever DE, Castolo AR. Nitrogen pollution by dairy cows and its mitigation. Nutrient Cycling in Agroecosystems 2001; 60 275-285.

[96] Eckard R J, Grainger C, and C. de Klein CAM. Options for the abatement of methane and nitrous oxide from ruminant production: A review. Livestock Science 2010; 130(1-3) 47-56.

[97] Boadi D, Benchaar C, Chiquette J, Massé D. Mitigation strategies to reduce enteric methane emissions from dairy cows: update review. Available at: http://classes.uleth.ca/200901/biol4500a/Readings/Beauchemin1.pdf; 2004 [accessed 5.20.12].

[98] Sejian V, Lakritz J, Ezeji T, Lal R. Forage and Flax seed impact on enteric methane emission in dairy cows. Research Journal of Veterinary Sciences 2011; 4(1) 1-8.

[99] Beauchemin KA, Kreuzer M, O'Mara F, McAllister TA. Nutritional management for enteric methane abatement: a review. Australian Journal of Experimental Agriculture 2008; 48(1-2) 21–27.

[100] Cahn TT, Sutton AL, Aarnink AJA, Verstegen MWA, Schrama JW, Bakker GCM. Dietary carbohydrates alter the fecal composition and pH and the ammonia emission from slurry of growing pigs. Journal Animal Science 1998; 76: 1887-95.

[101] Kim IB, Ferket PR, Powers WJ, Stein HH, Van Kempen TATG. Effects of different dietary acidifier sources of calcium and phosphorus on ammonia, methane and odorant emission from growing finishing pigs. Asian-austral as Journal Animal Science 2004; 17(8): 1131-8.

[102] Van Middelkoop JH, Van Harn J. The influence of reduced protein levels in broiler feed on NH3 emissions. Transl Silsoe Res Institute 1998; 66: 34.

[103] Elwinger K, Svensson L. Effect of dietary protein content, litter and drinker type on ammonia ammonia emission from broiler houses. J Agricultural Engineering Research 1996; 64(3) 197-208.

[104] Angel R, Powers W, Applegate T. Diet impacts for mitigating air emissions from poultry. In: Proceedings of the 8th International Livestock Symposium (ILES VIII), Iguassu Falls, August 31 – September 4, 2008.

[105] Van Duinkerken G, André G, Smits MCJ, Monteny GJ, Lebek LBJ. Effect of rumen-degradable protein balance and forage type on bulk milk urea concentration and emission of ammonia from dairy cow houses. J Dairy Science 2005; 88: 1099-112.

[106] Smits MCJ, Valk H, Elzing A, Keen A. Effect of protein nutrition on ammonia emission from a cubicle house for dairy cattle. Livest Prod Sci 1995; 44(6) 147-56.

[107] Velthof GL, Nelemans JA, Oenema O, Kuikman PJ. Gaseous nitrogen and carbon losses from pig manure derived from different diets. Journal Environmental Quality 2005; 34 698-706.

[108] VanderZaag, AC, Gordon RJ, Glass VM, Jamiesson RC. Floating covers to reduce gas emissions from liquid manure storage: A review. Transactions of the ASABE 2008; 24(5) 657-671.

[109] de Klein CAM. Mitigating N2O emissions from agriculture - an overview of the science. In: Proceedings of the Trace Gas Workshop, 18-19 March 2004, Wellington, New Zealand; 2004.

[110] Clemens J, Ahlgrimm HJ. Greenhouse gases from animal husbandry: mitigation options. Nutrient Cycling in Agroecosystems 2001; 60 287-300.

[111] Brink C, Kroeze C, Klimont Z. Ammonia abatement and its impact on emissions of nitrous oxide and methane: Part 2. Application for Europe. Atmospheric Environment 2001; 25 6313-25.

[112] Misselbrook T H, Chadwick DR, Pain BF, Headon DM. Dietary manipulation as a means of decreasing N losses and methane emissions and improving herbage N uptake following application of pig slurry to grassland. The Journal of Agricultural Science 1998; 130 183-191.

[113] Weier KL, Doran JW, Power JF, Walters DT. Denitrification and the dinitrogen nitrous-oxide ratio as affected by soil-water, available carbon, and nitrate. Soil Science Society of America Journal 1993; 57 66-72.

[114] Arnink AJA, Elzing A. Dynamic model for ammonia volatilization in housing with partially slatted floors, for fattening pigs. Livest Prod Sci 1998; 53(3) 153-69.

[115] Monteny GJ, Schulte DD, Elzing A, Lamaker EJJ. A conceptual mechanistic model for the ammonia emissions from free stall cubicle dairy cow houses. Transactions ASAE 1999; 41 193-201.

[116] Aarnink AJA, Schrama JW, Heetkamp MJW, Stefanowska J, Huynh TTT. Temperature and body weight affect fouling of pig pens. Journal Animal Science 2006; 84 2224-31.

[117] Hilhorst, MA, Mele RW, Willers HC, Groenestein CM, Monteny GJ. Effective strategies to reduce methane emissions from livestock. ASAE, Paper no 01-4070; 2001.

Chapter 14

Construction and Demolition Waste Management in Turkey

Hakan Arslan, Nilay Coşgun and Burcu Salgın

Additional information is available at the end of the chapter

1. Introduction

Natural resources generally were consumed by the construction sector in huge amounts and have produced significant quantity of construction and demolition (C&D) wastes. C&D waste constitutes the largest volume of all solid wastes. Statistics from various studies have reported the high amount of C&D waste generated. For instance U.S. construction industry generated over 100 million tons of C&D waste per year [1], and approximately 29% of solid waste stream in the U.S. created by the construction sector [2]. Also, C&D waste contributes more than 50% of all landfill volume in the UK, [3] and 70 million tons of C&D waste is discarded each year[4]. Craven et al. [5] have reported that construction activities generated approxiamety 20–30% of all waste in Australia which were entering to the landfills. The annual generation of C&D waste in Hong Kong between 1993 and 2004 doubled and have reached the amount of 20 million tons in 2004 [6]. Nearly 23% of the solid waste in Hong Kong comes from the construction sector activities [7, 8]. The huge amount of construction waste streams in different countries has revealed the importance of local actions in order to manage, recycle and re-use the wastes generated through the lifecycle of buildings.

C&D waste was generally defined as a mixture of inert and non-inert materials arising from construction, excavation, renovation, refurbishment, demolition, roadwork and other construction-related activities. Inert materials can be comprised of whether soft inert materials such as soil, earth and slurry or hard inert materials of rocks and broken concrete. Non-inert materials has also included wastes of metals, timber, plastics and packaging [6]. The negative environmental affects of C&D debris started by dumping them into forests, streams, ravines and empty land that has resulted and caused erosion; contaminates wells, water tables and surface waters. They also attracted pests and had a potential to create fire. All these large volumes of construction in many countries have contributed primarily in increasing waste strain landfill capacities and then have resulted to lead environmental concerns. It can also be pointed out that to disposal of construction waste is quite difficult because of its content of hazardous materials such as asbestos, heavy metals, persistent organic compounds and volatile organic compounds (VOCs). These wastes threaten the human health and the natural/artificial environment with various effects. The overall impact of the C&D waste not only had an impact on the environmental but also on the economical sustainability of the countries because of the fact that the construction sector as a base and crucial variable that was related all the other sectors.

The rise in the amount of C&D waste has caused serious problems both globally and locally. In this context, the management of C&D waste has become one of the major environmental issues in the construction industry. The environmental and economic effects of C&D wastes can be reduced by rational management. The aim of construction C&D waste management is waste minimization and appropriate disposal, both of which help to reduce negative environmental impacts. As C&D wastes are considered an important environmental problem in many countries and many regulations on these issues have been introduced. C&D waste management is adopted as part of state policy and specifications and guidance about these issues have been prepared. There are studies and documents on determining targets for the management and recovery of C&D waste management for member states of the European Union.

Construction sector in Turkey can be regarded as the main engine considering the whole economic activities. In the last decades, the role of contruction sector has drastically increased by the new housing and infrastructure invetsments countrywide. Also, due to high earthquake risks the intensive works such as retrofitting, reinforcement and demolition have continue for those which were under serious risk. 66% of Turkey's land is in the first and second level eathquake zone where this reflects nearly 71% of the countries population. When the existing building stock of Turkey is considered, it is obvious that most part of the housing stock must be transformed through demolition, retrofitting and reinforcement activities which should be applied in the short term due to high earthquake risk [9]. Thus, all these activities will increase the C&D waste and the need for efficient waste stream management will be more important during this transformation process to sustainable housing stock with legal building codes.

This study briefly focused on C&D waste generation and C&W management issues in Turkey by giving information about the general framework of C&D waste management with legal aspects. Management policy on C&D waste is discussed through the municipality actions and implementations. The case study of Istanbul Metropolitan Municipality aims to identify the current situation in construction and demolition waste management in Turkey by demonstrating a better understanding of the actual waste management strategy implementations in local level.

2. Construction and demolition waste generation

Waste materials have started to be generated during the construction phase and go on throughout the lifespan of the building such as usage, renovation and demolition. The generation of C&D waste differs depending on the type and function of the building in each phase of construction, usage, demolition and reconstruction stages. As an example it can be figured out that 10% of the materials used during the construction stage become waste. These were described as clean wastes that were relatively easy to classify for recycle and re-use. It is estimated that demolition and renovation operations produce ten times more waste than generated in construction phase [10]. The classification of these wastes generated during these stages were difficult as they were contaminated, complex.

The amount and type of waste materials varies during the production of construction materials, including wastage due to cutting or reformation during production and losses due to non-standard or defective products. Factors such as design defects, erroreous designation data, detail deficiencies, production preference defects, lack of communication between designers, lack of information during the design phase (management, preliminary works, pre-designing, application/detail design) may all result in wastage during production, usage and demolition stages.

During the construction stage, wastage is caused by many factors, including unused materials, incorrect materials; surplus stencils or nails, packages of construction materials or components; surplus concrete materials due to fractures or deformations due to improper storage or preservation of construction materials and components arriving at the construction site; necessary disassembly due to production errors; erroneous cuttings; improper equipment; bad weather conditions; relocating materials; and erroneous measurements [11]. The level of wastage caused during the construction stage depends on the construction system, the project and the variety of construction materials and components.

Due to renovations and refurbishments for reasons such as deterioration, corruption, alteration of needs, or fashion, many construction materials and components are changed and the old materials become wastes. These processes sometimes occur frequently. The usage stage is the longest stage of the construction lifecycle. Therefore, significant levels of wastes are generated during this stage. It is estimated that 30-50% of overall construction waste results from renovation activities [12]. A previous study determined that very frequent renovations are made in construction materials and components in Turkey, and that 74% of these materials are dumped [13].

During the demolition stage, when buildings have completed their operational life, the entire building, entire materials and components become wastes and thus the amount of waste material generated increases. Planned and systematic debris removal is of great importance. Structural problems of individual buildings may include collapsing entirely, due to earthquakes, illegal structuring, urban transformation processes etc. For example, after the Marmara Earthquake of 1999 in Turkey, the removal of approximately 13 million tons of debris created problems. These wastes were left in empty spaces and fill areas and some of them were dumped at sea [14]. Construction sector were aware of the waste management system during the intensive construction process and actively involved in a more efficient organization structure to reduce the negative impact of C&D wastes.

3. Construction and demolition waste management

The management of C&D waste has become one of the major environmental issue in the construction industry because of its long term affects. Uncontrolled dumping of C&D wastes not only represents a significant environmental burden but also a financial cost as well. Environmental and economic effects of C&D wastes can be reduced by a rational management policy. The aim of construction C&D waste management is based on waste minimization and appropriate disposal, which both two help to reduce negative environmental impacts. The specifications of European Union can be evaluated under three principles of waste management [15].

- **Waste Prevention:** This is a key factor in any waste management strategy. The aim is to minimize the waste before construction by detailed frame design and material use plans
- **Recovery:** This stage aims to reduce the environmental effects of unavoidable wastes, through reuse and recycle strategies.
- **Proper Storage:** This stage involves storage options in an appropriate way for non-recoverable wastes generated at registered sites. Easy access path will help to increase the efficiency and help the participation of building users.

Waste management is evaluated through its economic benefits but also the options of waste prevention not only have helped the reuse and recycling money savements but also generate broader environmental benefits such the conservation of natural resources. Reuse and waste prevention reduce the air and water pollution associated with materials manufacturing and transportation that indirectly is an asset for human health. This also saves energy and reduces attendant greenhouse gas production. The recycling of many materials requires less energy than production from virgin stock, and can also reduce transportation requirements and associated impacts.

3.1. Preventing and reducing C&D wastes

Waste prevention and reduction begins during the building materials production stage. Recuperation of wastes generated during the producing process has a crucial role in prevention of waste during subsequent stages of the construction lifespan. Design phase is very crucial part of the building production process in order to consider the generation of building waste and determination of the re-use and recycle options and features of building materials. The design phase prevention directly affects the volume of construction waste stream in construction, use and demolition phases. It is estimated in some international researches that nearly one third of the waste generated could be derived from the designers' inability to manage the waste reduction preventions [16]. The following precautions may then be taken during the design phase:

- Designs of buildings optimized to meet the needs of users,
- Rather than designing places at unnecessary scales, designing places of adequate sizes, based on the principle of "Less is More" and reducing the amounts of materials that will become C&D waste,

- Utilizing durable and repairable construction materials and components that reduces the need for frequent replacement during the usage stage,
- Creating flexible design solutions to reduce the amount of C&D waste when buildings are modified to meet the changing needs of occupants,
- Preferring standard modules at design phase to reduce material losses during construction phase,
- Selection of products based on lifecycle, production technology of the construction material and the effects of this technology on environment and health,
- Extending the economic lifespan of a structure by regarding the quality and resilience of construction materials,
- Utilizing reusable and/or recyclable materials,
- The use of local materials to reduce waste and energy consumption,
- Devising renovation project instead of demolition projects for old buildings in order to reduce wastes,
- Specifying water-based adhesives and paints.

Purchase and storage conditions should be carefully controlled during the construction stage in order to minimize wastage of surplus raw material and to provide financial savings. Precautions that can be taken during the construction stage include:

- Providing proper transportation of construction products,
- Using construction materials of appropriate sizes and amounts for the design,
- Specifying materials in appropriate sizes to prevent waste resulting from cutting and fracturing of materials,
- Purchasing materials in large quantities to reduce packaging waste associated with smaller quantities,
- Favoring material suppliers that retrieve product packaging,
- Favoring material suppliers that retrieve surplus materials from production,
- Purchasing materials with toxic constituents (when their use is necessary) in small amounts in order to reduce problems when disposing of surplus materials,
- Controlling material deliveries before use and/or storage to prevent on-site waste generation and the return of damaged materials,
- Storing season- sensitive materials appropriately,
- Taking security precautions against vandalism,
- Taking necessary precautions to prevent potential damage to natural habitats during excavation activities,
- Raising awareness among construction workers and making them responsible for the results of their task,
- Mentioning material and waste management issues in agreements with contractors.

The users and implementers should be informed about these issues in order to reduce changes/renovations during the usage stage. During renovation, disassembly of components and transportation of materials should be carried out carefully, to maximize the potential re-use of construction materials.

Before demolition, disassembly of materials to be recycled and reused should be carefully planned. This may prolong the demolition process. However, this is important to maximize economic and environmental benefits. The demolition should cover the following precautions:

- Reduce waste by suggesting appropriate demolition methods,
- Suggesting recycling or reusing the construction materials taken out of constructions completing their life,
- Determining reusable or recyclable materials before demolition (water and electrical system, doors, windows, sanitation systems, etc.),
- Instructing construction staff about the materials that will be disassembled and creating proper strategies to ensure the disassembly of undamaged elements,
- Determining the users and recyclers to whom disassembled materials will be given,
- Paying attention to materials that can contain hazardous constituents (lead-based paints, asbestos insulation materials etc.). Technical assistance should be sought for such issues [13].

3.2. Recovery of C&D wastes

Increasing the recycling and re-use of C&D waste within the industry will help to conserve the dwindling landfill resources. There are growing interests in many parts of the world in recycling and reusing C&D waste by the construction industry. Effective waste management has enabled recovery of 90% of C&D wastes [17].The Netherlands has the highest rate of construction recycling, at 90%, followed by Australia (87%) [18] and Denmark (82%) [19]. Total C&D waste for England was estimated at 86.9 million tonnes in 2008. 53 million tonnes were recycled and a further 11 million tonnes were spread on exempt sites (usually land reclamation, agricultural improvement or infrastructure projects) [20].

C&D waste management and recovery will enable:

- Reducing environmental effects of obtaining further raw materials, their transportation and processing,
- Reducing the dependence on natural resources,
- Reducing emission rates related to production and transportation of building materials,
- Reducing costs associated with new material purchases,
- Reducing the need for disposal sites for C&D wastes,
- Reducing the negative effects on environment through removal and disposal processes,
- Creating a source for the sector by recycling C&D materials to secondary salvaged materials,
- Creating new employment fields.

Reusing and/or recycling of concrete wastes, which represent the highest proportion of C&D wastes, offers both a solution to waste disposal problems and enables preservation of natural resources. While 40% of globally used rocks, pebble stones and sands are consumed in structuring activities each year, the availability of good quality aggregates decreases [21].

In addition, European Union members produce approximately 50 million tons of concrete waste each year, compared with 60 million tons in the United States 10-12 million tons in Japan. Japan has reduced the use of aggregates by 2.5 million m^3 by recycling concrete wastes in ready-mixed concrete plants [22]. Netherlands recycles 93% of concrete and Germany 18% [23].

In the recycling of steel, it is observed that 74% of the energy and 90% of the raw material are preserved, water consumption is reduced by 76% rate, air contamination is reduced by 86% and mining wastes by 97% [24]. Water consumption and emissions are reduced when recycling aluminum compared to extraction. Approximately 52% of steel wastes and 65% of aluminum wastes are recovered in the U.S. According to data from 1994, aluminum recovery was 85% in Switzerland and 91% in Sweden [25]. Doors and windows acquired as a result of demolition or renovation are ideal components for reuse. Even though kitchen and bathroom equipment if not be worn out, they are among reusable components.

3.3. Storage of C&D wastes

When recovery of C&D wastes is not possible, construction wastes should be stored in a controlled way. Storage facilities should not be established in primary farmland and areas of high agricultural production potential . Also these fields should not be close to drinking, irrigation and water reservoirs such as ponds. Distance to settlement areas is an important factor (at least 200 meters). Storage facilities should include units to accept arriving C&D wastes, an operation building and weighbridge, and the storage field should be separated by a proper divider.

An effective control, monitoring and tracking system should be created in the storage area, to enable segregated storage of different waste materials.

In order to prevent environmental contamination that can occur during transportation of C&D wastes to the storage location, vehicles should be covered with appropriate materials. Vehicles should not be overloaded and they should operate after cleaning mud or similar debris from their wheels. Vehicles should also carry signals indicating the C&D waste being transported [26].

Where the capacity of a storage area is exceeded, the area should be rehabilitated appropriately with the natural topographical features that allow for their reuse. For example, the Sidney Olympic Village was constructed on the site of a neglected industrial area in Homebush Gulf. The site was previously used for brick kilns, a military arsenal, state slaughterhouse and municipal garbage.

4. Research objectives and methodology

The research methodology of the study is based on the analysis of Turkey National C&D waste management general framework based on legal regulations and Istanbul Metropolitan City as a local case study. Brief information is given about the general framework of waste management in Turkey with legal aspects. The case study aims to identify the current situation in construction and demolition waste management in Turkey. The study also tries to highlight the actual implementation in local level.

5. The current state of construction and demolition waste management in Turkey

This section will give some brief information about the general framework of building waste management in Turkey with legal aspects. Management policy on construction and demolition waste will be discussed through municipality actions and implementations.

5.1. General national framework: Regulations regarding construction and demolition wastes in Turkey

Many developed countries have introduced legislation and strategies to reduce the environmental effects of C&D wastes. Turkey also has several legislations regarding the management of C&D waste.

In Turkey, the issue of debris produced by disasters is covered by Article 4 of the "Law Regarding Assistance and Precautions for Disasters Occurring in Public Life" (Number 7269, dated 1959) and Article 4, 14/b and 24 of the "Regulation Regarding Immediate Support Organization and Planning Principals of Disasters" (Number 88/12777, dated 1988) [27].

The issue of C&D wastes is briefly included in Article 23 of the "Regulation Regarding the Control of Solid Wastes" (Number 20814, dated March 14th, 1991 under the title "Storage of Excavation Soil" [28].

The Report of Sub-Commission of 8th Five- Year Development Plan Solid Waste Control suggests the regulation of "Regional Debris Management Plan and Waste Management System" under the title of Construction- Demolition and Excavation Wastes [14]. While 9th Five- Year Development Plan includes articles regarding reduction and disposal of wastes generally, in its Article 471, it involves a statement that "the generation of non-domestic wastes shall be reduced, collection, transportation, recovery and disposal systems shall be creates appropriate for country conditions".

- Laws and codes related with C&D waste in Turkey
- Environment Act (Number 2872 and dated 09/08/1983),
- Physical Development Planning Law (Number 3194 and dated 03/05/1985),
- Metropolitan Municipality Law (Number 5216 and dated 10/07/2004),
- Misdemeanors Law (Number 5326 and dated 30/03/2005),
- Regulation of Mining Activities with the Regain of the fields (Number 2747 and dated 23/01/2010),
- Regulation of "Pertaining to Regular Storage of Wastes" (Number 2753 and dated 26/03/2010),
- Forest Law article 16 (Number 27715 and dated 30/09/2010),

- Also, there are some regulations related with asbestos [29, 30] and after the regulations come into forde in 31/12/2010 date it was totally prohibited [31].

The most comprehensive regulation in Turkey regarding the control and recovery of C&D wastes is the "Regulation of the Control of Excavation Soil and Construction and Demolition Waste" number 25406 enacted by the Ministry of the Environment and Forestry which came into force on March 18th, 2004. This regulation includes general rules about administrative and technical subjects on the reduction, collection, temporary storage, recovery, evaluation and disposal of excavation soil and construction and demolition wastes [26]. C&D waste grouping were given according to the regulation (Figure 1).

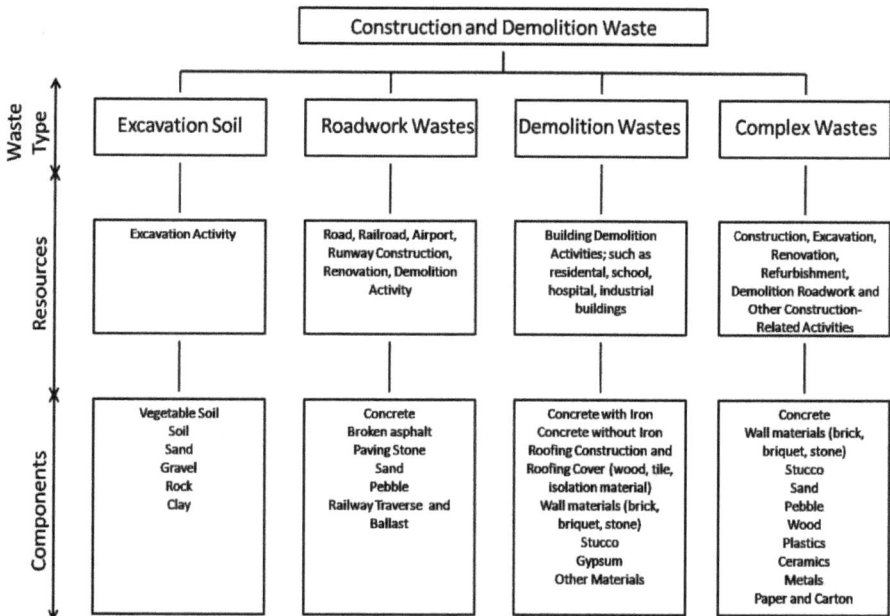

Figure 1. C&D Waste Grouping

5.2. Case study

The case study tries to show and gives a better understanding of the existing implementation in Istanbul Metroplitan Municipality in a local level and ongoing activities related the improvement of the waste management system and strategies.

- The study has evaluated solid waste management practices of the municipalities in Turkey in the context of the Istanbul Metropolitan Municipality works that were related and reflected the Istanbul city which connects the Asia and Europe has the highest population and had the highest population increase rate in Turkey. It is also city that has generated the highest C&D waste because of the high construction and demolition acitivties. Thus, the city had the comprehensive C&D waste management works. It is estimated that the annual increase/decrease of C&D waste generated from infrastructure, road, bridge, maintenance/repair, construction and demolition is 3% according to the population. The construction, excavation soil and demolition generation were determined as 1500 kg/person/year for excavated soil and 200 kg/person/year for C&D waste [32].

•   Until 2002 there were no regulation related with C&D waste management and all the waste generated were disposed randomly without any project and premission to the private or public land in different parts of the city. The landfill areas location and the amount of landfill was unknowable as it can not be prevented from unplanned construction in those landfill areas [33]. On 8th February 2002, in order to prevent from illegal spillation and espacillay to determine the landfill areas the central government – governorship the "Directive of Excavated Soil and Construction Debris Control" has put into action. But, short time iconsequences show that because the directive colu not be able to block the deficiencies in illegal disposal [34].

According to code issued in 2004 "Regulation of the Control of Excavation Soil and Conctruction and Demolition" Metropolitan Municipalities are responsible to take the actions such as preparing plans for excavation, construction and demolition waste/debris management. For example the "Building and Demolition Waste Mangement Plan for Istanbul City" has prepared by the municipality on March 2006. The municpalities are also responsible to determine, establish and operate all the recycle facilities and storage areas. So, it can be pointed our that the site activities related with all types of waste were managed by the local government. But the collection and transportation of the excavated soil, construction and demolition waste services can be private. The role of the municipality for the collection and transportation were limited to announce and inform the list,addresses and phone numbers of these private firms. Also the monitoring and controlling of these firms were the responsibilities of the municipality.

-   The responsibility for transportation of solid wastes including C&D, production of composted fertilizer, waste recycling, waste disposal in regular storage, electricity generation from storage areas, transportation of medical wastes and their disposal by incinerationi recycle centers management were given to the newly established municipality firm ISTAÇ A.Ş. (Istanbul Metropolitan Municipality Environmental Protection and Waste Materials Evaluation Joint-Stock Company) [35].

-   Totally 7 storage area with 54.343.676 m³ were determined for C&D waste generated in the city where the natural ground deformed such as quarries and open mine sites within the Istanbul province boundary. Additionally, excavated soil were reused as landfill for parks, gardens, open-space areas production, recreational areas, infrastructure and landscaping in district municipalities [33]. The C&D waste disposal and recycling amounts in Istanbul city between 2006-2011 can be seen in Table 1 [36].

	2006	2007	2008	2009	2010	2011	Total Amount
The amount of excavation soil disposed (gross ton)	2.999	18.354	16.796	16.257	24.100	47.709	126.215
The amount of C&D waste disposed (ton)	2.818	4.577	4.439	4.258	5.361	5.680	27.133
The amount of excavation soil recovered (ton)	-	1.389.986	102.421	175.540	34.356	35.996	1.738.299
The amount of C&D waste recovered (ton)	-	-	12.819	77.826	73.200	116.952	280.797

Table 1. Recycled and disposed waste generated from excavation, construction and demolition activities in Istanbul City (Population: 12.573.836 people in 2009) between 2006-2011) [36].

- 200 ton/hour the capacity Mobile C&D Waste Recycling Facility consisted of two sieves and one breaker were established in 2008 (Figure 2-3).
- Necessary measures have been taken for the factors related with these facilities such as dust and noise which may affect environment and human health.
- Excavation and debris were taken from the original place with an EPL (Environmental Phone Line). Building waste from renovation works divided into 40 kg package(s) and firstly sent to stopover stations then to the authorized disposal areas which belongs to ISTAÇ A.Ş. This service approximately cost 13 Euro per package between 1-20 packages and 0,65 Euro per extra package after 20 [37].

Waste Management Automation Project (WMAP) allow electronic web based control of all types of transportation vehicles that carry wastes (excavation, construction and debris, household, medical, industrial) within the İstanbul Metropolitan Municipality boundary (Figure 4). On the other hand, the works are going on in order to prevent illegal disposal with "Vehicle Control System" [35]. C&D waste collection and disposal processes illustrated as schemas in Figure 5-6 [35].

Figure 2. Construction Waste Recycle Facility [35]

Figure 3. Demolition Waste Recycle Facility [35]

Figure 4. WEB Based Control Unit [35]

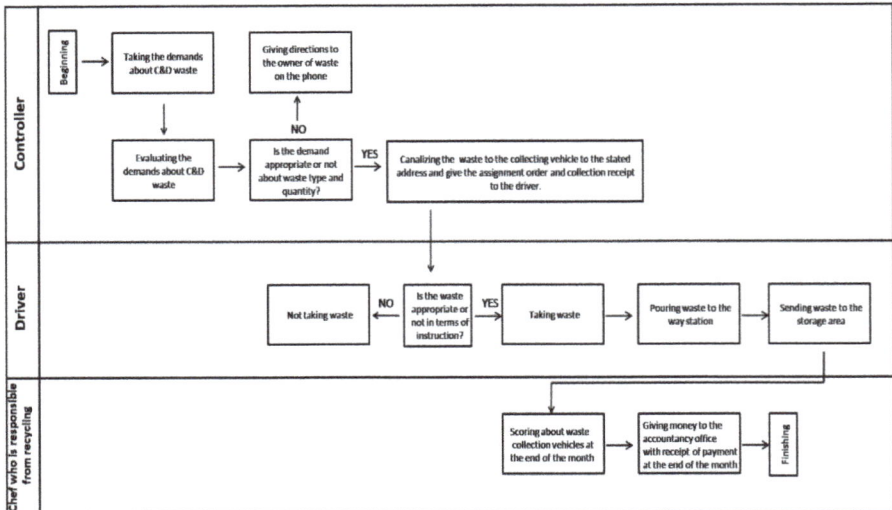

Figure 5. C&D Waste Collection Process

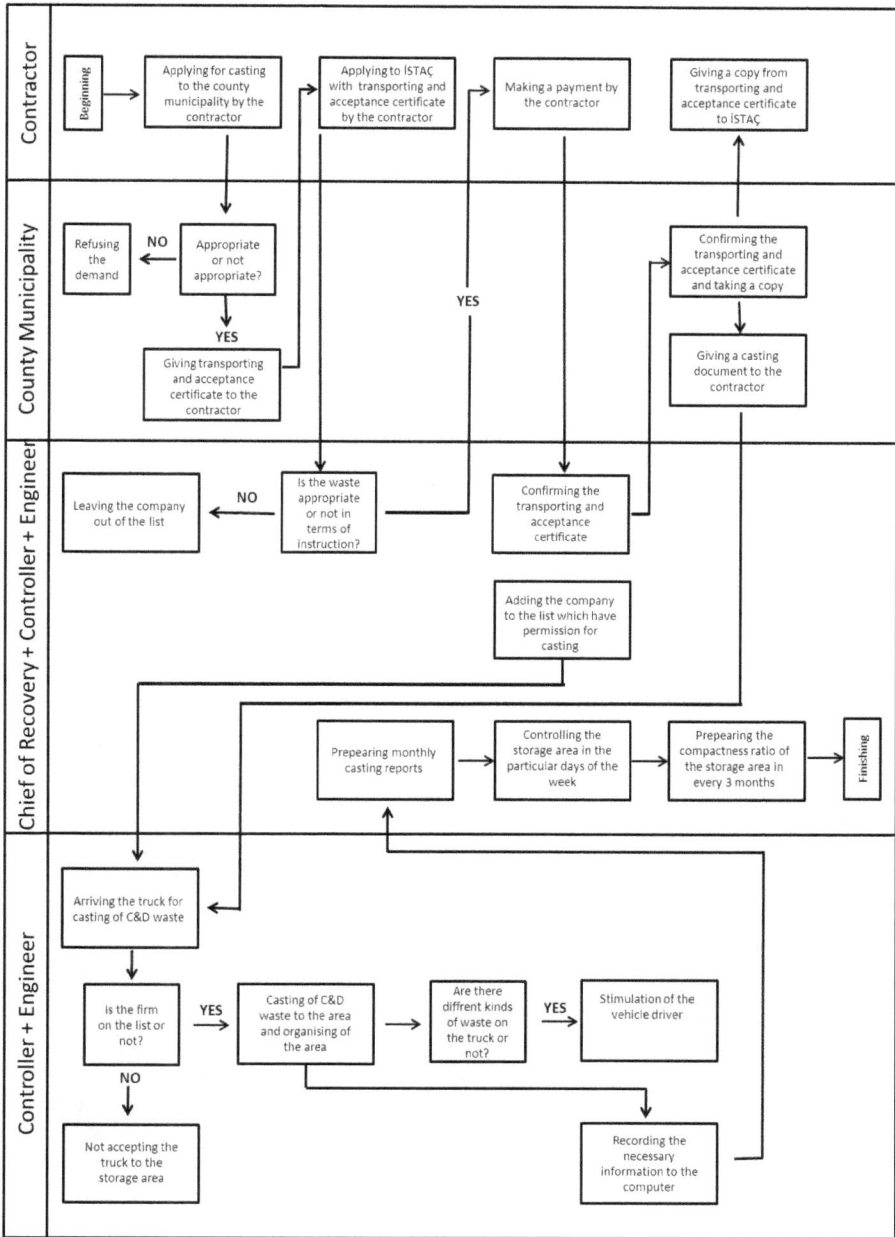

Figure 6. C&D Waste Disposal Process

The design phase has a very important role in reducing/preventing C&D waste during building production. In this context a research were conducted in order to understand and question the designers approaches to C&D wastes. Generally most of the respondents stated that the wastes occured in the construction site can be controlled in design phase but contrarily most of them did not characterize it as a responsibility of the designer to reduce. Also most of them stated that main waste produced during the construction. Most of the respondents also think that building waste management can be possible by waste materials recycle [38].

Because the necessary organizations that can recycle construction waste have not developed in Turkey, large economic losses occur. However, the reuse of construction waste does occur at certain levels albeit quite irregularly. Mostly in Istanbul's outlying suburbs there are building material collectors. The collected salvaged building materials are sold in open and semi-open markets that could be named as "salvaged building material outlets." At these outlets, wood and PVC doors and windows, kitchen and bathroom components (closet, wash-bowl, kitchen sink, kitchen counter and cupboard), strips, tiles, plastic pipes, asbestos roofing sheets, wooden lath, etc., various materials and items that are in good and bad shape that can be salvaged are sold mostly to low income wage earners or those building squatter homes [Figure 7-8]. While the collecting and selling of material from modifications by collectors is a positive application of reusing, it is not a system that functions at an efficient level [39].

Figure 7. Salvaged Building Materials Outlet

Figure 8. Salvaged Building Materials Outlet

6. Conclusion

Architect(s) and/or designer team(s) need to focus on understanding the importance of wastestream. Because their central role in design decision(s) determine the type of building materials. So, the higher they have knowledge about building materials and optimum sizing that may refer the lower the construction waste produced. Even the brief information they may give about wastes to their clients may boost recycle and re-use options of generated waste in the construction site. Also, use of modular systems, selecting nature friendly and high durability materials, considering salvaged and recycled building materials in material selection and detailing processes will result with a decrease in the wastestream. The role of the architect(s) and/or designer team(s) must not be limited with design and go further to construction phase and by means of providing a good coordination and communication with the contractor and/or the builder they may obviously contribute to reduce/prevent from building wastes generated during the building production process.

A "Construction and Demolition Waste Management" (C&DWM) plan should be created in every stage of the building production processes in order to take the necessary precautions to minimize the negative effects of construction wastes on the environment and human health. A C&DWM System can be created through the integration of waste management plans with construction production within lifecycle process in every stage. By creating a C&AWM System as part of the construction lifecycle, solutions at the national, local, institutional and individual scale will become clearer. The stakeholders role in each process can be summarized as;

-   **Legal Base for Waste Management Hierarchy:** Laws and regulations should be made at a national scale; effective monitoring and control should be introduced; a measurement and verification system should be developed to establish the potential of C&D wastes; integration potentials of C&DWM system and present implementations in Turkey should be determined; incentive policies should be developed, Regulations should be introduced to reduce the negative environmental effects of obtaining and using raw materials, their transportation and processing. Minimizing the raw material use will contribute to lower the global warming impacts through the production of these materials and their transportation to the construction site. National, regional and local standards and codes should be established for recycled and salvaged materials.

-   **Central and Local Government Initiatives and Role Distribution:** Altough the legal base will identify the role of government bodies it is important to implement and clarify the roles of the local and central government. Waste management need a governmental level approach for determining national goals but apart from that it is based on the local governments (municipalities) capacity and capability to manage the process. Decisions taken at local level should be implemented effectively, a database should be created to determine the potential of C&D wastes, and effective supervision and instruction should be implemented,

-   **Sustainable Construction and Demolition Waste Management Systems:** Recycling systems should be developed and supported at the national scale.The use of recycled and/or salvaged materials should be encouraged. Tax reductions should be used as a mechanism to encourage the use of recycled and salvaged materials. Reasonable price policies should be implemented for C&D wastes and classified waste materials should be charged with lower prices than the unclassified ones. Services for classifying and disassembling C&D waste should be created for better management of wastes. Quarries and open mines may be unsightly and unsuitable for immediate reuse, and should therefore be considered as sites for excavated wastes to also enable their rehabilitation. Dumping procedures of C&D wastes should be regulated and illegal dumping should be prevented.

-   **Institutional and Individual Action Plans:** At an institutional scale, especially in construction/demolition organizations (including local recyclers and haulers), functional waste management departments should be created, wastes should be classified and properly disposed of, and effective supervision and instruction should then be developed. At an individual scale, waste management plans should first involve reducing the waste with source control optimization. Latterly, classifying waste (for salvage, recycle or donation) and appropriate disposal of wastes is important in the construction site or in an existing building site by adhering to the decisions of local administration.

- **Clarifying Contractor(s)/Builder(s) Responsibilities:** Contractors' responsibilities for wastes should be clearly determined in pre-construction phase in contracts and supervised /controlled during construction. Contracts should require contractors responsibility to follow the waste management plan at least results with 50% recycle and salvage option during construction for sustainabile building(s) goals.
- **Educational and Research Activites:** Interdisciplinary training programs should be developed and encouraged to raise awareness of construction waste management at a national scale for all of the construction sector bodies. Research and development studies should be supported by the energy and construction related governmental organizations at a national scale in order to revise and sustain the existing waste management strategies.

Author details

Hakan Arslan
Department of Architecture, College of Engineering, Technology and Architecture,
Hartford University, West Hartford, USA

Nilay Coşgun
Department of Architecture, Faculty of Architecture, Gebze Institute of Technology, Kocaeli, Turkey

Burcu Salgın
Department of Architecture, Faculty of Architecture, Erciyes University, Turkey

7. References

[1] Mills T.H, Showalter E, Jarman D. (1999) A Cost Effective Waste Management Plan. Cost Engineering 41 (3): 35–43.

[2] Rogoff M.J, Williams J.F (1994) Approaches to Implementing Solid Waste Recycling Facilities. Noyes, Park Ridge, NJ.

[3] Ferguson J, Kermode N, Nash C.L, Sketch W.A.J, Huxford R.P (1995) Managing and Minimising Construction Waste: A Practical Guide. Institution of Civil Engineers, London.

[4] Sealey B.J, Phillips P.S, Hill G.J (2001) Waste Management Issues for the UK Readymixed Concrete Industry. Resources, Conservation and Recycling 32 (3–4): 321–331.

[5] Craven D.J, Okraglik H.M, Eilenberg I.M (1994) Construction Waste and a New Design Methodology. In: Kibert, C.J. (Ed.), Sustainable Construction. Center for Construction and Environment, Gainesville, FL, pp. 89–98.

[6] Poon C.S (2007) Reducing Construction Waste. Waste Management 27: 1715–1716.

[7] Environmental Protection Department (EPD) (2006) Available: www.info.gov.hk/wfbu/whatwehavedone/wms/cd01.htm. Accessed 2006 Nov 01.

[8] Lu W, Yuan H (2011) A Framework for Understanding Waste Management Studies in Construction. Waste Management 31: 1252–1260.

[9] Türkiye İnşaat Malzemeleri Sektör Görünüm Raporu (2011) Türkiye Odalar ve Borsalar Birliği. (In English: Report of Turkey Construction Material Sector Outlook, The Union of Chambers and Commodity Exchanges of Turkey) Available: http://www.tobb.org.tr/Documents/yayinlar/Türkiye%20İnşaat%20Malzemeleri%20Sek tör%20Görünüm%20Raporu.pdf. Accessed 2012 March 6.

[10] Higgins T.E (1995) Pollution Prevention Handbook. Lewis Publisher. A CRC Press Company.

[11] McGrath C (2001) Waste Minimisation in Practice. Resources, Conservation and Recycling 32 (2001): 227–238.

[12] Construction and Demolition Waste Practices and Their Economic Impact (1999) Report to DG XI, Symonds. Available: http://europa.eu.int/comm/environment/waste/report.htm. Accessed 1999 February 01.

[13] Esin T, Coşgun N (2005) Ecological Analysis of Reusability and Recyclability of Modified Building Materials and Components at Use Phase of Residential Buildings in Istanbul. UIA 2005 Istanbul XXII World Congress of Architecture - Cities: Grand Bazaar of Architectures, 3-10 July 2005, Istanbul.

[14] DPT, 8.Beş Yıllık Kalkınma Planı 2000–2005 (In English: State Planning Organization of Turkey, 8th 5 year Development Plan 2000–2005) (2000) Solid Waste Control Sub Commission Report –Ankara. Available: http://ekutup.dpt.gov.tr/icmesuyu/oik524.pdf. Accessed 2006 February 01.

[15] European Commission Environment (2011) Available: http://ec.europa.eu/environment/waste/index.htm. Accessed 2011 February 01.

[16] Osmani M, Glass J, Price A.D.F (2008) Architects' Perspectives on Construction Waste Reduction by Design. Waste Management. Volume 28. Issue 7: 1147-1158.

[17] Öztürk M (2005) İnşaat Yıkıntı Atıkları Yönetimi (In English: Construction Debris Management, Ministry of the Environment and Forestry). Ankara.

[18] Construction Waste Recyclıng Exceeds Target (2000) Available: http://eied.deh.gov.au/minister/env/2000/mr22jun00.html. Accessed 2000 February 01.

[19] European Commission, Directorate-General, Environment, Directorate E-Industry and Environment, ENV.E.3 - Waste Management, DG ENV.E.3 (2000) Management of Construction and Demolition Waste. Working Document N°1.

[20] Department for Environment, Food and Rural Affairs (2012) Available: http://www.defra.gov.uk/statistics/environment/waste/wrfg09-condem. Accessed 2012 April 6.

[21] Ngowi A.B (2001) Creating Competitive Advantage by Using Environment-Friendly Building Processes. Building and Environment 36: 291-298.

[22] Hansen T.C (1992) Recycling of Demolished Concrete and Masonry. Rilem Repot 6. Taylor&Francis Group London. New York.

[23] Corinaldesi V, Moriconi G (2004) Reusing and Recycling C&D Waste in Europe. Construction Demolition Waste. Ed. M.C.Limbachiya, J.J. Roberts. USA.

[24] Öztürk M (2004) Kullanılmış Çeliğin Geri Kazanılması (In Englsih: Recycle of Used Steel). Ankara.

[25] Öztürk M (2005) Kullanılmış Alüminyum Malzemelerin Geri Kazanılması (In English: Recycle of Aluminium Materials). Ankara.

[26] Hafriyat Toprağı, İnşaat ve Yıkıntı Atıklarının Kontrolü Yönetmeliği (in English: Regulation of the Control of Excavation Soil and Construction and Demolition Waste) (2004) Tarih: 18 Mart 2004. Sayı: 25406.

[27] Umumi Hayata Müessir Afetler Dolayısıyla Alınacak Tedbirlerle Yapılacak Yardımlara Dair Kanun (in English: Law Regarding Assistance and Precautions for Disasters Occurring in Public Life) (1959) Sayı: 7269. / Afetlere İlişkin Acil Yardım Teşkilatı ve Planlama Esaslarına Dair Yönetmelik (in English: Regulation Regarding Immediate Support Organization and Planning Principals of Disasters) (1988) Sayı: 88/12777.

[28] Katı Atıkların Kontrolü Yönetmeliği (in English: Regulation Regarding the Control of Solid Wastes) (1991) Resmi Gazete Tarih: 14 Mart 1991. Sayı: 20814.

[29] Asbestle Çalışmalarda Sağlık ve Güvenlik Önlemleri Hakkında Yönetmelik (2003) Resmi Gazete Tarih: 26/12/2003. Sayı: 25328. (In English: Regulation of "Health and Security Preventions Regarding Asbestos Works, Number: 25328 and dated 26/12/2003)

[30] Regulation of "Pertaining to Regular Storage of Wastes" (Number 2753 and dated 26/03/2010).

[31] Bazı Tehlikeli Maddelerin, Müstahzarların ve Eşyaların Üretimine, Piyasaya Arzına ve Kullanımına İlişkin Kısıtlamalar Hakkında Yönetmelikte Değişiklik Yapılmasına Dair Yönetmelik (2010) Resmi Gazete Tarih: 29/08/2010. Sayı: 27687. (In English: Regulation of "Changing the Regulation about Amendments and Restrictions on the use and placing on the market for certain dangerous substances, preparations and goods", Number 27687 and dated 29/08/2010).

[32] Istanbul Technical University, Faculty of Civil Engineering, Department of Environmental Engineering (2006) İstanbul içi AB Çevre Mevzuatı ile Uyumlu Entegre Katı Atık Yönetimi Stratejik Planı, İstanbul. (In English: Integrated strategic solid waste management plan with EU Environment legislation for İstanbul)

[33] Altındağ S (2011) İstanbul'da Hafriyat Toprağı, İnşaat ve Yıkıntı Atıklarının Tersine Lojistik Yöntemiyle Alternatif Yönetim Planı (in English: The Alternative Management Plan of Excavation Soil, Construction and Demolition Wastes with Reverse Logistics Method in Istanbul). Master of Science. Istanbul Technical University, Graduate School of Natural and Applied Sciences.

[34] İstanbul için AB Çevre Mevzuatı ile Uyumlu Entegre Katı Atık Yönetimi Stratejik Planı (2006) Istanbul Technical University, Faculty of Civil Engineering, Department of Environmental Engineering. Istanbul.

[35] İSTAÇ A.Ş. (2012) Available: www.istac.com.tr. Accessed 2012 April 6.

[36] Istanbul Metroplitan Municipality, 2012, Excavation soil, construction and demolition waste data, Istanbul

[37] Alo Çevre Hattı'ndan Serkan Öztürk ile yapılan görüşmede edinilen bilgiler (In English: Interview with Serkan Öztürk from EPL (Environmental Phone Line), 10th April 2012).

[38] Coşgun N, Güler T, Doğan B (2009) Yapısal Atıkların Önlenmesinde/ Azaltılmasında Tasarımcının Rolü (In English: The role of designer in reducing/preventing the construction waste). Mimarlık, Chamber of Architects Publications. Volume 348: 75-78.

[39] Esin T, Coşgun N (2007) A Study Conducted to Reduce Construction Waste Generation in Turkey. Building and Environment, 42(4): 1667–1674.

Chapter 15

E-Waste Disposal Challenges and Remedies: A Tanzanian Perspective

Daniel Koloseni and Faith Shimba

Additional information is available at the end of the chapter

1. Introduction

In recent years we have witnessed the rapid increase of number of mobile subscribers, mobile services providers, internet service providers, data operators and internet users. This is primarily caused by lifting of ban imposed by government on importation of computers and its peripherals in government entities, government decision to remove all taxes and duties on those electronic products and flexibility on regulations for establishment of media and telecommunication companies. All these steps have contributed much on inflow of electronic products particularly computers and its peripherals, mobile phones and television sets. This pose challenges on appropriate methods to dispose end of use electronic products without destroying environment, jeopardize people health and without loss of data and information stored in these products.

Additionally, it's unfortunate that this inflow of electronic products caught government entities, private organizations and the public in general unprepared on how to safely and economically dispose end of use electronic products. This has eventually left piles of unattended end of use electronic products both in streets and in office stores.

This chapter addresses the challenges on e- Wastes disposal steered by proliferations and usage of electronic devices both at homes and offices and propose solutions to the challenges. Also in this chapter we suggest appropriate measures that should be taken categorically, by Government entities, private organizations and public in general to thwart down the challenges.

2. Definition of some key terms

E-waste: Is defined as end-of-use or end-life of electronic products, components and peripherals such as: computers, fax machines, phones, Personal Digital Assistant (PDA), radios and TVs [1]

E-Waste recycling sites: This refers to the space allocated specifically for recycling e-waste or end-of-use ICT assets.

3. Current status of e- waste in Tanzania

The review of existing literature shows that Tanzania has no specific policy or regulation related to e-waste management. However, there are a number of policies and regulations which aim at protecting the environment and human settlements. Examples of these policies are: Environmental Policy [2], the Sustainable Industrial Policy [3], National ICT Policy, [4] among others. An overview of the different e-waste related policies is given in figure 1.

The review of these different policies reveals that there is a need for e-waste specific policies to address the different challenges and issues of e-waste management. There are also a number of regulations and laws that provides an institutional framework for a sustainable management of the environment in general. Among others, the Environmental Management Act (EMA) No. 20 of 2004 [7] is the cornerstone legislation in Tanzania. This legislation provides key principles for environment management, waste management, and impact and risk assessment among other things. A summary of the different regulations, their objectives and impact on waste management is given in figure 2.

4. E – waste management specific regulations

Tanzania lacks e-waste management specific legislation. Nonetheless, e-waste management is carried out through the Solid Waste and Hazardous Management Acts [5, 6] under the Environmental Management Act [7]. E- Waste, and specifically electronic waste, is addressed in section 4 of the Environmental Management Act [7].

Different regulations in the act provide principles and guidelines on how e-waste should be handled. For example, regulation 35 (1), requires that any persons in possession or in control of electrical or electronic goods, to separate or segregate e-waste from other types of waste, and dump them separately as prescribed by different national or local authorities [8, 9]. The separation also applies to collection of waste, its transportation and the final dumping or disposal (refer schedule eight of the regulations). Schedule eight (8) of the regulation also classifies electronic equipments such as large households appliances, IT and Telecommunication equipment, among others. Section 37 (1) of the same regulation allows for manufacturers and/or vendors to set-up and operate recycling systems for electrical and electronic equipments from consumers. The role of local government (local authorities) is detailed in section 39.

The government through the Prime Minister's Office, environment division, has developed a strategy and an action plan [10]. This strategy, among other things, addresses the issue of e-waste management. The plan aims at, as its goal in e-waste management, minimizing of environmental and health risks resulting from improper handling of e-waste through stakeholders' participation. The specific objectives of the strategy and action plan are; to review different policies and regulations related to e-waste management, awareness creation on e-waste management, and promotion of recycling and material recovery. However, the strategy is yet to be implemented.

5. The institutional framework

The institutional framework for e-waste management in Tanzania is not well developed. However, the existing framework is a result of a number of different policies and regulations which have resulted in the establishment of various bodies dealing with waste management but not directly related to e-waste. These policies and regulations are shown in figures 1 and 2 respectively, while; tables 1 summarizes the respective legal instrument with the corresponding responsible institutions/organization and table 2 summarizes the objectives and roles of the institutions/organization in waste management. As a result of these policies and regulations, the Tanzania government has established a number of institutions to deal with waste management. The key institution involved in waste management is the Vice President's office. In the Vice President's office there is a minister responsible for environment issues and the division for environment and the local government authorities are also responsible for environmental management.

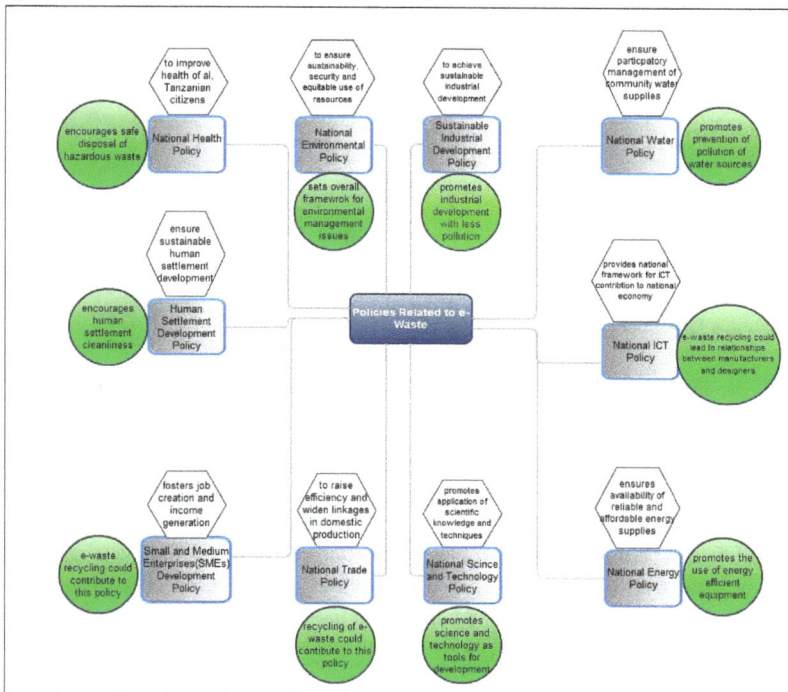

Figure 1. An Overview of Policies related to e-Waste Management in Tanzania: The Policy, Objective and Relevancy.

Figure 1shows a brief overview of different policies related to waste management. The figure summarizes the objectives and the relevancy of each particular policy as it relates to e-waste management. In the figure, the hexagon represents the objectives that a particular policy intends to fulfil, while the circle represents the relevancy of the policy to e-waste and the policy is represented by the smooth rounded rectangle.

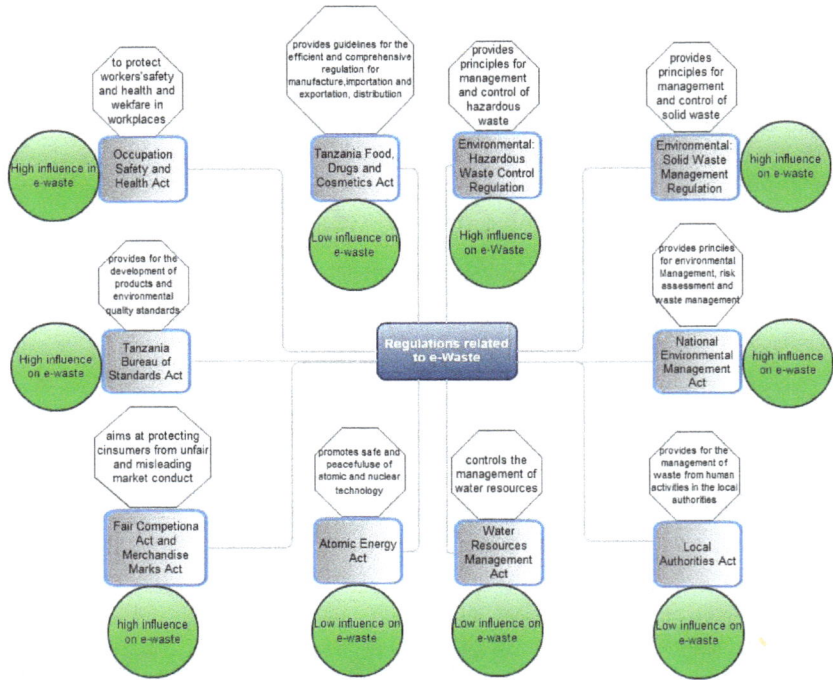

Figure 2. A summary of regulation related to e-waste management, their objectives and impact on e-waste management

Figure 2 shows the different regulations and act related to waste management and how they are related to e-waste and their impact or influence in e-waste management in Tanzania. The regulation or act id represented by a smooth rounded rectangle, the influence of the act is represented by the circle while the objective of the act is represented by octagon. The influence of the act is said to be low when it does not directly address e-waste or e-waste related issues, while a high influence signifies the importance of the act is regulating e-waste and e-waste management by addressing key issues related to e-waste.

These regulations and different bodies responsible for their implementations posses challenges in the management of e-waste. For example local government may decide to enact by laws but these are only applicable in the respective local authority and may be over ruled by a higher authority. However, this calls for more coordination efforts if e-waste management is to be effectively implemented in the country. From this example of how different institutions which have been given mandate by policy or regulation to deal with environmental management, can come into conflict in their carrying out of their responsibilities; it is clear that the existing policies, regulation and the institutional framework is inadequate. The policies and regulations are insufficient in addressing e-waste management issues and problems. These deficiencies in policies, regulation and institutional framework calls for the need of more effective and efficient framework that will adequately deal with e-waste management issues and problems.

Legal Instrument	Responsible Organ
National Environmental Management Act No. 20 of 2004	Vice President's Office National Environmental Management Council Local Government Authorities
The Environmental (Solid Waste Management) Regulations of 2009	Vice President's Office National Environmental Management Council Local Government Authorities
The Environmental (Hazardous Waste Control) Regulations of 2009	Vice President's Office National Environmental Management Council Local Government Authorities
Tanzania Foods, Drugs and Cosmetics Act of 2003	Tanzania Foods and Drugs Authority Ministry Health and Social Welfare
Occupational Safety and Health Act of 2003	Ministry of Labour, Youth and Culture Occupational Safety and Health Authority
Tanzania Bureau of Standards Acts No. 3 of 1975	Ministry of Industry and Trade Tanzania Bureau of Standards
Fair Competition Act of 2003 Merchandise Marks Act of 1963	Fair Competition Commission
Atomic Energy Act of 2003	Tanzania Atomic Energy Commission
Water Resources Management Act No. 11 of 2009	Ministry of Water Urban Water Authorities River Basins Authorities
Local Government (District Authorities) No. 7 of 1982 Local Government (Urban Authorities) No. 8 of 1982	Local Authorities

Table 1. A summary of regulations and responsible organs in implementation of the acts

Organization	Objective	Role
Vice President's Office: Environment Division	Coordination of all environmental Management issues	Policy formulation Advocacy and implementation Monitoring and evaluation Planning Legislation International cooperation
Local Government Authorities (LGAs)	Waste management and Control	To prevent or minimize e-waste in their jurisdictions
National Environmental Council (NEMC)	Regulatory authority	To oversee the implementation of the environmental management Act. Coordination, evaluation and implementation
Occupational Safety and Health Authority (OSHA)	Ensure occupational health and safety	Safe use and handling of hazardous waste
Ministry of Industry and Trade	Licensing	Registration and licensing of traders and equipment dealers
Tanzania Bureau of Standards	To develop products and environmental quality standards	To ensure that all manufacture or imported products meets the standards

Table 2. A summary of organs responsible for e-waste management, their objectives and responsibilities

6. E-waste, environment and health

E-waste comprises discarded electronic appliances of which computers and mobile phones make the great contribution due to their short life span [11,12] In addition to its damaging effects to the environment and the illegal importation(smuggling) to the developing countries, researches have shown that, e-waste has damaging impacts (effects) to human health [12]. The effects of e-waste to human health and well being includes: respiratory problems, oxidative stress, DNA damage and the possibility of causing cancer.

The reasons for the damaging effects to human health and environment of e-waste is caused by its chemical and physical characteristics which sets it apart from other forms of wastes that are produced by human activities or industrial wastes. These e-wastes contains both hazardous and valuable components that calls for specialized skills in handling, disposing and recycling in order to avoid contamination with the environment and safe guard the human health.

6.1. E-waste environmental contaminants

Table 3 below gives a summary of ingredients of e-waste that are potential environmental contaminants. Examples of the contaminants that can be found in e-waste are such as heavy metals like copper which are used in manufacture of electronic components. Other contaminants are those which are the results of disposal of e-waste through fire. An example is polycyclic aromatic hydrocarbons (PAHs) which are generated when electronic waste is burnt.

E-waste also contains some components which are distinct from other forms of wastes. An example of these are the batteries such as lithium batteries, contact materials and fire retardants [13], others include monitors (LCD) Chips [11].

Contaminant	Relationship with E waste	Typical E waste concentration (mg/kg)ᵃ
Polybrominated diphenyl ethers (PBDEs) polybrominated biphenyls (PBBs) tetrabromobisphenol A (TBBPA)	Flame retardants	
Polychlorinated biphenyls (PCB)	Condensers, transformers	14
Chlorofluorocarbon (CFC)	Cooling units, insulation foam	
Polycyclic aromatic hydrocarbons (PAHs)	Product of combustion	
Polyhalogenated aromatic hydrocarbons (PHAHs)	Product of low temperature combustion	
Polychlronated dibenzo p dioxins (PCDDs), polychlorinated dibenzofurans (PCDFs)	Product of low temperature combustion of PVCs and other plastics	
Americium (Am)	Smoke detectors	
Antimony	Flame retardants, plastics (Ernst et al., (2003))	1700
Arsenic (As)	Doping material for Si	
Barium (Ba)	Getters in cathode ray tubes (CRTs)	
Beryllium (Be)	Silicon controlled rectifiers	
Cadmium (Cd)	Batteries, toners, plastics	180
Chromium (Cr)	Data tapes and floppy disks	9900
Copper (Cu)	Wiring	41,000
Gallium (Ga)	Semiconductors	
Indium (In)	LCD displays	
Lead (Pb)	Solder (Kang and Schoenung, (2005)), CRTs, batteries	2900
Lithium (Li)	Batteries	
Mercury (Hg)	Fluorescent lamps, batteries, switches	0.68
Nickel (Ni)	Batteries	10,300
Selenium (Se)	Rectifiers	
Silver (Ag)	Wiring, switches	
Tin (Sn)	Solder (Kang and Schoenung, (2005)), LCD screens	2400
Zinc (Zn)		5100
Rare earth elements	CRT screens	

Table 3. Potential Environmental contaminants contained in e-Waste (adopted with modification from e-Waste 2009)

Therefore, the impact to the environment of e-waste and the concentration of the contaminants found in e-waste depends on the type of items that are discarded and the time that has elapsed since it was produced. Also the method used to dispose the e-waste impacts the effects that the disposed waste will have on the environment. For example the concentration of elements of e-waste such as Copper, Cadmium, Nickel, Lead and Zinc are of impact to the environment and human health if they were to be released as they pose risk to the ecosystem.

Taking into account that in Tanzania re-cycling is not a well established industry neither is it done properly, the amount of contaminants that could have been averted from leaching to the environment and endanger human health is increased as recycling could have reduced or removed some of the contaminants. Also, since most of the dumping sites in Tanzania either use fire or landfills large amount of these contaminants ends in landfills resulting in high concentration that may leach out into the environment and adversely affect the environment and human health.

Other contaminants are such as poly brominated dipheynl ethers (PBDEs). These are flame retardants which are mixed into the plastic components of electronics. However, these PBDEs have no chemical bond with the plastics and are very likely to escape to the environment from the surfaces of the plastics into the environment. Given the lipophilic characteristic of PBDEs, this causes bioaccumulation in organisms and biomagnifications in food chains [14]. Also, obsolete electronic products such as computers, refrigerators, and air conditioning units contain ozone depleting gases. These gases may escape to the environment from the improperly disposed items in the dumping sites.

6.2. E-waste health hazards

The impact of these toxic elements of e-waste most often impact on the health of humans through improper disposal. Table 4 summarizes the potential health and environmental hazards that are the results of e-waste.

e-Waste Component	Processing	Potential Health hazard	Potential Environmental Hazard
Cathode Ray Tubes (CRT), LCDs	Removal of copper, dumping, breaking	Silicosis, Cuts from the glass, Inhalation of phosphor 8 or contact	Water sources (ground water) contamination by phosphor
Printed Circuit Boards	Removal of chips, de-soldering of the board	Tin or Lead inhalation, Possibility of inhalation of mercury, beryllium and/or brominated dioxin	Air contamination by emitted gases
Chips and other gold plated components	Chemical stripping	Acid contact with skin and/or eyes resulting into permanent injury Inhalation of acid fumes resulting into respiratory problems	Acidification of water sources affecting the flora and fauna of the areas
Plastic components	Shredding, low temperature buring/melting	Exposure to hydrocarbons, brominated dioxins, PAH	Contamination of air by the emitted gases
Steel, Copper and precious metals	Recovering of copper/steel through fire – open burning/furnace	Exposure to dioxins and heavy metals	Contamination of air, water sources and the soil
Mother Board, dismantled printed circuit boards	Burning of the circuit board	Intoxication of dumping sites surrounding residents, workers in the dumping sites from Tin, Lead, Beryllium etc.	Contamination of surroundings and water sources-ground water

Table 4. A summary of potential health and environmental hazards of electronic components [14, 15]

Given the rise of importation of electronic products in Tanzania, and the nature of e-waste and how it is disposed in Tanzania, and the difficult in determining its mass and flux in the country, the health and environmental hazards that are the results of e-waste are likely to be considerable. The impact of this are degraded environment, and negatively affected human health.

7. Research approach

The study employed descriptive research design since the study is seeking for more detailed and accurate information on e- wastes management practices to uncover the challenges and come up with sound and workable solutions to the challenges. The aims of this study were: to establish if organizations have e-waste management plan disposal policy in place and whether asset disposal policy include electronic products, to understand how do organizations store or treat end of use electronic products to establish if there exist collaboration between organizations (consumer of electronic products) and recycling industry to recycle electronic products, to understand whether waste management practitioners are trained on e-waste management practices and understand methods used by waste management companies or organizations to destroy e- wastes in damp sites.

Target population of the study was people working in operational level and managers in companies which are major vendors of electronic products in Tanzania and firms which are major users (buyers) of electronic products. Sample size used in this study was fifteen (15) for major vendors of electronic products and fifteen (15) for major users of electronic products (as per Tanzania Revenue Authority records). The sample sizes were selected using simple random technique from a list of eighty one (81) and eighty (83) for major users and major vendors of electronics products respectively. The study took place in Dar es Salaam as all respondents companies (which were randomly selected from the list of major vendors and consumer of electronic products) have their headquarters in Dar es Salaam.

In line with that semi- structured interviews along with the questionnaires were used as data collection tools. We distributed two (2) sets of questionnaires. The first set of survey targeted companies dealing with waste management and the second set of questionnaire targeted companies which are major consumer of electronic products. Questionnaires were hand delivered to respondents and followed up. The survey was stopped after reaching 13 out of 15 targeted respondent companies for the first set of questionnaire and 10 respondent companies out of 15 targeted respondent companies due time constraint. Structured interviews were conducted to either Operation Managers or Estate Managers or Training Managers to gather their views on e-waste management practices. In addition, four (4) dumping sites and two (2) recycling industries were visited to observe e- waste management practices on site.

8. Results analysis and discussion

In this survey, twelve (12) organizations responded out of fifteen (15) yielding an effective respond rate of 87% for Waste management Organizations and ten (10) Organizations which are major consumers of electronic products out of fifteen (15) yielding an effective respond rate of 67% .

The study shows 75 % of respondents have been using electronic products in their Organizations for more than ten (10) years. This increases the reliability of study results because the majority of respondent are experienced consumers 'of electronic products and therefore are thought to be more familiar with electronic products. Additionally, 65.7% of respondents have more than ten (10) years experience in waste management.

The study found that, 37.5% of Companies that are major consumers of electronic products do not have electronic products disposal plan and policy that include electronic products. This is a challenge as electronic products disposal policies and procedures provide guidelines for disposing electronic products within the Company [16].These guidelines help determine assets to be disposed and reveal procedures for disposing of electronic products with value for money in mind. Without these policies and procedures, disposing of electronic products will be handled improperly, or the decision to dispose of the assets may be delayed and therefore this may keep the organization in stalemate of either to re-deploy or dispose the electronic products. Similarly, there is a possibility of disposing of electronic products which are still of value to the Organization.

Further, this study found that storage of e-waste in general needs special attention. 75 % of Organization surveyed had not special allocated rooms with enough space to store end of use electronic products. This increases the chance of leakage. Standard operating procedures for e-waste for companies in India for example require organizations to allocate sufficient storage space with each type of e-waste placed differently to ensure safety [17]. Some components of electronic products are made up of hazardous chemicals such as batteries of phones and cartridges which contain carbon, lithium and other chemical elements [18]. Likewise Cathode Ray Tubes (CRT's) found in televisions and computer monitors contain mercury, phosphorous, cadmium, barium and lead that may leak if stored carelessly. In this regard, sensing mechanisms to detect any leakage should be installed in storages of end-of-use electronic products.

Another challenge is lack of well trained personnel in waste management and allocation of dumping sites. It is shocking to find out that none of waste management Organizations surveyed has trained personnel in waste management. It should be noted that handling of e-waste is different from other wastes; therefore it requires well-trained personnel and specialized equipments. Further, it requires strategic allocation of recycling, landfills or dumping sites to decrease the effect on the environment. Unfortunately this is not the case in Tanzania. The study found, many recycling, landfills and dumping sites are found near residential areas. This poses a very dangerous situation to health and environment as most of Waste management Companies prefer fire (55.6%) to destroy wastes (including e-wastes).

Tanzania has few Institutions specialized in the recycling of waste products. Most of them target plastics (plastics containers), and metal related products. But there are no recycling centers specialized in recycling e-waste products. This is evidenced by the fact that we have not even managed to deal effectively with household waste. Therefore to deal with categorized waste such as e-waste will be more difficult. This situation could have been improved if consumers of electronic products could forge partnerships or collaboration with Waste management Companies, National Environmental Commission and other stakeholders to recycle end of use electronic products. In line with that, this study found that 100% of Organizations surveyed has no partnerships with recycling industry to recycle end of use electronic products. Another type of collaboration can be forged between vendors and users through an end of use take back programme.

In addition to that due to rapid increase in users of mobile phones, televisions, internet, internet hosts [1] increase in importation of computers, photocopiers, printers ,other computer peripherals, fridges, air conditioners and other electronic products to Tanzania, it is obvious that the existing recycling capacity is not sufficient to absorb the potential e-waste caused by all of this.

Finally the study observed that, privacy and confidentiality of information is in jeopardy as only 8.3 % of respondents sanitize storage devices such as hard disk to wipe out permanently data. If these devices hard drives are not wiped effectively or destroyed, privacy, confidentiality and security of information stored in may fall in wrong hands and cause disaster.

9. Suggested solutions

In this section we suggest solutions to the challenges identified in Section 3 above as follows, Companies should: craft electronic products disposal policies and procedures, improve of end of use electronic products storage facilities, provide training and improve disposing infrastructures and equipments, revisit legal framework, launch of awareness campaigns to the public, and properly clean hard drives of computers in order to secure privacy and confidentiality of information.

9.1. Formulation of disposal policies and procedures

As the research results have shown, most Companies lack e-wastes disposal policies and procedures.

Therefore e- waste disposal policies and procedures will lay down principles, guidelines and procedures in disposing of end of use electronic products in Companies and will clearly pave the way for the whole process in disposing of end-of-use electronic products. . In order to address this inadequacy; the authors suggests that, Tanzania government adopts the Durban declaration on e-waste management in Africa [19]. This declaration calls each country to develop its own process and define its own roadmap towards e-waste management.

The result of implementing the declaration; the government in collaboration with stakeholders should develop e-waste management guidelines for different stakeholders and organizations. The objectives of the guideline should be to:

a. enhance environmental and health protection from e-waste,
b. formulate a basis for policy and regulatory framework for e-waste management for the different players, and
c. to create/raise awareness towards sustainable e-waste management in Tanzania and both the national, organizational and household level.

The guidelines should address issues such as: approaches towards enhanced environmental and health protection, policy, regulatory and institutional framework for e-waste management, awareness creation, categories of e-waste, e-waste treatment and treatment methods and technologies, and disposal procedures for e-waste.

This will build a sense of responsibility and accountability related to management of electronic products within companies in order to make sure that value for money concept is realized during electronic products disposal exercise. Further, electronic products disposal policies and procedures will build a sense of responsibility and accountability related to management of electronic products within Companies in order to make sure that value for money concept is realized during electronic products disposal exercise. Formulation of these policies and procedures should involve all stakeholders in Companies and accommodate all electronic products used in the Company. To make them effective, the policies and procedures of disposing end of use electronic products should be enforced the same way as other policies. Since technology is changing very fast, periodic review of these policies and procedures is inevitable.

9.2. Improved storage facilities

Storage of end-of-use electronic products needs to be addressed seriously. Therefore to tackle this challenge, organizations and institutions need to make sure enough storage space is allocated for storing end-of-use electronic products. Storage facilities need to be equipped with sensing mechanisms to detect any leakage or emission of radioactive materials found within end-of-use electronic products. Meanwhile we recommend Companies to cut off time taken to keep in stores the end-of-use electronic products before they are taken for disposal. Reduction of delay-time will mean that organizations will save storage costs and space for storing end of use electronic products. This will ultimately help to mitigate all risks associated with storage for long term end-of–use and decommissioned electronic products including leakages and emissions of radioactive materials found in electronic products.

9.3. Training and improvement of disposing infrastructures

Training of people working in waste management sector and improvement of infrastructure for disposing e- wastes is a crucial step towards curbing e-wastes disposal challenges. The waste management sector is not getting the attention it deserves. Workers in this sector are not trained, equipped and are lowly-paid. Techniques related to personal protection, handling of hazardous products, first aid and combating fire and flames are essential for people working in this sector. E –wastes is still treated as any other waste. For example, currently, Tanzania has no specialized plant to destroy e- wastes. All wastes are disposed using the same method- fire. Partnerships or collaboration between Companies with electronic products major vendors or producers can help in improving e-waste disposing infrastructures.

Considering the electronic products growth trends in Tanzania there is no doubt that within the next few years these challenges will be realized, and therefore building and improving infrastructure is not an option any more, it is a necessity. Kenya, Nigeria, and South Africa, Egypt and many other West African countries have already started shaping their waste treatment infrastructure to accommodate e-waste[20],[21], and [22].

9.4. Deliberate support for recycling initiatives

Due to rapid increase in importation of electronic products and usage at homes and offices, it is clear that e-waste is becoming a "time ticking bomb". Recycling activities in Tanzania are mainly focusing on plastic and metal scrap recycling only. Van de Brink and Szrimai's study of the Tanzania scrap recycling showed that Tanzania have large surplus of scrap and the players in metal scrap industries are not capable of utilizing all available scrap. [23].Additionally little has been done to support plastic recycling. This can be evidenced by the fact that few plastic products producers are engaging in recycling activities. From Van de Brink and Szrimai's study we can realize that deliberate support for recycling initiatives and recycling activities is urgently needed to extend recycling initiatives and activities to accommodate e-waste products in order to curb this alarming problem before is too late. The Government through National Environmental Commission (NEMC), Municipal and City Councils, Non- Governmental Organizations and other stake holders has to take charge in launching recycling campaigns and initiatives.

9.5. Proper cleaning of hard drives and memory

Proper hard drives cleaning is essential for guaranteeing privacy and confidentiality of information stored in end of use electronic products. Deleting a file or formatting a hard drive does not sterilize completely computer hard drives. When a hard drive is formatted, information stored in is still alive and can easily be recovered using data recovery software. There is a lot of software available that are capable of deleting a file permanently. They clean hard drives without destroying the sectors and leave no trace of information. This makes it impossible for information stored to be recovered. Therefore in order to safeguard privacy and confidentiality of information we recommend organizations to apply secure data cleaner software that conforms to DoD 5220.22-M standard, which requires overwriting of all addressable locations with a character, its complement, then a random character and verify[24]. Secure wiping of data and information should be conducted regardless of whether the electronic device is decommissioned or destroyed. This is crucial because data and information stored in hard drives or memory of these devices can be accessed and used illegally by unauthorized people.

9.6. Enforcement of e-waste related legislation and e- waste awareness campaigns

Having legislation without enforcing it is useless. It is therefore the responsibility of the Government through its law enforcement units to make sure these e- waste legislations however do not in full adhere to the Basel Convection Basel Convection on the control of trans-boundary movements of hazardous wastes and their disposal are enforced. This will largely help to stop few culprits to transport and dispose irresponsibly e- wastes.

The study conducted by UNIDO e- waste initiatives for Tanzania in 2011 indicated that, 80% of people interviewed are unaware of e-wastes and its hazardous impact on the environment [25]. This suggests the need of e- waste awareness campaigns to the public in order to safeguard the environment and public health in general.

10. Conclusion

This chapter has discussed various challenges on e- Wastes disposal in Tanzania and proposed remedy to the challenges. These challenges are:

a. Lack of ICT asset disposal policy
b. Lack of storage facilities for end of use electronic products
c. Lack of trained personnel in e- waste management
d. Lack of proper re-cycling initiatives and partnership and
e. Privacy and confidentiality of data and information in end of use electronic products such as phones and computers.
f. Lack of enforcement of e- waste related legislation.

We suggest that, Government and non government organizations should set and enforce electronic products disposal policy and work in collaboration with recycling industries to recycle end of use electronic products safely. Further, the main importers of electronic products should be involved in raising awareness to users of electronic products regarding safe disposal of electronic products and be involved in e- waste recycling initiatives.

Our findings illustrate that disposing of end of use electronic products is alarming situation and therefore needs special attention. It is essential that all key players be involved in thwarting down e-waste issue. Different regulations provide principles on handling of e- wastes and provide room for producers and vendors of electronic products to set up recycling systems of electronic products from consumers. These regulations are not enforced and therefore not followed. The intention of these regulations is to involve consumers of electronic products in different levels in disposing e- waste safely. The existence of these regulations is seen as important and crucial, but is not executed. Even the strategy and action plan on e- waste has not been implemented while things are getting out control.

Tanzania is one of the signatories of Basel Convection on the control of trans-boundary movements of hazardous wastes and their disposal which was ratified on 7 April, 1993. Unfortunately, Tanzania has not dealt seriously with the problem of importation of sub-standard electronic products which are donated to majority of African countries as means of off-loading e-waste from developed countries.

It is now high time for the Government of United Republic of Tanzania to amend and enforce Environment Law to accommodate the Basel Convection. This will prevent Tanzania from becoming a dump site for end-of-use electronic products. Amendment of environmental law will make Tanzania part of world-wide team who prevents illegal e-waste trade and at the same time help hold responsible people importing sub-standard electronic products and disposing of them irresponsibly. Our desire to stimulate development of ICT industry in Tanzania by slashing all taxes and duties on computers and their peripherals needs to be revisited in order to save our environment for present and future generations.

Author details

Daniel Koloseni and Faith Shimba

The Institute of Finance Management,Faculty of Computing, Information Systems and Mathematics,Department of Information Technology, Dar Es salaam, Tanzania

11. References

[1] e-Waste Guide (2009). E- Waste definition. Available: http://india.ewasteguide.info/e_waste_definition. Accessed on December12, 2009)

[2] URT (1997) National Environmental policy, Government of Tanzania.

[3] URT (1996) The Sustainable Industrial Development Policy (1996-2020).Government of Tanzania

[4] URT (2003) National ICT Policy. Government of Tanzania

[5] URT (2009) The Environmental (Solid Waste Management) Regulations. Government of Tanzania.

[6] URT (2009) Environmental Management (Hazardous Waste Control). Government of Tanzania

[7] URT (2004) National Environmental Management Act No. 20. Government of Tanzania

[8] URT (1982) Local Government (Urban Authorities) Act No. 8. Government of Tanzania

[9] URT (1982) Local Government (District Authorities) Act No. 8. Government of Tanzania

[10] URT (2009) National Waste Management Strategy and Action Plan. Government of Tanzania

[11] Ladou, J. and Lovegrove, S (2008) Export of electronics equipment waste. International Journal of Occupational and Environmental Health, 14(1): 1-10.

[12] Robinson, B.H., E-waste: An assessment of global production and environmental impacts. Science of the total environment, 2009. 408(2): p. 183-191.

[13] Ernst, T., et al. (2003), Analysis of eco-relevant elements and noble metals in printed wiring boards using AAS, ICP-AES and EDXRF. Analytical and bio-analytical chemistry, 375(6): 805-814.

[14] Deng, W.J., et al (2006) Atmospheric levels and cytotoxicity of PAHs and heavy metals in TSP and PM at an electronic waste recycling site in southeast China. Atmospheric Environment, 40(36): 6945-6955.

[15] Wath, S.B., P.S. Dutt, and T. Chakrabarti, E-waste scenario in India, its management and implications. Environmental monitoring and assessment. 172(1): p. 249-262.

[16] London Metropolitan University(2009). Disposal of ICT Assets Policy. Available: http://www.londonmet.ac.uk/londonmet/library/o49519_3.pdf. Accessed December 12, 2009)

[17] e-Waste Guide(2009). Standard Operating Procedures for E- waste. Available http://india.ewasteguide.info/standard_opreating_procedure_for_e_waste_managemen t. Accessed December 12, 2009

[18] Brigden, K., Santillo, D (2006) Determining the presence of hazardous substances in five brands of laptop computers. Greenpeace Research Laboratories Technical Note p. 20

[19] WasteCon2008 (2008), The Durban Declaration on e-Waste Management in Africa. Available: http://ewasteguide.info/Durban_declaration Accessed May 28, 2012.

[20] Wanjiku, R (2008) Kenya opens first e-waste management plant. In: IDG News Service

[21] Finlay,(2005) E-waste Challenges in Developing Countries: South Africa Case Study. APC "Issue Papers" Series Association for Progressive Communication, 1–22

[22] Moubasher, H (2009) BCRC-Egypt E-waste Activities

[23] Brink, J., Szirmai, A (2002) The Tanzanian scrap recycling cycle. Technovation 22(3), 187

[24] DoD., (1995) National Industrial Security Program. Operating Manual, Defense Technical Information Center

[25] Magashi, A. and M. Schluep (2011) e-Waste Assessment Tanzania.

Permissions

The contributors of this book come from diverse backgrounds, making this book a truly international effort. This book will bring forth new frontiers with its revolutionizing research information and detailed analysis of the nascent developments around the world.

We would like to thank Luis Fernando Marmolejo Rebellón, for lending his expertise to make the book truly unique. He has played a crucial role in the development of this book. Without his invaluable contribution this book wouldn't have been possible. He has made vital efforts to compile up to date information on the varied aspects of this subject to make this book a valuable addition to the collection of many professionals and students.

This book was conceptualized with the vision of imparting up-to-date information and advanced data in this field. To ensure the same, a matchless editorial board was set up. Every individual on the board went through rigorous rounds of assessment to prove their worth. After which they invested a large part of their time researching and compiling the most relevant data for our readers. Conferences and sessions were held from time to time between the editorial board and the contributing authors to present the data in the most comprehensible form. The editorial team has worked tirelessly to provide valuable and valid information to help people across the globe.

Every chapter published in this book has been scrutinized by our experts. Their significance has been extensively debated. The topics covered herein carry significant findings which will fuel the growth of the discipline. They may even be implemented as practical applications or may be referred to as a beginning point for another development. Chapters in this book were first published by InTech; hereby published with permission under the Creative Commons Attribution License or equivalent.

The editorial board has been involved in producing this book since its inception. They have spent rigorous hours researching and exploring the diverse topics which have resulted in the successful publishing of this book. They have passed on their knowledge of decades through this book. To expedite this challenging task, the publisher supported the team at every step. A small team of assistant editors was also appointed to further simplify the editing procedure and attain best results for the readers.

Our editorial team has been hand-picked from every corner of the world. Their multi-ethnicity adds dynamic inputs to the discussions which result in innovative outcomes.

These outcomes are then further discussed with the researchers and contributors who give their valuable feedback and opinion regarding the same. The feedback is then collaborated with the researches and they are edited in a comprehensive manner to aid the understanding of the subject.

Apart from the editorial board, the designing team has also invested a significant amount of their time in understanding the subject and creating the most relevant covers. They scrutinized every image to scout for the most suitable representation of the subject and create an appropriate cover for the book.

The publishing team has been involved in this book since its early stages. They were actively engaged in every process, be it collecting the data, connecting with the contributors or procuring relevant information. The team has been an ardent support to the editorial, designing and production team. Their endless efforts to recruit the best for this project, has resulted in the accomplishment of this book. They are a veteran in the field of academics and their pool of knowledge is as vast as their experience in printing. Their expertise and guidance has proved useful at every step. Their uncompromising quality standards have made this book an exceptional effort. Their encouragement from time to time has been an inspiration for everyone.

The publisher and the editorial board hope that this book will prove to be a valuable piece of knowledge for researchers, students, practitioners and scholars across the globe.

List of Contributors

Elias Mazhindu
Ethiopian Civil Service University, Addis Ababa, Ethiopia

Trynos Gumbo
African Doctoral Academy, Stellenbosch University, South Africa

Tendayi Gondo
Department of Urban and Regional Planning, University of Venda, Thohoyandou, South Africa

James Okot-Okumu
Department of Environmental Management; School of Forestry Environmental and Geographical Sciences; College of Agricultural and Environmental Sciences; Makerere University Kampala, Uganda

Jayashree Sreenivasan, Marthandan Govindan, Malarvizhi Chinnasami and Indrakaran Kadiresu
Multimedia University, Malaysia

Aretha Aprilia and Tetsuo Tezuka
Department of Socio-Environmental Energy Science, Graduate School of Energy Science, Kyoto University, Yoshida-honmachi, Sakyo-ku, Kyoto, Japan

Gert Spaargaren
Department of Environmental Policy, Wageningen University, Wageningen, The Netherlands

Asmawati Desa, Nor Ba'yah Abd Kadir and Fatimah Yusooff
School of Psychology and Human Development, the National University of Malaysia, Bangi, Selangor, Malaysia

Manfred Fehr
Institute of Geography, Federal University at Uberlândia, Uberlândia, Brazil

Mariana Matos-Moreira and Emilio Carral
Department of Cellular Biology and Ecology, University of Santiago de Compostela, Lugo, Galiza, Spain

Mario Cunha
Department of Mathematics-Research Center for Geo-spatial Sciences, Porto University, Porto, Portugal

M. Elvira López-Mosquera
Department of Plant Production, University of Santiago de Compostela, Lugo, Galiza, Spain

Teresa Rodríguez
Department of Zoology and Physic Anthropology, University of Santiago de Compostela, Lugo, Galiza, Spain

Luis F. Marmolejo, Patricia Torres and Mariela García
Facultad de Ingeniería, Universidad del Valle, Cali, Colombia

Luis F. Diaz
CalRecovery Inc., Stanwell Drive, Concord, CA, USA

Tiew Kian-Ghee
Faculty of Engineering and Built Environment, University Kebangsaan, Malaysia

Noor Ezlin Ahmad Basri, Hassan Basri and Shahrom Md. Zain
Faculty of Engineering & Built Environment, Department of Civil and Structural Engineering, University Kebangsaan, Malaysia

Sarifah Yaakob
Alam Flora Sdn Bhd., Malaysia

Christia Meidiana
Faculty of Engineering, Brawijaya University, Malang, Indonesia

Yoichi Kodera
Research Institute for Environmental Management Technology, National Institute of Advanced Industrial Science & Technology (AIST), Tsukuba, Japan

John J. Classen
Biological and Agricultural Engineering Department, North Carolina State University, Raleigh, NC, USA

Harbans Lal
National Water Quality and Quantity Team, NRCS/USDA, Portland, OR, USA

Md Saidul Borhan and Shafiqur Rahman
Agriculture and Biosystems Engineering Department, North Dakota State University, Fargo, ND, USA

Saqib Mukhtar and Sergio Capareda
Biological and Agricultural Engineering Department, Texas A&M University, TAMU College station, TX, USA

Hakan Arslan
Department of Architecture, College of Engineering, Technology and Architecture, Hartford University, West Hartford, USA

Nilay Coşgun
Department of Architecture, Faculty of Architecture, Gebze Institute of Technology, Kocaeli, Turkey

Burcu Salgin
Department of Architecture, Faculty of Architecture, Erciyes University, Turkey

Daniel Koloseni and Faith Shimba
The Institute of Finance Management, Faculty of Computing, Information Systems and Mathematics, Department of Information Technology, Dar es Salaam, Tanzania